EVOLUTION AT THE MOLECULAR LEVEL

WITHDRAWN
FROM
UNIVERSITY OF PLYMOUTH

90 0077591 3

TELEPEN

2
1
2

This book is s
B
CHARGES W

EVOLUTION AT THE MOLECULAR LEVEL

Edited by Robert K. Selander,
Andrew G. Clark,
and Thomas S. Whittam

INSTITUTE OF MOLECULAR EVOLUTIONARY GENETICS
PENNSYLVANIA STATE UNIVERSITY

SINAUER ASSOCIATES INC. • PUBLISHERS
Sunderland, Massachusetts 01375

POLYTECHNIC SOUTH WEST
LIBRARY SERVICES

Item No.	9000775913
Class No.	574.873281 EVO
Contl No.	0878938184

EVOLUTION AT THE MOLECULAR LEVEL

Copyright © 1991 by Sinauer Associates, Inc.
All rights reserved. This book may not be reproduced in whole or in part by any means without permission from the publisher. For information address Sinauer Associates, Inc., Sunderland, MA 01375

Library of Congress Cataloging-in-Publication Data

Evolution at the molecular level / edited by Robert K. Selander,
    Andrew G. Clark, and Thomas S. Whittam.
        p.   cm.
    Includes bibliographical references and index.
    ISBN 0-87893-818-4   —ISBN 0-87893-819-2 (pbk.)
    1. Chemical evolution.   2. Molecular genetics.   I. Selander,
    Robert K.   II. Clark, Andrew G., 1954–   III. Whittam, Thomas S.
    QH325.E96   1990   90-43140
    574.87'3282—dc20   CIP

Printed in U.S.A.

6 5 4 3 2 1

Contents

Preface

Fundamental research in molecular biology continues to reveal novel aspects of variation in the structure of genes and genomes for study and interpretation by evolutionary population geneticists. And technical innovations, such as development of the polymerase chain reaction method of directly amplifying DNA, are providing powerful tools for addressing classical questions concerning the mechanisms of evolutionary change, the genetic structure of populations, and the phylogenetic relationships of organisms.

The intent of this book is to present the content and capture the excitement of recent advances in the study of evolution that have been achieved through the integration of molecular biology and evolutionary genetics. In this volume, which has its roots in a symposium sponsored by the Society for the Study of Evolution and the American Society of Naturalists at Pennsylvania State University in June of 1989, 26 scientists have written 13 chapters on a variety of topics ranging from the population genetics of transposable elements and organelle genomes to the mechanisms of divergence of multigene families, with examples drawn from retroviruses, bacteria, plants, and animals. The contents of the chapters should be accessible to advanced biology students, professional biologists, and other scientists interested in the recent developments in the study of evolution at the molecular level. The book may serve as a text for a graduate-level course in molecular population genetics or molecular evolution and as supplemental reading for courses in evolutionary biology.

In the initial chapter, Carl Woese reviews the development of evolutionary concepts in microbiology and the recent impact of the analysis of ribosomal RNA sequences on interpretations of bacterial relationships. His discovery of the *Archaea* domain has revolutionized our understanding of genomic diversity and phylogenetic relationships among microorganisms and higher organisms and has forced a major revision of ideas relating to early stages in the radiation of life forms.

The growing field of bacterial population genetics is represented by two chapters. Selander and his colleagues illustrate how a population genetic approach is yielding insight concerning the genetic structure and evolution of populations of pathogenic bacteria of the genus *Salmonella*. Comparative sequencing of flagellin genes in diverse clonal lineages within a phylogenetic framework provided by multilocus enzyme genotypes indicates that horizontal transfer and recombination is a primary mechanism for the origin of new serotypes. DuBose and Hartl combine comparative sequence analysis with an experimental approach to elucidate the role of selective forces operating on alkaline phosphatase in *Escherichia coli*. Their approach heralds a major experimental advance in the study of evolutionary genetics in which deliberate manipulations of molecular sequences afford tests of evolutionary hypotheses relating to the structure and function of proteins.

The power of comparative molecular analysis is exemplified by the study of the tryptophan synthetic pathway presented by Crawford and Milkman. By tracing the modifications of the *trp* gene complex in light of bacterial phylogeny, they illustrate how complex enzyme structures have been produced through a series of paralogous and orthologous events.

Interest in the structure, function, and evolution of retroviruses has increased dramatically as a result of the AIDS epidemic. Yokoyama's chapter recounts the recent history of the study of human retroviruses and presents some of the basic evolutionary features of HIV and related viruses. He concludes that comparative studies will provide valuable clues to the origin of pathogenicity and host specificity in these viruses.

The intricacies of the theory of the evolution of organelle genomes are discussed by Birky, who explains how levels of variation in genes of organelles are influenced by forces of genetic drift and selection operating at cellular, individual, and population levels. And Clegg and colleagues review the growing body of information on the structure of chloroplast genomes, with special attention to the implications of this new type of data for our understanding of the phylogenetic relationships among plant taxa.

Two chapters focus on "selfish" genetic elements and their effects on the reproductive fitness of their carriers. Charlesworth and Langley combine mathematical theory with experimental results to assess the parameters influencing the dynamics of transposable elements in *Drosophila* populations. These analyses explain how transposable elements are maintained in populations and why they tend to accumulate in chromosomal regions of restricted recombination. New answers to old questions about segregation distortion are provided by Wu and Hammer's detailed account of the selfish genes underlying systems of meiotic drive in *Drosophila*.

The last four chapters focus on eukaryotic genes and gene complexes that are model systems for evolutionary research. Kreitman gives an overview of the molecular variation underlying the classical polymorphism of alcohol dehydrogenase in *Drosophila melanogaster*. The extensive sequence data generated by Kreitman have permitted identification of the evolutionary forces acting on and around the coding region of the *Adh* gene and have also stimulated the development of new mathematical models and statistical tools for the analysis of molecular data. The major histocompatibility complex, which includes some of the most polymorphic genes in humans and other mammals, provides an exemplary system for studying evolutionary processes. Two chapters, with complementary approaches, assess the relative roles of natural selection and nonselective forces in maintaining polymorphism of these highly diverse genes. Nei and Hughes explain how statistical analysis of the sequences of MHC genes has revealed unusual patterns of nucleotide substitution and has identified ancient polymorphisms shared among species. Strong evidence that these genes have been subject to balancing selection is marshalled. Hedrick and colleagues apply techniques of classical population genetics to investigate the bases for maintenance of antigenic variation in MHC molecules. In the final chapter, Hardison summarizes an extensive molecular analysis of the globin gene cluster that has yielded remarkable insights concerning the phylogenetic history and mechanisms of divergence of a multigene family. His chapter highlights the roles of duplications and transpositions in the radiation of globin genes.

ROBERT K. SELANDER
ANDREW G. CLARK
THOMAS S. WHITTAM

UNIVERSITY PARK, PA
July 1990

Contributors

PILAR BELTRAN Department of Biology, Pennsylvania State University, University Park

C. WILLIAM BIRKY, JR. Genetics Department, Ohio State University, Columbus

BRIAN CHARLESWORTH Department of Ecology and Evolution, University of Chicago, Chicago, IL

MICHAEL T. CLEGG Department of Botany and Plant Sciences, University of California, Riverside

IRVING P. CRAWFORD Late, Department of Microbiology, University of Iowa, Iowa City

ROBERT F. DUBOSE Institute of Molecular Biology, University of Oregon, Eugene

EDWARD M. GOLENBERG Department of Botany and Plant Sciences, University of California, Riverside

MICHAEL F. HAMMER Museum of Comparative Zoology, Harvard University, Cambridge, MA

ROSS C. HARDISON Department of Molecular and Cell Biology and Institute of Molecular Evolutionary Genetics, Pennsylvania State University, University Park

DANIEL L. HARTL Department of Genetics, Washington University School of Medicine, St. Louis, MO

PHILIP W. HEDRICK Department of Biology and Institute of Molecular Evolutionary Genetics, Pennsylvania State University, University Park

AUSTIN L. HUGHES Institute of Molecular Evolutionary Genetics, Pennsylvania State University, University Park

WILLIAM KLITZ Department of Integrative Biology, University of California, Berkeley

MARTIN E. KREITMAN Department of Biology, Princeton University, Princeton, NJ

MARY K. KUHNER Department of Integrative Biology, University of California, Berkeley

CHARLES H. LANGLEY Department of Genetics, University of California, Davis

GERALD H. LEARN Department of Botany and Plant Sciences, University of California, Riverside

ROGER MILKMAN Department of Biology, University of Iowa, Iowa City

MASATOSHI NEI Institute of Molecular Evolutionary Genetics, Pennsylvania State University, University Park

WENDY P. ROBINSON Department of Integrative Biology, University of California, Berkeley

ROBERT K. SELANDER Institute of Molecular Evolutionary Genetics, Pennsylvania State University, University Park

NOEL H. SMITH Institute of Molecular Evolutionary Genetics, Pennsylvania State University, University Park

GLENYS THOMSON Department of Integrative Biology, University of California, Berkeley

CARL R. WOESE Department of Microbiology, University of Illinois, Urbana-Champaign

CHUNG-I WU Department of Biology, University of Rochester, Rochester, NY

SHOZO YOKOYAMA Department of Ecology, Ethology, and Evolution, University of Illinois, Urbana-Champaign

EVOLUTION AT THE MOLECULAR LEVEL

THE USE OF RIBOSOMAL RNA IN RECONSTRUCTING EVOLUTIONARY RELATIONSHIPS AMONG BACTERIA

Carl R. Woese

The technologies that permit the manipulation and sequencing of genetic material are revolutionizing biology. To the layman and to the biologist in general, the revolution lies in applying this technology to various problems that face society. New strains of agriculturally important plants and animals can now be engineered, as can organisms to help remove the vast quantities of pollutants our race inflicts on this planet. New, faster, and more accurate means of diagnosing diseases are being developed, and powerful, specific treatments for some incurable diseases seem in the offing. The culmination of this revolution is seen as that remarkable undertaking, the sequencing of the human genome in its entirety—sometimes referred to as the Holy Grail of biology. What medical miracles we could then perform; *Homo sapiens* would come to know himself to the very molecular essence of his being! The great American dream of a brighter tomorrow through science and technology has been recast by the biologist in images fit for the twenty-first century.

However, this is not the revolution I see. The real revolution is a far quieter one, and generally goes unrecognized by biologist and layman alike. Its implications are in basic, not applied biology. The revolution we are witnessing is in evolutionary biology. Organisms, in a very important sense, are historical documents; their structure at all levels reflects (is determined by and records) their evolutionary history. Therefore, knowledge of evolution (and evolutionary relationships) is an integral part of explanation and understanding at virtually

1

all levels of biology. To know the sequence of every gene in the genomes of a number of prokaryotes is to possess wisdom of incredible antiquity and enormous value—something that was unimaginable to the biologist even as recently as the 1970s.

The classical view of evolution, whose images still shape our view today, is morphocentric and strongly founded in paleontology. Its scope is restricted to two relatively small groups of organisms, the metazoan animals and plants; its time span is barely more than 10% of our planet's history. The coming revolution widens the study of evolution to include all living systems, and increases the time span to near the full 4.5 billion year scope of Earth's history. Evolution's focus is shifting from the morphological to the metabolic, from a concern with the ancestors of various species to a concern with the ancestor common to all species. The history of the planet and the history of life thereon now need to be viewed as very much the same problem; primitive geochemistry encompasses primitive biochemistry.

The starting point for this biological revolution (once nucleic acids could be sequenced) was the use of ribosomal RNA sequence comparisons to open the door of bacterial phylogeny. It is this historic transition in microbiology and what led up to it that is the starting point of this chapter.

HISTORICAL BACKGROUND

In the early part of this century microbiologists were as keenly aware of and as vitally concerned with evolutionary considerations as were other biologists. Like botanists and zoologists, early microbiologists set out to construct a phylogenetic classification for microorganisms. In the words of Kluyver and van Niel:

[T]he only truly scientific foundation of classification is to be found in appreciation of the available facts from a phylogenetic point of view. Only in this way can the natural interrelationships of the various bacteria be properly understood. . . . the studies in comparative morphology made by botanists and zoologists have made phylogeny a reality. Under these circumstances it seems appropriate to accept the phylogenetic principle also in bacteriological classification. (Kluyver and van Niel, 1936)

A crucial difference between microbiology and its sister sciences, botany and zoology, however, lies in the morphological simplicity of bacteria. Bacterial shapes are phylogenetically uninformative; bacteria have no meaningful developmental stages, and (except for the dating it provides) the bacterial fossil record is of little value. Consequently, developing a microbial classification system along the same lines as zoologists and botanists had done was a foredoomed enterprise. However, in their zeal to know the "natural relationships" among bacteria early microbiologists tended to overlook this.

The history of microbial taxonomy, of the search for the evolutionary relationships among prokaryotes, we can now see was a trail that began in deception, then led to disillusionment and finally to despair. The deception, of course, was the self-deception that a phylogenetic system of microbial classifi-cation could be reliably based on morphological characteristics, as had been done so successfully in zoology and botany. Kluyver and van Niel (1936) felt it to be "self-evident that the shape of the cells is of outstanding importance for determining the place of a bacterium in any phylogenetic system." Elaborate schemes were constructed on this premise (Pringsheim, 1923; Buchanan, 1925; Kluyver and van Niel, 1936; Prévot, 1940; Stanier and van Niel, 1941). Bacterial evolution was seen as a progression from the simplest forms (cocci) to the most complex (for example, spore formers and actinomycetes)—i.e., just like the general metazoan progression (Kluyver and van Niel, 1936).

Disillusionment came with the realization that these taxonomies, regardless of what premises they were based on, were getting nowhere (van Niel, 1946). New organisms and new characteristics did not fit comfortably into existing categories (the systems showed no tendency to refine and develop themselves from within); old disputes remained unresolved; what passed for progress was usually cleverly rationalized prejudice. In frustration van Niel, once the leading exponent of a natural bacterial system, dismissed the whole venture:

[It] must be obvious to those who recognize in . . . the systematics of plants and animals . . . an increasingly successful attempt at reconstructing a phylogenetic his-tory . . . that comparable efforts in the realm of the bacteria (and bluegreen algae) are doomed to failure because it does not appear likely that criteria of truly phylogenetic significance can be devised for these organisms. (van Niel, 1955)

The despair that followed can be seen in the ready resignation (possibly implicit in the quote from van Niel) to the "fact" that since existing methods could not determine microbial phylogenies, they will *never* be known; and in the ensuing disdain for them, seen in the attitude that, determinable or not, there is no value in knowing microbial phylogenetic relationships.

For [most] major biological groups [such as the bacteria], the general course of evolution will probably never be known, and there is simply not enough objective evidence to base their classification on phylogenetic grounds. For these and other reasons, most modern taxonomists have explicitly abandoned the phylogenetic approach . . . (Stanier et al., 1970, and previous editions thereof)

Having created this impasse, microbiologists sought to circumvent it by redefining the basic problem of their science, replacing concern with phyloge-netic relationships by a concern with "the nature of the bacterial cell":

[T]he biological nature and relationships of the bacteria have been subjects of perennial discussion. Why have these questions obsessed some members of each succeeding generation of microbiologists? There can be no doubt about the principal reason. Any good biologist finds it intellectually distressing to devote his life to the study of a group that cannot be readily and satisfactorily defined in biological terms; and the abiding intellectual scandal of bacteriology has been the absence of a clear concept of a bacter-ium. . . . Our first joint attempt to deal with this problem . . . 20 years ago . . . was framed in an elaborate taxonomic proposal, which neither of us cares any longer to defend. But even though we have become sceptical about the value of developing formal taxonomic systems for bacteria . . ., the problem of *defining these organisms . . . in terms of their biological organization* is clearly still of great importance . . . [italics added] (Stanier and van Niel, 1962)

Although great progress in defining the organization of the prokaryotic cell subsequently occurred, the fundamental problem posed by Stanier and van Niel was not solved. Nor could it have been. If organisms are the products of their history, their complete description necessarily involves comparative phylogenetic analysis; and phylogeny is precisely what the microbiology community now eschewed.

The refocusing of microbiology in terms of the submicroscopic and molecular organization of bacteria had the unfortunate consequence of contributing to (if not causing) the general acceptance and dogmatization of the notion that there exists a fundamental dichotomous (phylogenetic) division of living systems into prokaryotes and eukaryotes.

Early microbiologists did not perceive the bacteria as any monolithic, monophyletic assemblage; quite the contrary. The blue-green algae (and their presumed nonphotosynthetic relatives) were traditionally separated from other bacteria. Even the division *Schizomycetes,* which comprised the three classes Eubacteriae, Myxobacteriae, and Spirochaetae, was considered to be "of polyphyletic origin" (Stanier and van Niel, 1941). Microbiologists had good reason to be suspicious of this prokaryote–eukaryote dichotomy, of its assertion that prokaryotes are monophyletic:

The entirely negative characteristics on which this group [prokaryotes] is based should be noted, and the possibility of . . . convergent evolution . . . be seriously considered. (Pringsheim, 1949)

[T]he bacteria and blue-green algae encompass a number of distinct major groups, which do not now appear to be closely related to one another; their only common character is that they are procaryotic. It thus appears that the procaryotic cell has provided a structural framework for the evolutionary development of a wide variety of microorganisms. (Stanier and van Niel, 1962)

[T]here are remarkably few *comparative* studies [of prokaryotes] . . . the application of the newer adjuncts of morphology for taxonomic purposes entails generalization from limited cases. (Murray, 1962)

Yet despite these earlier reservations, microbiologists ultimately did come to accept prokaryote–eukaryote as the fundamental phylogenetic distinction.

All organisms . . . can be assigned to one of two primary groups [prokaryotes and eukaryotes], readily distinguishable by differences in cellular organization . . . These differences are now so widely recognized that descriptions of them can be found in the better textbooks of general biology . . . , a sure indication that they have acquired the status of truisms. (Stanier, 1970)

There is little doubt . . . that biologists can accept the division of cellular life . . . into two groupings at the highest level expressing the encompassing characters of procaryotic and eucaryotic cellular organization. (Murray, 1974)

[Though not stated explicitly, the contexts from which these last two quotes were taken make it apparent that the distinction is a phylogenetic one.]

This turn of events, microbiology's uncritical acceptance of the monophyletic nature of prokaryotes, has no rational explanation. It must somehow have grown out of the disaffection, distrust, and disregard with which microbiologists came to view phylogenetic considerations. While phylogenetic studies had revealed nothing about the nature of bacteria, submicroscopic and molecular characterizations did provide something both tangible and unique. The problem was that the characterization of the prokaryote was based essentially on one organism, *Escherichia coli*. And the absurdity is that microbiologists, despite their earlier misgivings, now became party to the unsubstantiated generalization of *E. coli*'s properties to all prokaryotes!

The impact of the prokaryote–eukaryote dichotomy on biology has been enormous. It has shaped the concept of a bacterium, defined the relationship between bacteria and eukaryotes, and strongly influenced our notions about the early evolutionary course—particularly the origin of the eukaryotic cell. Advances over the past decade have drastically changed the concept of a prokaryote, have provided the historical dimension so badly needed if we are to understand prokaryotes. Yet "prokaryote" today still means to most biologists (connotations and all) what it meant a decade ago—continuing to influence strongly both scientific thought and action (e.g., defining the scope of textbooks, defining university departments and the courses they offer, defining the structure and policies of funding agencies). The time is past when the biologist can productively operate within the framework of such an outmoded concept. We are now in a position to, and must, put "prokaryote–eukaryote" into proper perspective, and thereby move beyond its stultifying confines.

As dogma the prokaryote–eukaryote dichotomy has a ring of truth, understanding, and completeness that are totally unwarranted. The principal positive effect the concept has had was in leading the biologist to discover how very different organisms can be on the molecular level when they represent different (primary) kingdoms, different domains (Woese et al., 1990a). And it also gave microbiologists confidence that they dealt with a real, definable entity in a prokaryote. Beyond that the dogma's appealing simplicity served, however, only to impede scientific progress. It created complacency—a sense that we had grasped a fundamental phylogenetic truth when we had not. It lulled microbiologists into a "seen one, seen 'em all" attitude toward bacteria. It justified those who felt there to be little value in understanding the natural relationships among microorganisms. It almost certainly delayed discovery of the archaebacteria by at least a decade. And, the connotations of the prefix "pro-" convinced us that prokaryotes were more primitive than eukaryotes, were older than eukaryotes, and gave rise to eukaryotes—all of which is uncertain at best, and definitely misleading in where it led us (or prevented us from going).

This strong condemnation of the prokaryote–eukaryote dichotomy is both justified and necessary—as a counterperspective from which the biologist can understand that archaebacteria and eubacteria, though both "prokaryotes" [in the original, *negative* cytological definition thereof (Chatton, 1937)], are not necessarily related to one another (Woese and Fox, 1977; Fox et al., 1980). More than a decade after their discovery many biologists still view archaebacteria as "only prokaryotes," as "just bacteria," and thereby fail to comprehend their real biological meaning. In an attempt to rectify this situation a new formal system of organisms has been proposed (Woese et al., 1990a), in which all life is organized at the highest level into three "domains": the *Archaea* (archaebacteria), the *Bacteria* (eubacteria) and the *Eucarya* (eukaryotes). Hopefully, the new sytem, the new terminology, will catalyze the full realization of the uniqueness of the archaea (as I will from here on call them).

One of the grand ironies in this story is that as microbiology moved away from and disavowed microbial evolution, techniques for validly determining microbial evolutionary relationships (molecular techniques) were being developed. The reason why molecular approaches would succeed where classical microbial characterizations had failed is obvious: At the molecular level prokaryotes are just as complex as are eukaryotes, and within that complexity lies an enormous amount of historical (phylogenetic) information.

"Molecules as documents of evolutionary history," Zuckerkandl and Pauling's seminal paper (1965), put the matter clearly. Molecular sequences are linear strings of hundreds to thousands of quasi-independent characters (amino acids or nucleotides). The number of *possible* sequences is enormous. In such a vast space of possibilities convergence ceases to be a significant consideration, and (extensive) similarity must mean evolutionary relatedness. In other words, "a recognition of many differences between two [sequences] does not preclude the recognition of their similarity" (Zuckerkandl and Pauling, 1965). Some molecular sequences, such as ribosomal RNA, change so slowly over time that they retain the traces of ancestral patterns billions of years old. From the extent (and nature) of the differences among a set of homologous sequences one can then reconstruct molecular genealogies, evolutionary trees of organisms (Fitch and Margoliash, 1967; Fitch 1971; Felsenstein, 1982; Olsen, 1988).

MOLECULAR SEQUENCES AS CHRONOMETERS

The relationship between molecular sequence and molecular function is by no means a one-to-one correspondence. A very large number of different sequences all correspond to the exact same molecular function. Consequently, the sequence of a molecule can be changed in any of a large number of ways without altering the overlying function. This realization has led to the concept of the "evolutionary clock," arguably the most important insight into the evolutionary process in this century: Evolution, in effect, occurs on two different levels, in two different ways. On the level of the genotype (in terms of molecular sequences) changes occur more or less continuously (on an evolutionary time scale). The majority of these changes, however, are either deleterious (so we don't encounter them) or selectively neutral, and so do not produce significant change at the overlying phenotypic (functional) level (Kimura, 1983). On the other hand, changes at the phenotypic level (ostensibly selected changes) are relatively rare and far more sporadic in occurrence. The neutral genotypic changes in effect mark time. This quasi-independent "evolutionary clock" embedded in the genotype is what gives the biologist the power to infer evolutionary histories and relationships in all living forms.

The molecular chronometer is one thing in principle, another in practice. To read any chronometer, its state has to be known at two or more times. Since the genetic sequences with which we deal exist only in one state (their present state), past states cannot be known directly. However, two different versions of the same molecule (from two different organisms) have at some past time shared a common ancestor. Therefore, knowing the differences between the two is almost as good as knowing the difference between the present state of either one and the earlier, ancestral state; in other words, the difference between two extant (homologous) sequences is a (relative) measure of evolutionary time, the time since the two sequences shared a common ancestor. (This difference is, of course, akin to distance, not time. Thus, any time interpretation given it requires knowledge or assumptions regarding relative rates of change in both lineages.)

In practice, not all molecular chronometers are equally useful for reconstructing phylogenies, and no method of phylogenetic inference in use today is completely satisfactory. The majority of molecules change in sequence too rapidly to retain sufficient information to be used in detecting the deeper phylogenetic branchings. Relatively few remain strictly constant in function over time, which adds selected changes to the neutral (time marking) ones, and so interferes with the analysis. And, many functions are not universally distributed, which restricts their utility as molecular chronometers.

The two most commonly used methods for reconstructing phylogenetic trees from sequence information are distance matrix analysis and maximum parsimony analysis. The first looks only at the distance between two homologous (aligned) sequences, i.e., the *extent* to which, the fraction of positions in which, they differ. From a matrix of such pairwise distances the phylogenetic relationships among a set of sequences are inferred, by selecting that tree whose branching order and branch lengths best accommodate the measured distances (Fitch and Margoliash, 1967; Felsenstein, 1982). The second method, maximum parsimony analysis, is more ambitious. It looks at the actual changes between pairs of sequences, not merely their number, and attempts to reconstruct the course of events that converted an assumed ancestral sequence into its various descendant lineages, on the assumption that evolution proceeds by some least action principle, i.e., makes the fewest changes possible in going from a given ancestral sequence to any given descendant sequence (Fitch, 1971; Felsenstein, 1982).

Both treeing methods work well under "ideal" conditions, when sequences are closely related and various parameters change at the same rate. However, when any or all of the following situations obtain (which is usually the case in the real world), both methods tend to produce incorrect branching orders (not to mention branch lengths): (1) Some lineages are only distantly related to the others, (2) some lineages are rapidly evolving, (3) the different positions in a sequence change composition at vastly different rates, and (4) sequences have significantly different compositions (base ratios).

As the number of differences between two sequences increases (as they become increasingly distantly related) so does the possibility that positions in a sequence alignment will have undergone more than one change, which in a fraction of such cases will appear to be no change at all. Thus, the observed difference between two related sequences always underestimates the actual number of changes that separate them—the more distant their relationship, the more pronounced the underestimate. As a result the most distantly related sequences in any set tend to be artificially drawn toward one another by both parsimony and (uncorrected) distance matrix analyses, to the extent in some cases that they appear specifically related when actually they are not (Felsenstein, 1982; Olsen, 1988). This same artifact is seen for (relatively) rapidly evolving lineages; i.e., they are drawn toward outgroup (distantly related) lineages, and so appear to branch more deeply in a tree than they actually do. (Other factors contribute to the artificial exclusion of rapidly evolving lineages as well.)

The rates at which the different positions in a (ribosomal RNA) sequence change can vary by at least two orders of magnitude (Woese, 1987). This means that the amount of phylogenetic information that the various (rate) classes of positions carry will change as the phylogenetic distance among species changes. The positions that are the most useful in ordering closely related lineages, for example, the enteric bacteria, have changed in composition so frequently among more distantly related species that they carry little phylogenetic information; and if such positions are retained in the analysis of more distantly related species, they can only adversely affect the outcome.

The first two problems (distantly related and rapidly evolving lineages) have received the most attention from theoreticians. The third has only relatively recently come in for serious treatment (Olsen, 1988). The final problem, disparity in compositions, is far more troublesome than is generally recognized, and has received almost no attention to date. A number of prokaryotic groups contain both thermophilic and mesophilic organisms. The rRNAs of thermophilic prokaryotes tend to have considerably higher $G + C$ contents than do those of the mesophilic prokaryotes. In prokaryote phylogenetic trees the thermophiles generally cluster near the base, and their lineages tend to be appreciably shorter than those of mesophiles. It is important to understand the extent to which this general pattern reflects rRNA compositional disparity rather than the true phylogeny.

With distance matrix analysis it is possible to correct statistically (at least to some extent) for the underestimation of the number of observed vs. actual changes that separate two sequences (Jukes and Cantor, 1969; Olsen, 1988). This is not possible with parsimony analysis, which consequently is more sensitive to the distortions in analysis caused by distantly related lineages and the like (Olsen, 1988).

A third method of analysis, called "evolutionary parsimony," has recently been introduced (Lake, 1987). Its strength lies in its apparent insensitivity to "unequal rate effects" (artifacts caused by distantly related/rapidly evolving lineages, to which parsimony analysis in particular is sensitive). The method has produced some topologies, most importantly the primary branching order of the universal phylogenetic tree, that are at total variance with the topologies produced by the more traditional methods. While these aberrant topologies (Lake, 1988) have been heralded as the truth revealed when "unequal rate" artifacts are overcome, it has become apparent that they themselves are undoubtedly artificial.

A recent study by Gouy and Li (1989) demonstrates that the evolutionary parsimony method gives the so-called eocyte topology (Lake, 1988) for the universal phylogenetic tree only with small subunit rRNAs, not large subunit rRNAs. And Olsen and Woese (1989) have shown that even with small subunit rRNAs the method gives that topology only when particular sequences are used in the analysis. Most telling are the simulations of Jin and Nei (1990), which demonstrate that evolutionary parsimony has difficulty recovering the correct topology when transversion rates are unequal; and the alignment used to produce the universal topology has precisely this characteristic (Olsen and Woese, unpublished analysis).

THE CURRENT STATE OF MICROBIAL PHYLOGENY

Almost all that is known about microbial phylogeny (above the genus level) has been determined within the last 10 or so years, and comes from analysis of small subunit rRNA sequences (Woese, 1987). There can be little doubt that the small subunit rRNA is as reliable a molecular measure of phylogenetic relationships among organisms as is now available (for prokaryotes, at least). Ribosomal RNA is universal in distribution; its function was established at an early stage in the evolutionary process, and appears extremely constant and highly constrained; and it is not affected by most changes in an organism's environment (except for changes in basic physical parameters such as intracellular pH and temperature). Being large molecules, the ribosomal RNAs contain a great deal of information. For this and other reasons RNAs are less erratic chronometers than smaller molecules (Woese, 1987). Moreover, they appear not to be subject to lateral gene transfer (Woese et al., 1980). Nevertheless, the fact that the whole microbial phylogenetic superstructure rests almost solely on this single molecular species is disquieting. It is, therefore, reassuring to 'see that as other molecules become characterized, they give results in agreement with the rRNA-defined taxa—for example, the unexpected grouping of (anaerobic) bacteroides and (aerobic) flavobacteria and their relatives predicted by rRNA similarities (Paster et al., 1985; Weisburg et al., 1985a) is confirmed by common idiosyncrasies in their ATPase sequences (Amann et al., 1988). However, it is also clear that this testing of the phylogenies produced from rRNAs by using other molecules needs to be done to a much greater extent than it now is.

At this writing roughly 500 (90–100% complete) 16 S rRNA sequences are known for prokaryotes, about 90% of which are (eu)bacterial sequences. In addition about 50 small subunit rRNA sequences exist for eukaryotes. The prokaryotic sequence collection appears representative of the bacteria (that have been cultured). The single most striking general characteristic of the universal phylogeny to emerge is that the higher taxa are clearly and easily defined in rRNA terms. In these cases no sophisticated analyses are required; positions of conserved composition, characteristic stretches of sequence, and idiosyncrasies in rRNA higher order structure are common to and readily define and distinguish the major taxa. However, when it comes to defining the branching order among major groups or to defining many of the lesser taxa within them, one is at the mercy of powerful and sophisticated tree construction algorithms, and can consequently fall prey to their vagaries.

THE DOMAINS AND BRANCHING ORDER OF THE UNIVERSAL TREE

The primary and strongest division of living systems is into the domains *Archaea*, *Bacteria* and *Eucarya* (Woese et al., 1990a). Each is readily recognized by its phenotypic characteristics and molecular (sequence) idiosyncrasies. In terms of the small subunit rRNA, gross secondary structural features readily distinguish among the three (Woese, 1987). Two particularly good examples of the distinction between bacteria and archaea are the regions between position 500–545 and 991–1045 in the small subunit rRNA. In the first of these (500–545), a bulge loop protrudes from the stalk of an otherwise normal stalk/loop structure (Woese et al., 1983). In all bacterial examples the bulge loop comprises six nucleotides and is situated between the fifth and sixth base pair of the stalk (of 12 pairs); however, in all archaeal sequences the loop comprises seven nucleotides and is situated between the sixth and seventh pair in the (12-pair) stalk (Woese et al., 1983; Woese, 1987). In the second example, all bacterial sequences (except the mycoplasmas and their relatives, where its form is idiosyncratic) show a bifurcating, Y-shaped (compound), secondary structure; two helices arising out of an underlying, third helix (Gutell et al., 1985; Woese,

1987). All archaeal sequences, on the other hand, show only a (complex) linear helical structure in this region (Gutell et al., 1985; Woese, 1987).

At a somewhat more refined level, the two domains can be distinguished by a number of lesser structural irregularities, such as individual "bulged" bases in helical regions (Woese, 1987). Examples here are the two adjoining helices 27–37·547–556 and 39–47·394–403. In every bacterial small subunit rRNA a "bulged" (unpaired) nucleotide occurs in each helix, at positions 31 and 397, respectively. However, the archaeal versions of these helices show no bulged bases (Woese, 1987).

Finally, the three domains are well defined and separated at the (linear) sequence level, in terms of the compositions of individual positions. In the small subunit rRNA approximately 300 positions show an invariant composition (or nearly so) in each of the three domains. [Composition is the same (and completely invariant) in all three domains for approximately 200 of these.] Table 1 shows a subset of 49 of these positions that defines and distinguishes among the three domains. The bacteria, archaea, and eucarya can be defined by 24, 10, and 20 of these positions, respectively. Assuming that the root of the universal tree is not *within* any of the three groups (see below), the table provides strong evidence for the monophyletic nature of the bacteria and eucarya; there are 23 and 17 positions, respectively, in which the domain in question shows one characteristic composition while the other two domains exhibit a common but different composition. Less, but still significant support exists for a monophyletic grouping of the archaea (8 positions). It is interesting that by the measure of Table 1, the archaea resemble the eucarya at least as much as they do the bacteria. This stands in sharp contrast to the impression one gains from overall evolutionary distances derived from the small subunit RNA as a whole, by which measure the archaea and bacteria are appreciably closer to one another than either is to the eucarya (Woese, 1987).

Given the sum total of comparative rRNA evidence, it can be said with assurance that each of the three domains has arisen from an ancestor whose rRNA was also characteristic of that domain. Since this same general statement can be made for many other molecular functions as well, none of the domains can be considered polyphyletic; each must have arisen from an ancestor whose molecular phenotype would have placed it within that taxon.

Recently it has become possible to determine the root of the universal tree, by comparing sequences from pairs of paralogous genes that duplicated when all extant life was still represented by a common lineage. [This rooting strategy (Schwartz and Dayhoff, 1980) in effect uses the one set of (aboriginally duplicated) genes as an outgroup for the other, thereby allowing a determination of the root of the universal tree.] Using either two (related) ATPase subunits or the related translation factors EFTu vs. EFG, the root of the universal tree has been fixed on the (eu)bacterial branch (Iwabe et al., 1989). If these results using two pairs of paralogous molecules truly represent what happened as the universal ancestor "broke up" into the three primary lineages, then the initial bifurcation

TABLE 1. Small subunit rRNA sequence signatures defining the three domains, archaea, bacteria, and eucarya[a]

Position(s) of base or pair[b]	Composition[c] in		
	Bacteria	Archaea	Eucarya
9:25	G:C†	C:G	C:G†
10:24	R:U	Y:R	Y:R*
32:552	A:U	A:U†	U:A†
33:551	A:U	Y:R	A:U
42:400	G:C†	G:C†	U:A†
53:358	R:Y	C:G	C:G
115:312	G:C†	C:G*	C:G
125:236	U:R†	U:R	G:C†
338	A	G	A
339:350	C:G	G:C†	C:G†
361	R	C	C*
365	U	A*	A
367:393	U:A	C:G†·*	U:A†
377:386	G:C†	C:G†	Y:R*
514:537	Y:R	G:C	G:C†
516:535	U:A	U:A	U:U*
523	A*	C*	A
549	C	U*	Y†
563	A	A	G†
567:883	G:C	G:C†	C:G†
575	G†	G	A*
585:756	R:Y	C:G†	U:A
675	A	U	U*
684:706	U:A†	G:C†	G:Y*
716	A	C	Y
783:799	Y:R†	C:G*	G:C†
785:797	G:C	G:C	Y:R

[a]Except where indicated, the compositions shown are invariant in each domain. Analysis based upon approximately 380 bacterial, 40 archaeal, and 50 eucaryal sequences. R=purine; Y=pyrimidine.
[b]Numbering follows E. coli 16 S rRNA convention (Brosius et al., 1978) and base pairing is according to Woese et al. (1983).
[c]Exceptions to strict constancy are noted as follows: † denotes one or a few (phylogenetically independent) exceptions to given composition, but a composition that differs from that characteristic of the other two domains; * denotes a single (phylogenetically independent) exception whose composition is the same as that characteristic of one of the other two domains; all positions not so marked show no exceptions to the given composition.

TABLE 1. *(Continued)*

Position(s) of base or pair[b]	Composition[c] in		
	Bacteria	Archaea	Eucarya
884	U	U	G*
912	C*	U*	U
923:1393	A:U**	G:C	A:U
927:1390	G:U	G:U	A:U
930:1387	Y:R	A:U*	G:C†
931:1386	C:G†	G:C	G:C†
940:1343	Y:R	R:Y*	R:Y
962:973	Y:G*,d	G:C	Y:G^d
966	G†	U	U
1063:1193	C:G	G:G	U:A
1064	G	G	Y
1094	G	G	U
1109	C*	A	A
1110	A†	G	G
1194	U†	G†	R
1211	U†	G	U†
1212	U†	A	A†
1381	U†	C	C
1413	A	A	U†
1419	G†,*	G†	U†
1430:1470	Y:R*	R:Y	R:Y*
1516	G†,*	G	U†
—	—	—	—

[d]The eubacterial version is always C:G except for one phylogenetically independent example of U:G and one of G:C, while the eukaryotic version is U:G except for one case of C:G.

of the universal tree split the bacterial lineage from the (then) common archaeal/eucaryal lineage. Thus, although the archaea are prokaryotes (cytologically), they bear no specific relationship to the other class of prokaryotes, the bacteria.

A specific (but distant) relationship between the archaea and the eucarya is actually not all that surprising. Studies of the archaea over the past decade have revealed an increasing number of properties they share specifically with the eucarya. And, for functions common to all three domains, the archaeal version is usually closer in sequence to its eucaryal homolog than either is to their bacterial homolog. One of the most impressive examples of this is the recent report by Reeve and coworkers of an archaeal histone (from *Methanothermus fervidus*) with a close relationship to eucaryal histones (Sandman et al., 1990). Not only is the archaeal histone specifically related to the eucaryal versions, but its sequence is closer to the (consensus) sequences of the *four* eucaryal histones (H2A, H2B, H3, and H4) than these four are to one another (Sandman et al., 1990).

The archaeal domain

The archaea are known to comprise four general phenotypes, the methanogens, the extreme halophiles, the sulfate-reducing archaea, and the extreme thermophiles (in whose metabolism sulfur and sulfur compounds play important roles; hence the name "sulfur-dependent archaebacteria" formerly used for them). However, these four distinct phenotypes do not define four separate phylogenetic groupings (of equivalent taxonomic rank). Rather, the four are phylogenetically intermixed in a way suggestive of particular evolutionary relationships among them.

Figure 1 shows an archaeal phylogenetic tree. Its root divides the domain into two major lineages, two kingdoms. One of these, *Crenarchaeota*, comprises a relatively phenotypically pure group of hyperthermophiles; the other, *Euryarchaeota*, represents a diverse collection of phenotypes and broad range of niches [which accounts for the prefix "eury-" in their name (Woese et al., 1990a)]. The euryarchaeotes exhibit examples of all the major archaeal phenotypes—three distinct clusters of methanogens (Balch et al., 1979), a group of (rather closely related) halophiles, and the lineages for *Thermoplasma* (a thermophilic mycoplasma), *Archaeoglobus* (a sulfate-reducing phenotype) and the *Thermococcus–Pyrococcus* cluster (typical hyperthermophiles). The *Thermococcus–Pyrococcus* unit represents the deepest branching among the euryarchaeotes (Woese, 1987; Achenbach-Richter et al., 1988); *Archaeoglobus* branches next (Achenbach-Richter et al., 1987a); followed by the three methanogen groups (Woese, 1987). The extreme halophiles are specifically related to (are a sister group of) one of the three methanogen clades, the *Methanomicrobiales* (Woese and Olsen, 1986). This combined grouping is then related at a deeper level to a second group of methanogens, the *Methanobacteriales*, with the *Thermoplasma* lineage (probably) arising somewhere within this combined unit (Woese and Olsen, 1986). In general form the archaeal tree has the property, mentioned above, that thermophilic lineages tend to cluster near the base and are much shorter than the lineages of their mesophilic counterparts.

Those aspects of the archaeal tree that are considered firmly established are (1) that all methanogens so far characterized (on the order of thirty species in all) fall into one of three distinct groups, (2) that the extreme halophiles are a sister group of one of them, i.e., the *Methanomicrobiales*, and (3) that the *Thermococcus–Pyrococcus* cluster is a member of the *Euryarchaeota* rather than the *Crenarchaeota* (in which their *phenotypic* counterparts reside). These rela-

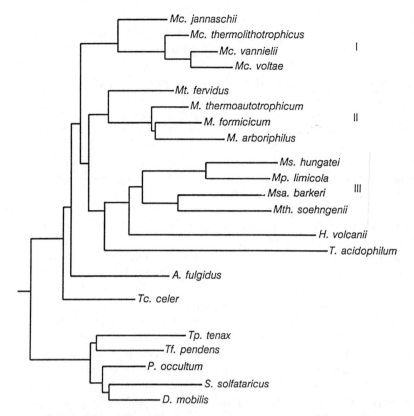

FIGURE 1. Archaeal phylogenetic tree. The tree was produced from an evolutionary distance matrix derived from an alignment of archaeal small subunit rRNA sequences. Only those positions represented by a known base in all sequences in the alignment were used in the calculation. Detailed descriptions of these procedures can be found in Woese (1987) or references cited therein. The root of the tree was established by the use of bacterial sequences as an outgroup. The *Crenarchaeota* lie below the root as shown, the *Euryarchaeota* above it. The three major methanogen groups are indicated by Roman numerals. Generic abbreviations: A., *Archaeoglobus*; D., *Desulfurococcus*; H., *Halobacterium*; M., *Methanobacterium*; Mc., *Methanococcus*; Mp., *Methanoplanus*; Ms., *Methanospirillum*; Msa., *Methanosarcina*; Mt., *Methanothermus*; Mth., *Methanothrix*; P., *Pyrodictium*; S., *Sulfolobus*; T., *Thermoplasma*; Tc., *Thermococcus*; Tf., *Thermofilum*; and Tp., *Thermoproteus*.

tionships are all defined by strong rRNA sequence signatures in addition to being given by the standard types of tree inference methods (Woese and Olsen, 1986; Woese, 1987; Achenbach-Richter et al., 1987a, 1988). For example, the placement of *Thermococcus* is demanded not only by the position of the root of the archaeal tree (Achenbach-Richter et al., 1988), but can be further rationalized by the relatively high overall similarity between its (16 S) rRNA sequence and those of the methanogens, by the possession of a general euryarchaeotal rRNA signature, and by other characteristic euryarchaeotal-specific properties, such as the presence of a tRNA gene in the spacer region of the rRNA operon (Woese, 1987; Achenbach-Richter and Woese, 1988).

Less firmly established is the specific relationship between the *Methanobacteriales* and the *Methanomicrobiales* (and extreme halophiles), shown in Figure 1. This branching is almost always seen with distance matrix analysis, and is supported by a small but significant signature (Woese and Olsen, 1986; Woese, 1987). However, parsimony analysis usually groups the *Methanobacteriales* specifically with the *Methanococcales* rather than with the *Methanomicrobiales* (unpublished analysis)—probably because this last lineage is relatively rapidly evolving, and is, therefore, artificially excluded. While the general position of the *Thermoplasma* lineage appears to be correct (Woese and Olsen, 1986), its exact location on the methanogen branch of the tree is somewhat problematic (its lineage too is rapidly evolving).

The rooting of the archaeal tree (determined by outgroups) has been extensively investigated (Achenbach-Richter et al., 1988). Bacterial and eucaryal lineages, taken separately or together, all intersect the unrooted archaeal tree on the same line segment (i.e., that shown in Figure 1). Moreover, it has been demonstrated that this root position is not the "random" root positioning, where distance or parsimony analysis would artificially tend to position a totally unrelated lineage on an unrooted archaeal tree. In fact, as noise (randomness) is progressively introduced into an outgroup sequence, its intersection with the unrooted archaeal tree shifts from the position shown in Figure 1 toward the dominant "random attractor," i.e., the *Methanomicrobiales*–extreme halophile cluster (Achenbach-Richter et al., 1988). Thus, the general area of the root of the archaeal tree appears firmly fixed.

The position of the *Archaeoglobus* lineage (Achenbach-Richter et al., 1987a) is not considered certain, however. Overall the A. *fulgidus* small subunit rRNA sequence is most similar to those of *Tc. celer* and its relatives, and vice versa (Achenbach-Richter et al., 1987a). Evolutionary distance analysis generally places the lineage where it is shown in Figure 1, but occasionally (with different makeups of the alignment) positions it specifically with the *Thermococcus* lineage or immediately above the *Methanococcus* lineage (unpublished analysis). Parsimony analysis places the lineage either as shown in Figure 1 or immediately above the *Methanococcus* lineage (unpublished analysis). The A. *fulgidus* small subunit rRNA exhibits an uncomfortably large number of the signature features characteristic of the *Methanomicrobiales* (P. Rouviere et al., in preparation). [The best example here is the possession of a U residue at position 1520. This particular composition is found in all members of the *Methanomicrobiales* so far characterized, but is found *nowhere else* among the archaea, eucarya, or bacteria (including mitochondrial sequences) (Rouviere et al., in preparation).] The small subunit rRNA of *Archaeoglobus* is relatively rich in G + C content (64%)—an additional reason to suspect the placement of its lineage given by distance and parsimony analyses. For comparison, the rRNAs of mesophilic methanogens are all less than 58% G + C, while that of *Thermococcus* is 66% G + C.

As mentioned above, the influence of composition on the outcome of phylogenetic analyses has not been adequately examined, and so good corrections for compositional effects have not been developed. However, rRNA sequences can be analyzed in a way that should be insensitive to these compositional differences. Although relatively broad spreads in G + C content can be seen among (prokaryotic) rRNAs, their total purine content remains almost constant (varying in archaeal examples between 55% and 57% A + G). Therefore, by confining analysis to the generic two-state situation (purines vs. pyrimidines), compositional differences should not be a significant factor.

Figure 2 is an archaeal tree based on these transversion distances. To a first

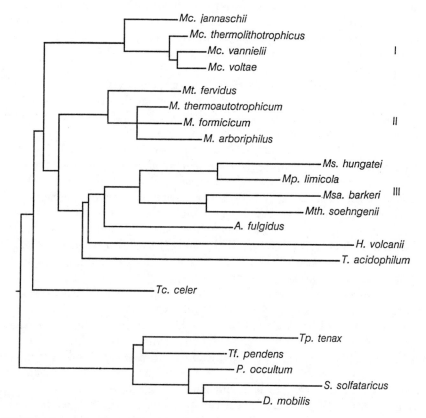

FIGURE 2. Archaeal tree based on transversion distances. Analysis was done as in Figure 1 except that A and G were treated as equivalent, as were C and U. The Olsen distance correction, which takes base ratio into account, was used (as it was in producing Figure 1; see Weisburg et al., 1989b for description).

approximation the tree resembles that of Figure 1; all the above-mentioned well-established features of the latter are characteristic of the former. The prominent difference between them, however, lies in the placement of the A. *fulgidus* lineage, which in Figure 2 shows the association with the *Methanomicrobiales*–extreme halophile group suggested by signature analysis (P. Rouviere et al., in preparation). Parsimony analysis of this same (two state) alignment also places the *Archaeoglobus* lineage on a common stem with the *Methanomicrobiales* and extreme halophile lineages (P. Rouviere et al., in preparation). Note also that in the transversion distance tree of Figure 2 the thermophilic lineages no longer appear as short relative to mesophilic lineages as they do in Figure 1, and that the firmly established groupings shown in Figure 1 are, if anything, even more distinct in Figure 2.

The distribution of phenotypes on the archaeal tree (Figure 2) suggests that the archaeal domain had a thermophilic, anaerobic, sulfur-metabolizing ancestor—i.e., only this phenotype is found in both archaeal kingdoms (Woese, 1987). Thermophilia is universal among the crenarchaeotes and common among the euryarchaeotes. It is noteworthy that mesophilic crenarchaeotes have yet to be found, although they have been extensively sought.

The name *Crenarchaeota* is meant to suggest that organisms of this general phenotype most resemble the common archaeal ancestor (Woese et al., 1990a). If so, it follows that crenarchaeotes should also resemble eukaryotes more than euryarchaeotes do; and this would seem the case. Take, for example, the ribosome and the organization of the rRNA genes. The crenarchaeotal translation apparatus shows a number "eukaryote-like" characteristics not seen in euryarchaeotes (except in some cases among the deep branching euryarchaeotes), such as (1) relatively high levels of modified bases in rRNA and tRNA (Woese et al., 1984), (2) ribosomes having relatively high protein:RNA ratios—found also among the deeper branching euryarchaeotes (Cammarano et al., 1986), (3) lack of a tRNA gene in the spacer region of the rRNA operon (Achenbach-Richter and Woese, 1988), and (4) no 5 S rRNA gene linked to the rRNA operon at its 3′ end (Leffers et al., 1987)—found also among the deeper branching euryarchaeotes (Noll, 1989).

A number of other crenarchaeal or general archaeal features can be cited to support a specific relationship between the archaea and eucarya: for example, the subunit structure and sequence(s) of DNA-dependent RNA polymerase (Schnabel et al., 1983), the nature of transcription signals (Hudepohl et al., 1990), a common sensitivity to diphtheria toxin (Kessel and Klink, 1980), and (as mentioned above) the presence of a true eukaryotic histone in archaea (Sandman et al., 1990). In both this last case and for the RNA polymerases (Puhler et al., 1989), the archaeal version appears closer to their presumed common ancestral version than are its eucaryal counterparts. This, of course, raises the interesting possibility that the archaeal phenotype, particularly the general crenarchaeal phenotype, resembles the phenotype of the common eukaryotic ancestor, and, therefore, archaea may prove of considerable value in trying to understand the nature and evolution of the eukaryotic cell.

The eubacterial domain

Figure 3 shows a representative (eu)bacterial phylogenetic tree; Table 2 lists the main bacterial groups, their major subdivisions, and representative genera therein. Although the tree has been produced by evolutionary distance matrix analysis, it is important to reiterate that in almost all cases the major bacterial groupings and their higher subdivisions can be defined simply by shared derived characters (signatures), either at the level of individual sequence positions or at the higher order structural level.

Of the 11 main bacterial groupings (together with their subdivisions) listed in Table 2, three—the purple bacteria, the Gram-positive bacteria, and the flavobacteria and relatives—account for the vast majority of (cultured) bacterial species. None of the names for the main bacterial groupings is properly descrip-

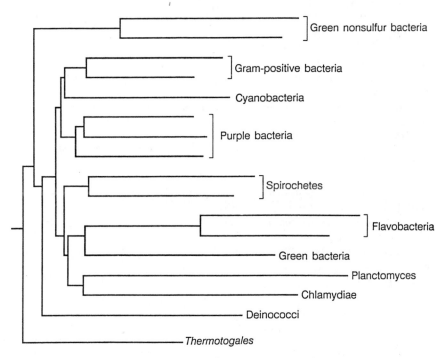

FIGURE 3. Bacterial phylogenetic tree. The tree was produced in the manner described in the Figure 1 caption. The alignment comprised 1–3 representative small subunit rRNA sequences from each of the 11 main recognized (eu)bacterial groups (Woese, 1987).

TABLE 2.   Partial listing of bacterial genera encompassed by the main bacterial taxa

---

Purple bacteria

α Subdivision
Azospirillum, Agrobacterium, Beijerinkia, Caulobacter, Ehrlichia, Rhodospirillum, Rhodomicrobium, Rhodopseudomonas, Rickettsia

β Subdivision
Alcaligenes, Chromobacterium, Eikenella, Neisseria, Nitrolobus, Nitrosomonas, Rhodocyclus, Spirillum, Vitreoscilla

γ Subdivision
Acinetobacter, Aeromonas, Chromatium, Coxiella, Escherichia, Ectothiorhodospira, Legionella, Pseudomonas, Wolbachia

δ Subdivision
Bdellovibrio, Cystobacter, Desulfotobacter, Desulfuromonas, Desulfosarcina, Desulfovibrio, Myxocoxxus, Nanocystis, Stigmatella

Gram-positive bacteria

Bacillus–Lactobacillus subdivision
Bacillus, Kurthia, Lactobacillus, Leuconostoc, Mycoplasma, Streptococcus

Gram-negative walled organisms
Sporomusa, Megasphaera, Selenomonas

High G + C subdivision
Arthobacter, Bifidobacterium, Corynebacterium, Frankia, Mycobacterium, Nocardia

True clostridia
Clostridium (most), *Eubacterium, Acetomaculum*

Flavobacteria and relatives
Bacteroides, Cytophaga, Flavobacterium, Flexibacter, Microscilla, Saprospira, Spirosoma

Cyanobacteria
Synecococcus, Oscillatoria, Spirulina, Plectonema, Fischerella, Nostoc

Spirochetes
Borrelia, Spirochaeta, Treponema, Leptospira

Green sulfur bacteria
Chlorobium, Prosthecochloris

Planctomyces and relatives
Planctomyces, Isocystis

Chlamydiae
Chlamydia

Radio-resistant micrococci and relatives
Deinococcus, Thermus

Green nonsulfur bacteria and relatives
Chloroflexus, Herpetosiphon, Thermomicrobium

Thermotogales
Fervidobacterium, Thermosipho, Thermotoga

---

tive: Purple (photosynthetic) bacteria constitute less than half the species of "purple bacteria"; a number of species without Gram-positive walls are encountered in the "Gram-positive" group (in addition to the wall-less mycoplasmas); and few members of "flavobacteria" fit that phenotype. In fact, no unique phenotypic description can adequately cover all (or even the vast majority of) species in most of the main bacterial groups. The (provisional) names used here and previously have been chosen to reflect the presumed ancestral phenotype of the group, which criterion, strangely, is not generally accepted by microbial taxonomists as a valid basis for naming. What follows is a brief characterization of each of the main bacterial groups.

The purple bacterial group has four major subdivisions, α, β, γ, and δ, with *Wolinella* and its relatives (the camphylobacters) making a possible fifth (Woese, 1987). The branching order among the four (or five) remains uncertain, except for the readily demonstrable specific relationship between the β and γ subdivision (Oyaizu and Woese, 1985; Woese, 1987). (However, evidence is accumulating to the effect that β and γ cluster with α, at a deeper level, to the exclusion of the others.) These three subdivisions all contain purple photosynthetic species, from which fact the ancestral phenotype of the group has been inferred.

The α subdivision contains many photosynthetic species (*Rhodospirillum, Rhodospeudomonas*, etc.) in addition to numerous nonphotosynthetic derivatives of these (Woese, 1987). The subdivision is notable for its many species that have intracellular (or close) relationships to eukaryotic cells—e.g., *Agrobacterium* and *Rhizobium* with plants (Yang et al., 1985), and the rickettsias and ehrlichias with animals (Weisburg et al., 1985b; 1989a). At some very early time it was a member(s) of this subdivision that entered the evolving eukaryotic cell to become the protomitochondrion (Yang et al., 1985). Photosynthetic representation in the β subdivision is confined to a few species of *Rhodocyclus*. (However, as now defined, the genus *Rhodocyclus* is polyphyletic.) Its mainly nonphotosynthetic representation includes species of *Alcaligenes, Chromobacterium, Nitrobacter, Nitrosolobus, Vitreoscilla, Spirillum, Neisseria*, and others (Woese, 1987). Some thiobacilli are peripherally associated with the β subdivision. At present γ has the largest representation of the four main purple bacterial subdivisions. Photosynthetic species are relatively minor in occurrence in γ, being confined to *Chromatium* and its relatives and the ectothiorhodospiras (Woese, 1987). The subdivision comprises a variegated collection of metabolisms and niches. It includes the enteric bacteria and their relatives, the fluorescent pseudomonads and their relatives, the oceanospirilla, some beggiatoas and others. The δ subdivision comprises an unexpected grouping of phenotypes, none photosynthesic, i.e., various sulfur and sulfate-reducing species (e.g., *Desulfovibrio* and *Desulfuromonas*), the myxobacteria and relatives, and the bdellovibrios (Hespell et al., 1984; Woese, 1987).

The Gram-positive bacteria do not structure along classical taxonomic lines, into high and low G + C groups. The former is a coherent phylogenetic unit, the branching within which is remarkably shallow (recent), especially given the great phenotypic variety the subdivision contains (Woese, 1987). However, the latter is not; it encompasses a number of subdivisions, each as deep or deeper (more ancient) than the high G + C Gram-positive subdivision. Among these are a clostridial lineage, which contains many but not all of the *Clostridium* species, in addition to a number of nonclostridial (but Gram-positive) phenotypes; a bacillus–lactobacillus lineage, which contains, in addition to species of these two types, the streptococci and the mycoplasmas [the latter a group that also includes a collection of misnamed specific walled relatives (Weisburg et al., 1989b)]; a lineage comprising various species such as *Sporomusa, Megasphaera*, and others, which have non-Gram-positive walls (Stackebrandt et al., 1985); and a lineage of photosynthetic bacteria of the *Heliobacterium* type, also having non-Gram-positive walls (Woese et al., 1985). A little known lineage of anaerobic halophiles, e.g., *Haloanaerobium* (Oren et al., 1984; Woese, 1987), would qualify as a peripheral member of the group, as may the fusobacteria and their relatives (Woese et al., unpublished).

The third most extensive bacterial group, the so-called flavobacteria and relatives, is made up of species whose classification has proven problematic by traditional methods. In the main the group is drawn from the "genera" *Flavobacterium, Flexibacter, Cytophaga*, and *Bacteroides* (Paster et al., 1985). The relationship of the first three (all aerobic) to the anaerobic *Bacteroides* went totally unrecognized until fortuitously detected by rRNA analysis. Arguments as to whether various species are flavobacteria, flexibacteria, or cytophagas have been going on for decades. It is not surprising that all three "genera" turn out (on rRNA analysis) to be polyphyletic assemblages; their various species intermix phylogenetically (Woese et al., 1990b). Although no photosynthetic members of the group are known, the group as a whole bears a sister relationship to the (photosynthetic) green sulfur bacteria (Woese et al., 1990c), the two together making what will probably be defined ultimately as a kingdom within the domain *Bacteria*. (The purple bacteria and Gram-positive bacteria should also be accorded kingdom status.)

The fourth largest group, the cyanobacteria, contains only photosynthetic members, all of the same type (Giovannoni et al., 1988). This is somewhat surprising in that classical bacteriologists felt certain that the morphological resemblances between the blue-green algae (cyanobacteria) and certain bacteria such as the beggiatoas and thiobacilli (which are now known to be purple bacteria) made the latter types colorless variants of the former—one more example of the problems that result from taking morphology as the primary indicator of microbial phylogenetic relationships. Strong, but not completely convincing evidence exists that the cyanobacteria are specifically related at a deep level to the Gram-positive bacteria (Woese, 1987). This is consistent with the fact that chlorophyll a (in cyanobacteria) is most similar in structure to heliobacterial chlorophyll g (Gest and Favinger, 1983).

The only other main bacterial group that has appreciable phylogenetic depth

and representation is the spirochetes. The unit comprises spirochetes, treponemes, borrelias, and leptospiras. However, as classically defined these units
are not phylogenetically proper taxa; some *Spirochaeta* species, for example,
cluster with the treponemes, and one species of treponeme, *T. hyodysenteriae*,
represents a lineage distinct from the combined spirochete–treponeme cluster
(Woese, 1987; B. Paster et al., in preparation). The leptospiras (which cover
the genera *Leptospira* and *Leptonema*) are the deepest branching lineage in the
group. The spirochetes is one of the few major bacterial groups wherein all or
almost all members share some common phenotypic trait (Woese, 1987); the
other is the cyanobacteria.

The remaining six major bacterial lineages are only poorly represented by
cultured (characterized) species, so the groups are not necessarily well understood. The green sulfur bacteria comprises only photosynthetic species (of the
Chlorobium type) and so far seems a phylogenetically shallow unit (Gibson et
al., 1985; Woese, 1987). As mentioned above, the unit is related at a deep
level to the flavobacteria and relatives. The green nonsulfur bacteria contain
only three known genera, *Chloroflexus*, *Herpetosiphon*, and *Thermomicrobium*
(Oyaizu et al., 1986). The phylogenetic position and ancestral phenotype of the
group are of particular importance, for they bear on whether or not photosynthesis evolved ancestrally (or very early) among the bacteria. The deinococcus
group comprises various cocci and noncocci that share a characteristic type of
cell wall (Brooks et al., 1980), in addition to the genus *Thermus*. A phenotypic
connection between these two very different phenotypes is not evident at present.
The planctomyces group contains in addition to *Planctomyces* and related
genera, the genus *Isosphaera*, a thermophile (Woese, 1987). The chlamydia
group is so far confined to a set of closely related species of medical interest
(Weisburg et al., 1986). At a deep level this group appears to show a sister
relationship to the planctomyces group, which can perhaps be phenotypically
rationalized by the fact that all show unusual types of cell walls (Weisburg et
al., 1986).

The base of the bacterial tree is at present defined by a lineage known to
produce only thermophilic offshoots, called the *Thermotogales* (Achenbach-
Richter et al., 1987b; Huber et al., 1989). The lineage is relatively short, and
its branching in the overall bacterial tree does not seem to be an artifact resulting
from a rapid evolutionary rate (Achenbach-Richter et al., 1987b). There is at
present a debate (as yet largely unpublished) among workers in bacterial phylogeny as to whether the planctomyces lineage branches more deeply even than
the thermotoga lineage. Although the matter remains to be definitively resolved,
I am convinced of the approximate placement of the planctomyces lineage
where shown in Figure 3 for two reasons: (1) The planctomyces lineage gives
evidence of being rapidly evolving, which would tend to force its placement
artificially deep in the tree; and (2) when the planctomyces lineage appears
deeply in the tree its apparent sister relationship to *Chlamydia* (Weisburg et al.,
1986) disappears. Thus, resolution of this problem would seem to lie in estab-

lishing beyond a reasonable doubt that the chlamydial and planctomyces lineages are sister groups. If so, the former (which appears the less rapidly evolving of the two, and which alone does not vary greatly in its position in the bacterial tree) is the more reliable indicator of the phylogenetic position of both.

CONCLUSION

Evolutionary development in the two prokaryotic domains is similar in some respects, though not in all. In both cases, thermophilic species predominate in the lower reaches of the tree. Among the bacteria, however, there is also a broad distribution of photosynthetic phenotypes, whereas archaeal photosynthesis is strictly confined to the halobacteria, a relatively superficial branching. Five of the 11 major bacterial taxa contain photosynthetic species, each of a different type. Although the deepest bacterial branching (the *Thermotogales*) is not known to have photosynthetic representation, the next deepest one (*Chloroflexus*) does. Autotrophic phenotypes appear mixed with heterotrophs at all levels of the bacterial and archaeal trees.

Given these facts and the realization that the first branching in the tree of life (the first speciation event) involves the bacteria on the one hand, and the common archaea–eucarya lineage on the other, it is reasonable to suggest that life arose under thermophilic conditions, that (chlorophyll-based) photosynthesis arose early in the bacterial line of descent, and that perhaps life arose autotrophically. In any case it can be said that the long-standing prejudice that heterotrophy precedes autotrophy in evolution (the so-called "Oparin-ocean" view that life arose heterotrophically) receives no support from what we now know of microbial phylogeny.

CHAPTER 2

EVOLUTIONARY GENETICS OF

SALMONELLA

Robert K. Selander, Pilar Beltran, and Noel H. Smith

In these days of transformism it need hardly be pointed out that any classification which does not express an evolutionary idea is nothing but an arbitrary if convenient method of pigeon-holing facts.
—P. B. WHITE (1926)

Research on the biochemistry, physiology, ecology, and pathogenicity of bacteria has, with a few notable exceptions (Carlile and Skehel, 1974), been conducted with relatively little concern for population genetics or evolutionary processes. Although molecular methods of measuring genetic variation in populations, analyzing population structure, and inferring phylogenies have been routinely applied to eukaryotes since the 1960s, microbiologists have continued to characterize bacterial strains and infer their genetic relationships largely on the basis of physiological and serological characters (Goodfellow and Board, 1980; Krieg and Holt, 1984; Ewing, 1986; Barker and Old, 1989; Threlfall and Frost, 1990). However, in most cases, variation in these phenotypic characters cannot presently be interpreted in terms of allelic substitutions at individual gene loci. Moreover, some of these characters are controlled by plasmid-borne genes rather than chromosomal genes, and many of them are highly variable in expression, evolve rapidly, and are subject to convergence under the pressure of natural selection in response to environmental conditions. In particular, the cell-surface characters (polysaccharide and protein antigens, fimbriae, outer-membrane proteins, and phage attachment sites) that are widely used to classify strains of bacteria are, for adaptive reasons, likely to be among the most labile features of the organism (Silverman and Simon, 1980; Riley and Sanderson, 1990).

The critical shortcoming of conventional bacteriological studies of phenotypic characters is that they do not provide data for population genetic analysis, which requires information on allelic variation at specific gene loci. For this reason, there has long been a critical need for application of the laboratory methods, theory, and statistics of molecular population genetics to bacteria to achieve an understanding of genetic diversity, population structure, and evolutionary processes and relationships within and among species.

Bacterial population genetics has recently emerged as an area of considerable interest and activity, largely through the efforts of workers who have come to microbiology from backgrounds in the evolutionary genetics of eukaryotes (Milkman, 1973; Levin, 1981; Whittam et al., 1983; Hartl and Dykhuizen, 1984; Selander et al., 1987; Young, 1989; Selander and Musser, 1990). By reason of their great phenotypic and genetic diversity, haploid genomes, adjunct episomal genetic systems, and short generation periods, bacteria are attractive subjects for both descriptive and experimental studies of evolutionary processes. Since 1980, our laboratory has been studying the molecular population genetics of the major human pathogenic bacteria, including species of *Escherichia*, *Salmonella*, *Legionella*, *Neisseria*, *Haemophilus*, *Bordetella*, *Streptococcus*, and *Staphylococcus* (reviews in Selander et al., 1987; Selander and Musser, 1990; Musser et al., 1990; Selander and Smith, 1990), as well as soil bacteria of the genera *Pseudomonas* (Denny et al., 1988) and *Rhizobium* (Pinero et al., 1988; Eardly et al., 1990).

This chapter reviews current research on the genetic diversity, population structure, evolutionary dynamics, and phylogenetic relationships of bacteria of the genus *Salmonella*, and particularly those forms causing human typhoid and other enteric fevers. This work illustrates the ways in which a population genetic approach is contributing information and insights of value to students of both molecular evolution and bacterial pathogenesis.

PATHOGENICITY AND EPIDEMIOLOGY

Salmonella belongs to the large, medically important eubacterial family Enterobacteriaceae and is related to *Escherichia coli* (Krieg and Holt, 1984; Ochman and Wilson, 1987). But unlike that familiar species, most strains of which are harmless commensals of mammals, the salmonellae are ubiquitous pathogens of both warm-blooded and cold-blooded vertebrates (Rubin and Weinstein, 1977).

Diseases caused by *Salmonella*

Salmonellae cause four disease syndromes in humans, mammals, and birds (Hook, 1979; Timoney et al., 1988). Gastroenteritis occurs when bacteria invade the epithelium of the small intestine and stimulate extensive fluid secretion.

The resulting diarrhea is normally self-limiting in a few days but may progress to bacteremia (septicemia), particularly in adults with carcinoma or other serious underlying illness. About 15% of cases of bacteremia are followed by focal infections (localized tissue infections), such as osteomyelitis, meningitis, and pneumonia. And finally, some strains produce the serious invasive diseases known as enteric fevers (typhoid and paratyphoid fevers).

Typhoid fever is caused by *Salmonella typhi*, which infects only humans and is spread by water and food contaminated with human feces. Bacteria penetrate the intestinal mucosa and are taken up by mononuclear phagocytes in Peyer's patches in the ileum. But instead of being killed by these cells, the bacteria multiply within them and, subsequently, reach the mesenteric lymph nodes, liver, and spleen. Then, escaping into the bloodstream, they cause a sustained secondary bacteremia, which ushers in the long febrile phase of the disease, with varied other symptoms such as abnormal liver function and bradycardia. This stage may be followed by necrosis of the gallbladder or Peyer's patches, which causes hemorrhage and perforation of the bowel. Mortality is about 20% if untreated with antibiotics, but only 1% if treated. Following typhoid fever, 2% of patients become asymptomatic chronic carriers, as bacteria living in the gallbladder are secreted into the intestine with the bile. Carriers may excrete 10^6 cells per gram of feces. In Santiago, Chile, where *S. typhi* is endemic, there are an estimated 30,000 chronic carriers.

Paratyphoid fevers are essentially the same disease syndrome as typhoid fever but with generally milder symptoms. They are caused in humans by *S. paratyphi* A, *S. paratyphi* B, *S. paratyphi* C, and *S. sendai*, and in animals by other host-adapted forms of *Salmonella*.

Frequency of disease

Typhoid fever was once a common disease in the United States. For example, 1000 people died of it in Philadelphia in 1899 as a result of drinking river water. And through the early part of this century, typhoid continued to touch the lives of many people. In 1912, a promising young songwriter named Irving Berlin married Miss Dorothy Goetz, who contracted the disease while honeymooning in Cuba and died. In her memory, Berlin wrote "When You Left Me" and some of his other great songs.

As a result of improvements in sanitation and the treatment of chronic carriers, typhoid fever has steadily declined in frequency in the United States since 1900, to a current level of about 500 cases per year (Chalker and Blaser, 1988), 75% of which can be traced to international travel (Ryan et al., 1989). Still, there are occasional outbreaks, such as 222 cases in a migrant farm-worker camp in Dade County, Florida, in 1973 (Hoffman et al., 1975).

In many developing countries, typhoid fever is endemic and epidemic (Hornick, 1985), and the recent emergence of antibiotic resistant strains of *S.*

typhi has made treatment and control increasingly difficult (Goldstein et al., 1986; Robbins and Robbins, 1984). Excluding China, for which data are not available, there are 13 million cases of typhoid fever each year, with a frequency of 5.4 cases per 1000 persons in developing countries (Edelman and Levine, 1986). In Indonesia, the incidence is twofold greater and typhoid is among the five major causes of death. Closer to home, there was a severe epidemic in Mexico in the early 1970s, which resulted in 80 cases in travelers from the United States (Baine et al., 1977).

For *Salmonella* organisms causing nontyphoidal human disease, domesticated animals are the primary reservoir, although infection is also transmitted person to person by the fecal-oral route. Contaminated poultry and poultry products are responsible for 50% of all outbreaks and epidemics of gastroenteritis in the United States, and contamination of beef and pork causes 13% of outbreaks.

The frequency of gastroenteritis has actually increased in the United States in recent years to a current estimated 3 million cases per year, 67% of which involve children (Chalker and Blaser, 1988). This higher incidence of human disease reflects a corresponding increase in the frequency of infection among domesticated animals. Animal feeds are supplemented with unsterilized bonemeal and other by-products of the meat packing industry, and crowding of animals in feeding lots and holding pens before slaughter spreads infection. Consequently, processing equipment often becomes contaminated from the carcasses of infected animals. *Salmonella* has frequently been in the news in recent years (Table 1).

SEROLOGICAL CLASSIFICATION OF STRAINS

For more than half a century, the salmonellae have been classified and identified largely on the basis of their cell-surface antigenic properties according to a serotyping scheme developed by F. Kauffmann and P. B. White (Kauffmann, 1966; Le Minor, 1984; Ewing, 1986). The primary antigens used are those of the somatic lipopolysaccharide (O antigens) and the flagellin protein (H1 and H2 antigens). Each distinctive combination of O, H1, and H2 antigens is recognized as a serotype or serovar (Table 2), of which more that 2200 have been distinguished. Most serovars have been assigned Latin binomial species names (e.g., *Salmonella enteritidis* and *S. dublin*), but some are designated simply by antigen formula. In general, each serovar has a unique set of antigens; but in several cases, two or three serovars with the same antigen formula have been recognized because of apparent differences in host range or disease syndrome produced (e.g., see *S. sendai* and *S. miami* in Table 2) or because of biochemical differences or minor antigenic variation not detected in routine serotyping (e.g., *S. choleraesuis*, *S. typhisuis*, and *S. decatur*).

TABLE 1. Representative recent large outbreaks and epidemics of salmonellosis

Region	Year	Serovar	Contaminated vehicle	Reference
International	1985–present	*S. enteritidis*	Grade A chicken eggs[a]	St. Louis et al. (1988)
United States and Canada	1973–74	*S. eastbourni*	Chocolate[b]	Craven et al. (1975)
Norway and Finland	1987	*S. typhimurium*	Chocolate[c]	Kapperud et al. (1989)
California	1985	*S. dublin*	Unpasteurized milk[d]	Fierer (1983)
Arizona	1983	*S. typhimurium*	Unpasteurized milk	Tacket et al. (1985)
New Jersey and Pennsylvania	1981	*S. newport* and *S. typhimurium*	Precooked roast beef[e]	Riley et al. (1983)
California	1985	*S. newport*	Hamburger[f]	Spika et al. (1987)
United States	1985	*S. muenchen*	Marijuana[g]	Taylor et al. (1982)

[a]Bacteria present inside intact eggs as a result of ovarian transmission.
[b]Traced to a chocolate factory in Canada.
[c]Traced to a chocolate factory in Norway. Although only 349 bacteriologically verified cases of diseases were recorded, it is estimated that the outbreak involved 40,000 cases, mostly in children (G. Kapperud, personal communication).
[d]Caused a number of deaths among AIDS patients.
[e]Traced to precooked roast beef distributed by a meat processing plant in Philadelphia.
[f]Caused a number of deaths among persons receiving antibiotic therapy before becoming infected. The contaminating strain was traced to several meat processors and slaughter houses back to several dairy farms from which animals had been sent for slaughter.
[g]Traced to marijuana plants fertilized with contaminated cow manure.

TABLE 2. Antigenic formulas of selected *Salmonella* serovars

Serotypic name	Somatic (O)[a]	Antigens Phase-1[b] (H1)	Phase-2[b] (H2)
S. typhi[c]	9,12,[Vi]	d	—
S. paratyphi A	1,2,12	a	[1,5]
S. sendai	1,9,12	a	1,5
S. miami	1,9,12	a	1,5
S. panama	1,9,12	l,v	1,5
S. paratyphi B	1,4,[5],12	b	[1,2]
S. java[d]	1,4,[5],12	b	[1,2]
S. typhimurium	1,4,[5],12	i	1,2
S. heidelberg	1,4,[5],12	r	1,2
S. rubislaw	11	r	e,n,x
S. saintpaul	1,4,[5],12	e,h	1,2
S. derby	1,4,[12]	f,g	—
S. enteritidis	1,9,12	g,m	—
S. dublin	1,9,12,[Vi]	g,p	—
S. paratyphi C	6,7,[Vi]	c	[1,5]
S. choleraesuis	6,7	[c]	1,5
S. typhisuis	6,7	[c]	1,5
S. decatur	6,7	c	1,5
S. newport	6,8	e,h	1,2
S. muenchen	6,8	d	1,2
S. saliantis[e]	4,12	d,e,h	d,e,n,z_{15}
S. sandiego	4,12	e,h	e,n,z_{15}
"Arizona" serovar in subspecies IIIb	$48_{1,3,4}$	i	z

[a]Underlining of the O antigen 1 indicates that its expression is connected with phage conversion. Antigenic factors in brackets may be absent. Vi is the capsular virulence antigen.

[b]The genes encoding the phase 1 and phase 2 flagellins, long known as H1 and H2, were recently redesignated as *fliC* and *fliB*, respectively.

[c]Strains can sometimes be induced to express a phase 1 flagellar antigen j by growth in anti-d serum; and strains normally expressing the j antigen occur in natural populations in Indonesia. Some Indonesian isolates also are biphasic, expressing a z_{66} flagellar antigen that presumably is encoded by an H2 locus.

[d]Sometimes classified as *S. paratyphi B* variety *java* or combined with *S. paratyphi B*.

[e]A triphasic serovar, so-called because growth in medium containing anti-d serum leads to permanent loss of the ability to express the d antigen.

Somatic O antigens

The O antigen is a polysaccharide that covers much of the surface of the cell. It is typically a linear polymer, with a repeating oligosaccharide unit of three to six sugar residues, that is linked through an oligosaccharide core to lipid A, the whole comprising the lipopolysaccharide (LPS). Antigenic type is determined by the character and arrangement of the sugar residues involved, and about 40 types have been recognized (Jann and Jann, 1984). Variation in the O antigen is largely mediated by genes of the *rfb* cluster, which is located at 42 minutes on the *Salmonella* chromosome (Makela and Stocker, 1984; Sanderson and Roth, 1988). (See "Evolution of somatic polysaccharide antigens.") However, the structure of at least one antigen, O54, which occurs in eight serovars, is determined by plasmid-borne genes (Popoff and Le Minor, 1985).

Some O antigens, notably antigen O1, are subject to lysogenic conversion, a process in which infection with certain phages causes changes in the antigens expressed (see discussion in Ewing, 1986). As long as the cell remains lysogenic (i.e., while the converting phage is present), the particular O antigen is expressed; but when the phage is lost, the antigen cannot be detected.

Flagellar antigens and phase variation

Most strains of *Salmonella* have two flagellin genes (H1 and H2) that encode proteins with different antigenic factors (diphasic condition). But many strains have only one active gene (usually H1) and consequently, produce only one type of flagellin (monophasic condition); and strains of two serovars, *S. gallinarum* and *S. pullorum*, are aflagellate and, hence, nonmotile.

In diphasic strains of *Salmonella*, the alternate expression of the H1 and H2 structural genes results in an oscillation of phenotype known as phase variation (Lederberg and Iino, 1956; Silverman and Simon, 1980; Simon et al., 1980; Kutsukake and Iino, 1980), which is a classic example of the regulation of gene expression by site-specific recombination (Smith, 1985). The frequency of phase transition ranges in different strains of *Salmonella* from 10^{-3} to 10^{-5} per bacterium per generation (Stocker, 1949). The mechanism involves recombinational inversion of a 900-base pair (bp) controlling DNA segment located adjacent to the H2 gene (Figure 1). The invertible segment carries a promoter sequence close to one of the two crossover points, and phase transition is achieved as inversion alternately associates and disassociates this promoter and the H2 structural gene. The controlling segment also contains a gene (*hin*) encoding a substance that greatly increases the frequency with which inversion events occur (Figure 2). [Remarkably, this gene shares structural and functional homology with the *gin* gene of phage Mu and the *tnpR* gene of transposon Tn3, which strongly suggests derivation from a common ancestor (Simon et al., 1980)]. Still another gene (*rh1*), which is linked to and coordinately ex-

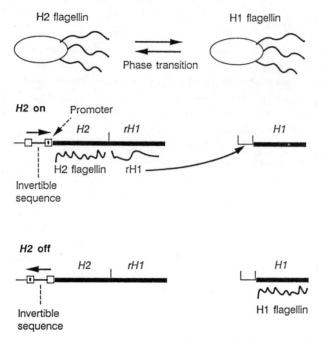

FIGURE 1. The mechanism of flagellar phase transition in diphasic serovars of *Salmonella*. Alternate expression of the H1 and H2 flagellin genes is achieved by site-specific inversion of a 900-bp controlling segment that carries a promoter at one end. In the "H2 on" configuration, both the H2 flagellin gene and the rH1 repressor gene are expressed; expression of the H1 flagellin gene is blocked by the product of the rH1 gene. In the "H2 off" configuration, inversion of the controlling segment gene dissociates the promoter from the H2 and rH1 loci (see Figure 2), leading to expression of the H1 flagellin gene. After Silverman and Simon (1980).

pressed with the H2 gene, codes for a repressor that prevents expression of the H1 gene (Figure 1).

Vi capsular antigen

The virulence antigen (Vi antigen), a polymer of galactosaminuronic acid that forms a capsule on the surface of the bacterial cell (covering the LPS), occurs in isolates of *S. typhi*, *S. paratyphi* C, and *S. dublin*, and, curiously, in some isolates of *Citrobacter freundii*, another member of the Enterobacteriaceae (Ewing, 1986). The capsule is presumed to provide protection against the immune defenses of the host and has been implicated as a virulence factor in *Salmonella typhi* (Robbins and Robbins, 1984). Expression of the Vi antigen is mediated by two widely separated chromosomal gene clusters, *viaA* and *viaB* (Johnson et al., 1965; Snellings et al., 1981). (See "Distribution and expression of Vi capsular antigen.")

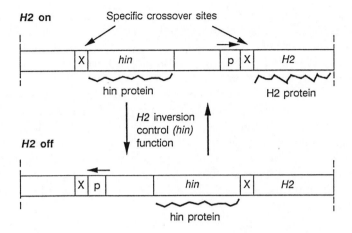

FIGURE 2. Structure and orientation of the 900-bp invertible sequence controlling flagellar phase variation in biphasic serovars of *Salmonella*. p is the promoter sequence for the H2 gene, and the two X's mark crossover points involved in site-specific recombination. The *hin* gene codes for a substance that increases the frequency with which inversion events occur by at least three orders of magnitude. After Silverman and Simon (1980).

METHODS OF STUDYING BACTERIAL POPULATION GENETICS

Multilocus enzyme electrophoresis

There are 3.5 million bp—enough for 3000 genes—in the chromosome of a bacterium such as *Salmonella* (Sanderson and Roth, 1988). Population genetic research requires a method for rapidly indexing allelic variation in a representative sample of these genes among hundreds or thousands of strains. The method of choice has been multilocus enzyme electrophoresis (MLEE) (Selander et al., 1986), which in practice has been used to measure allelic diversity in 20 to 30 randomly selected structural genes of the chromosomal genome by indexing variation in the primary structure (amino acid sequence) of gene products—polypeptides. The key concepts underlying application of this technique are that electromorphs (mobility variants) of a protein can be equated with alleles at the corresponding structural gene and that electromorph profiles for multiple proteins (electrophoretic types, ETs) can be equated with multilocus genotypes. For *Salmonella* and other bacteria, this technique has opened the door to population genetics by providing a basis for estimating levels of single-locus and multilocus genotypic variation in populations and species and for deducing the genetic structure of natural populations by measuring the extent of linkage disequilibrium (nonrandom associations of alleles). Multilocus enzyme electrophoresis also yields estimates of genetic distance between strains and makes possible the reconstruction of the phylogenies of cell lineages and species. A special advantage of the method is that much of the allelic variation detected is selectively neutral or nearly so (Whittam et al., 1983; Kimura, 1983; Hartl et al., 1986) and, hence, minimally subject to convergent evolution in separate phylogenetic lineages.

With the recent advent of the polymerase chain reaction (PCR) technique and continuing improvement in the methodology of nucleotide sequencing, the day is rapidly approaching when the complete assessment of allelic variation in multiple genes in natural populations of bacteria will be possible. (See "Evolution of flagellin genes.") However, multilocus enzyme electrophoresis is presently the most efficient and productive way of indexing sequence variation in the large numbers of genes and isolates required for population genetic analyses.

Other methods of assessing genetic variation

Other methods of studying genetic variation that have been useful in epidemiology but to date have had little or no application in bacterial population genetics include restriction fragment length polymorphism (RFLP) analysis of chromosomal and plasmid DNA (Wachsmuth, 1986; Mayer, 1988; Stull et al., 1988; Altwegg et al., 1989), pulsed-field electrophoresis of DNA (Arbeit et al., 1990), and assessment of number and distribution of insertion sequences (Sawyer et al., 1987a; Lawrence et al., 1989). [For a recent population genetic analysis of the evolution of transposable elements, see Condit (1990).]

GENETIC VARIATION AND POPULATION STRUCTURE IN SOME COMMON SEROVARS

Our first population genetic study of *Salmonella* (Beltran et al., 1988) addressed the following questions: Genetically, what is a serovar? Does serotypic identity of strains indicate overall genetic identity or similarity? Is the population structure of *Salmonella* clonal, with most cases of disease being caused by a small proportion of existing clones, as we have found for other types of pathogenic bacteria (Selander and Musser, 1990)?

In this study, 1527 isolates of eight common serovars from worldwide sources were analyzed for allelic variation at 23 enzyme loci (Table 3), and 71 distinctive multilocus genotypes (ETs) were recognized (Figure 3). The analysis demonstrated that populations of all serovars are genetically variable, being represented by multiple ETs. For some serovars (e.g., *S. typhimurium* and *S. choleraesuis*, all ETs are closely related (monophyletic serovars). But for other serovars, notably *S. derby* and *S. newport*, the same antigen profile occurs in strains of ETs belonging to two or more branches of the dendrogram (polyphyletic serovars). And organisms that are closely similar in multilocus genotype may have

TABLE 3. Number of ETs represented by isolates of eight common *Salmonella* serovars

Serovar	Antigen formula	Number of		Percentage of isolates of commonest ET
		Isolates	ETs	
S. *choleraesuis*	6,7,:c:1,5	85	6	88
S. *dublin*	1,9,12;g,p:–	115	4	89
S. *enteritidis*	1,9,12;g,m:–	257	14	93
S. *typhimurium*	1,4,5,12:i:1,2	299	17	83
S. *heidelberg*	1,4,5,12:r:1,2	204	8	87
S. *infantis*	6,7:4:1,5	113	4	96
S. *newport*	6,8:e,h:1,2	105	13	
Division I		72	7	38
Division II		32	3	84
Unassigned		1	1	
S. *derby*	1,4,12:f,g:–	349	6	
Division I		71	2	93
Division II		267	3	61
Division III		11	1	100

Source: Beltran et al. (1988).

different serotypes, as, for example, S. *typhimurium* and S. *heidelberg*, which have different phase 1 flagellar antigens (Table 3).

Our discovery that identity of serotype does not necessarily indicate genetic identity, or even similarity, or strains clearly invalidates the Kauffmann-White scheme as a method of biological classification. To explain the occurrence of the same serotype in distantly related phylogenetic lineages, we invoked the horizontal transfer and recombination of antigen-determining genes (Beltran et al., 1988), but in the absence of gene sequence data, phenotypic convergence in antigenic factors could not be ruled out. (See "Evolution of flagellin genes.")

The occurence of strong linkage disequilibrium among alleles of metabolic enzyme loci, the association of each antigen profile with a small number of lineages (Figure 3), and the global distribution of the common ETs clearly indicate that the genetic structure of *Salmonella* populations is clonal; and for each serovar, most disease is caused by one or a few clones of global distribution (Table 3). These findings explain why the Kauffmann-White scheme has been useful in epidemiology. Clonality means that recombination of chromosomal genes among cell lineages is very infrequent, occurring at a rate well below that required to randomize genes in chromosomal genomes.

Among the polyphyletic serovars, strains in different divisions (major lin-

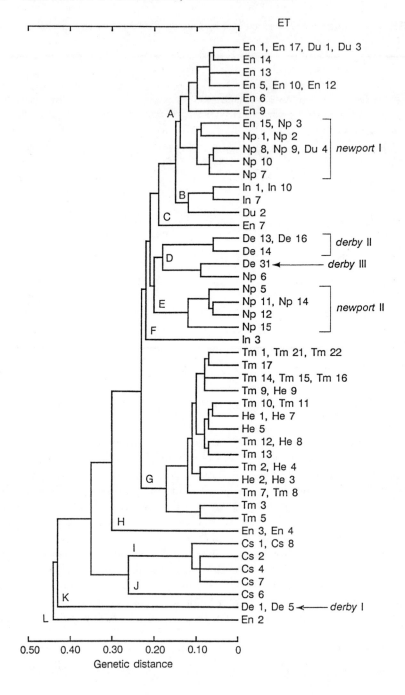

◀ **FIGURE 3.** Dendrogram showing estimated genetic relationships among 71 ETs, representing clones, of eight common serovars of *Salmonella*, based on electrophoretically demonstrable allelic variation at 23 chromosomal enzyme loci. The dendrogram was generated by the average-linkage method from a matrix of pairwise genetic distances between ETs; genetic distance between a pair of ETs is the proportion of enzyme loci at which dissimilar alleles occur (mismatches). The dendrogram was truncated at a genetic distance of 0.04, which corresponds to a single locus difference between ETs. On the assumption that the rate of evolution is constant over lineages, the dendrogram may be interpreted as a phylogenetic tree. Major lineages and clusters are designated by letters, and ETs of the eight serovars are designated as follows: Cs (*S. choleraesuis*), De (*S. derby*), Du (*S. dublin*), En (*S. enteritidis*), He (*S. heidelberg*), In (*S. infantis*), Np (*S. newport*), and Tm (*S. typhimurium*). Note that ETs of *S. newport* occur in clusters A (labeled *newport* I) and E (*newport* II) and that ETs of *S. derby* occur in lineage K (*derby* I) and in two branches of cluster D (*derby* II and *derby* III). ETs of *S. enteritidis* are in cluster A and lineages C, H, and L. After Beltran et al. (1988).

eages) usually show distinctive patterns of host occurrence (Table 4) and vary geographically in relative frequency (Table 5). These are aspects of *Salmonella* biology that could not be detected until the clones were identified and genetically defined. Obviously, the proper unit of analysis for studies of pathogenicity is the clone, not the serovar.

EVOLUTIONARY RELATIONSHIPS OF SUBSPECIES

Seven taxonomic groups of the salmonellae have been distinguished on the basis of studies of DNA–DNA hybridization and variation in biochemical traits such as the ability to ferment certain sugars (Crosa et al., 1973; Le Minor et al., 1986). Sixty percent of the serovars, including those responsible for 99% of

TABLE 4. Host distribution of isolates of *Salmonella derby* and *S. newport*

Serotype and division	Number of isolates	Isolates from indicated host (%)		
		Birds	Humans	Swine and other mammals
S. derby[a]				9.4
I	53	20.8	69.8	32.0
II	206	7.3	60.7	
S. newport[b]				
I	72	6.9	75.0	18.1
II	30	6.7	30.0	63.3

Source: Beltran et al. (1988) and Selander (unpublished data).
[a]$\chi^2_{(2)} = 16.1$ ($p < 0.001$).
[b]$\chi^2_{(2)} = 20.8$ ($p < 0.001$).

TABLE 5. Geographic variation in frequency of recovery of isolates of several ETs of *Salmonella derby*

Division and ET	Number (%) of isolates from indicated region[a]		
	United States	Mexico	France/Switzerland
I			
De 1	67(19)	9(11)	7(7)
De 5	4(1)	0	1(1)
II			
De 13	163(47)	36(45)	10(10)
De 14	103(30)	25(32)	53(55)
De 14[b]	0	0	25(26)
III			
De 31	11(3)	9(11)	0
Total	348	79	96

Source: Beltran et al. (1988) and Selander et al. (unpublished data).
[a]$\chi^2_{(10)} = 176$ ($p < 0.001$).
[b]De 14a is combined with De 14 in Figure 1. De 14a differs from De 14 at one of 24 enzyme loci assayed.

cases of enteric fever and salmonellosis in humans, belong to group I. All seven groups have generally been regarded as closely enough related to be considered members of a single genospecies that is comparable in total span of genetic diversity to *E. coli* (Crosa et al., 1973; Le Minor et al., 1986), but some workers regard group V, which is the most divergent, as specifically distinct from the others (Reeves et al., 1989).

Multilocus enzyme electrophoresis (24 gene loci) of 179 strains distinguished 94 ETs and placed them in eight groups (Figure 4), seven of which are precisely those identified by DNA hybridization, thus demonstrating a strong correlation of results of the two methods of assessing overall genetic relationships (see also Ochman et al., 1983; Gilmour et al., 1987). The eighth group (labeled "undesignated" in Figure 4) is known from only five isolates (of two serovars) that formerly were assigned to group IV on the basis of biochemical characteristics. Although these isolates are indistinguishable in 99 biotype and other phenotypic characters from strains of group IV, DNA hybridization experiments have confirmed our finding that they are genetically distinct (Selander et al., unpublished data).

EVOLUTION OF THE HUMAN-ADAPTED CLONES THAT CAUSE ENTERIC FEVER

Only a few of the more than 2200 serovars of *Salmonella* are known to be primarily or exclusively limited in host range (host-adapted) to humans (Table

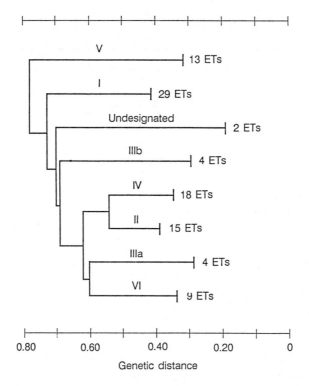

FIGURE 4. Genetic relationships of the subspecies of *Salmonella*. Ninety-four ETs (represented by 179 isolates) were clustered on the basis of estimates of genetic distance derived from an analysis of allelic variation at 24 enzyme loci. Each lineage is truncated at the genetic distance level at which branching occurs. After Selander et al. (in preparation).

6). Medically the most important of these is *S. typhi*, the agent of human typhoid fever; others are *S. paratyphi* A, *S. paratyphi* C, and *S. sendai*, all of which cause typhoid-like enteric fevers. Additionally, certain strains of *S. paratyphi* B cause human enteric fever, while others (often designated as *S. java*) produce gastroenteritis in both humans and animals. Finally, *S. miami*, which is serologically the same as *S. sendai*, is largely limited to humans but causes gastroenteritis rather than enteric fever.

Despite the considerable effort of microbiologists to differentiate and classify strains of *Salmonella* by serological, biochemical, and various other conventional methods, the evolutionary relationships of the human-adapted organisms to one another and to strains of other serovars have remained largely unknown. Consequently, there has been no systematic framework for studying the evolution of host adaptation and pathogenicity in this group of bacteria. Many questions remain to be answered: Are the strains causing human enteric fever

TABLE 6. Mean genetic diversity per locus (*H*) among clones (ETs) of *Salmonella* serovars in relation to host range

Serovar	Number of Isolates	Number of Clones	*H*
Host-adapted			
S. *typhi*	334	2	0.083
S. *paratyphi* A	135	4	0.042
S. *sendai*	5	4	0.083
S. *paratyphi* C	96	6	0.074
S. *choleraesuis*	159	9	0.089
S. *dublin*	206	2	0.042
S. *gallinarum*	50	4	0.055
Mean	·	4.4	0.067
Non-host-adapted			
S. *muenchen*	73	4	0.326
S. *panama*	96	13	0.126
S. *saintpaul*	34	4	0.188
S. *typhimurium*	340	17	0.119
S. *heidelberg*	204	8	0.092
S. *derby*	349	6	0.258
S. *newport*	105	13	0.149
S. *enteritidis*	257	14	0.176
S. *infantis*	113	4	0.152
Mean		9.2	0.176

Source: Selander et al. (1990b) and Selander (unpublished data); estimates of *H* based on 23 or 24 enzyme loci.

related to one another or has there been convergent evolution in multiple phylogenetic lineages? Where did the human-adapted clones come from; what are their closest relatives? What processes are involved in the evolutionary origin of new pathogenic strains?

To address some of these questions, we studied 761 isolates of the seven human-adapted serovars and thousands of isolates of 22 other group I serovars (Selander et al., 1990b). A dendrogram summarizing estimates of genetic relationships among the clones of the human-adapted serovars and those of seven other serovars that are genetically similar to one or more of the human-adapted serovars is presented in two parts in Figures 5 and 6. Among the total of 1482 isolates of the 14 serovars analyzed for allelic variation at 24 enzyme loci, 106 ETs were distinguished. On the assumption that rates of evolution of metabolic enzyme genes are more or less constant in all lineages, the dendrogram may be interpreted as a phylogeny of chromosomal genomes (Nei, 1987; Ochman and Wilson, 1987), and inferences can be drawn regarding the evolution of clones that are largely or entirely adapted to humans and the development of the propensity to cause enteric fever.

The dendrogram shows that, except in the case of S. *paratyphi* A and S. *sendai*, there is no close phylogenetic relationship between clones of different human-adapted serovars, which implies that both host restriction to humans and the ability to cause enteric fever have evolved several times independently, either through mutation or the horizontal transfer of genes among distantly related cell lineages.

S. *paratyphi* A, S. *sendai*, and S. *miami*

The human-adapted serovars S. *paratyphi* A and S. *sendai* apparently evolved from a common ancestral lineage that also gave rise to the clones of S. *panama*, which are agents of gastroenteritis in a broad range of hosts, including humans, and to S. *miami*. With regard to serological characters, the simplest hypothesis is that the a antigen of the phase 1 flagella is the ancestral condition and was retained in S. *paratyphi* A, S. *sendai*, and S. *miami* while it was replaced by the l and v antigens in S. *panama* (see Table 2). Additionally, in S. *paratyphi* A the O antigen 9 was lost and the phase 2 flagellin gene (H2) was silenced. The derivation of S. *paratyphi* A and S. *sendai* from an ancestor with a broad host range would have involved both a restriction in host range to humans and the development of invasive ability. These changes presumably occurred in an ancestral clone that very recently differentiated into the clones of S. *paratyphi* A and S. *sendai*.

The serovar S. *miami*, which is predominantly North American in distribution, is genotypically heterogeneous and may have arisen several times from various clones of an S. *panama*-like ancestor. The predominant S. *miami* clone is Mi 1; isolates of this clone and of Mi 2 and Mi 2a were recovered from humans, whereas isolates of clones Mi 3, Mi 4, and Mi 5, which are on a different evolutionary branch (Figure 5), were cultured from a frog, a fish, and guano; and Mi 6, which is still another branch, was isolated from a blackbird. These data suggest that only the first group of clones has become adapted to humans, without, however, acquiring characters facilitating invasiveness. The alternative possibility that host restriction to humans is the ancestral condition, with S. *panama* and some clones of S. *miami* secondarily acquiring broad host ranges, seems less likely. (Note that Se 5 is actually a clone of S. *miami*.)

S. *paratyphi* B

S. *paratyphi* B (including the d-tartrate-positive strains long designated as S. *java*) is a genotypically heterogeneous complex of clones related to clones of S. *typhimurium*, S. *heidelberg*, and S. *saintpaul* (Figure 6); these four serovars

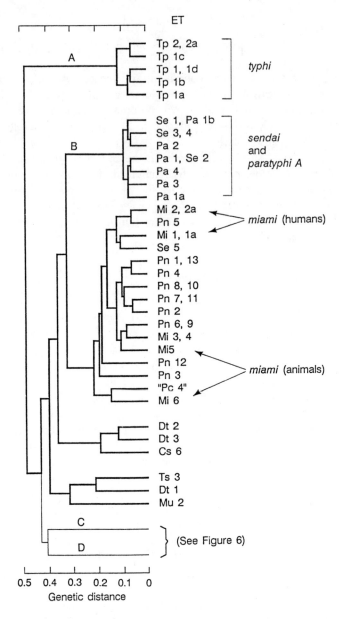

FIGURE 5. Part 1 of a dendrogram showing estimated evolutionary genetic relationships (based on 24 enzyme loci) among ETs, representing clones, of serovars of *Salmonella* causing human typhoid and other enteric fevers and certain other serovars with which they are phylogenetically allied. The lineage of the cluster of ETs of *S. typhi* (Tp) is labeled A, and that of the cluster containing ETs of *S. paratyphi* A (Pa) and four of the five ETs of *S. sendai* (Se) is labeled B. Other serovars represented by ETs in this part of the dendrogram are *S. miami* (Mi), *S. panama* (Pn), *S. paratyphi* C (Pc), *S. decatur* (Dt), *S. typhisuis* (Ts), and *S. muenchen* (Mu). Lineages labeled C and D are shown in part 2 of the dendrogram (Figure 6). After Selander et al. (1990b).

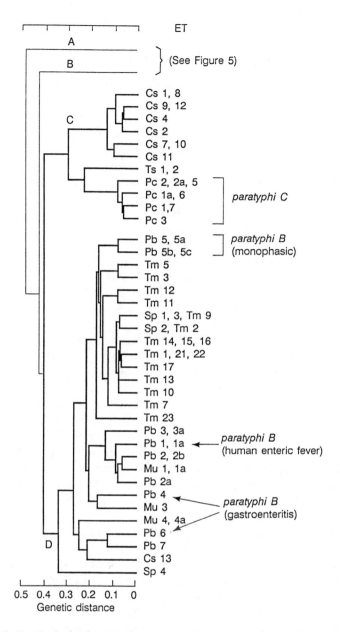

FIGURE 6. Part 2 of a dendrogram showing estimated evolutionary genetic relationships (based on 24 enzyme loci) among ETs, representing clones, of serovars of *Salmonella* causing human typhoid and other enteric fevers and certain other serovars with which they are phylogenetically allied. The lineage of the cluster containing nine ETs of *S. choleraesuis* (Cs), two ETs of *S. typhisuis* (Ts), and eight ETs of *S. paratyphi* C (Pc) is labeled C; and D marks the lineage of the large *S. typhimurium* complex of clones (see text), which consists of ETs of *S. typhimurium* (Tm), *S. heidelberg* (He) (not shown here), *S. paratyphi* B (including *S. java*) (Pb), *S. saintpaul* (Sp), and *S. muenchen* (Mu), together with one ET of *S. choleraesuis* (Cs 13). Note that one ET of *S. typhisuis* (Ts 3) is a member of a cluster shown in part 1 of the dendrogram (Figure 5). Lineages labeled A and B are shown in part 1 of the dendrogram (Figure 5). After Selander et al. (1990b).

differ only in their phase 1 flagellar antigens (Table 2). *S. muenchen* is also part of this complex, although it is distinctive in both somatic and phase 1 antigens (Table 2).

An analysis of 24 enzyme loci distinguished 14 ETs and suggested that *S. paratyphi B* is polyphyletic, with the monophasic clones being closely related to *S. typhimurium, S. heidelberg*, and the main group of *S. saintpaul* clones (Selander et al., 1990a). Human adaptation and the ability to cause enteric fever evolved only in the globally distributed clone Pb 1, presumably rather recently, since Pb 1 is only weakly differentiated from certain other clones that have broad host ranges and do not cause enteric fever. No such changes occurred in the other lineages of *S. paratyphi B*, which, together with *S. typhimurium, S. heidelberg, S. saintpaul*, and *S. muenchen*, have retained the presumed ancestral ecological and pathogenic characteristics of the complex.

The strains of *S. paratyphi B* studied had earlier been characterized for variation in 13 biotype traits (e.g., ability to ferment *m*-inositol), phage type (profile of sensitivity to 12 standard test phages), sensitivity to colicin M and phage ES18, and electrophoretic pattern of ribosomal RNA (Barker et al., 1988). A classification of strains had been proposed on the basis of these characters, but our analysis revealed that, individually or in combination, they fail to mark clones or other meaningful phylogenetic subdivisions. This finding is illustrated in Figure 7, where the coefficients of correlation (r) indicate the relative value of individual characters in indexing the overall genetic relationships among isolates estimated by multilocus enzyme electrophoresis. The coefficient for a given character (e.g., the ability to ferment rhamnose) is the product-moment correlation between two matrices of genetic distances between all pairs of isolates, one a matrix based on allelic variation at 11 polymorphic enzyme loci (13 of the 24 loci studied were monomorphic), and the other a matrix based solely on variation in the character in question.

Apart from the polymorphic enzymes, there are few characters for which r is greater than 0.300. Five of the characters, including three enzyme loci (CAT, ACO, and AC2), *m*-tartrate utilization, and sensitivity to the Dundee test phage are both highly polymorphic and moderately to strongly correlated with estimates of overall genetic relatedness between pairs of isolates. (These characters are enclosed in an oval in Figure 7.) Four of the five primary biotype characters, all eight secondary biotype characters, the rRNA pattern, and the response to colicin M and phage ES18 are of little or no value in indexing overall genetic relationships among strains. The rRNA pattern character involves the presence or absence of intact 23 S rRNA; fragmentation of this RNA has recently been shown to result from the excision (without religation) of an intervening sequence of about 90 bp in the course of rRNA maturation (Burgin et al., 1990). Because these intervening sequences are highly volatile evolutionarily, sometimes occurring in only some of the multiple rRNA operons of a particular strain, it is not surprising that their distribution is phylogenetically uninformative.

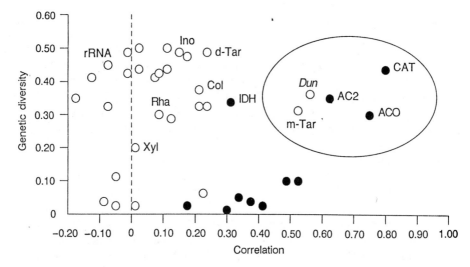

FIGURE 7. Relationship between the diversity of a character and the correlation between pairwise distances between ETs (for enzyme loci) or isolates (for other characters) of *Salmonella paratyphi* B (including *S. java*), based on the character in question and on 24 enzyme loci. Dots represent enzyme loci; circles represent biotype and other characters. Characters identified by symbols are IDH, isocitrate dehydrogenase; AC2, acid phosphatase-2; ACO, aconitase; CAT, catalase; m-Tar, *m*-tartrate reaction; d-Tar, *d*-tartrate reaction; Rha, L-rhamnose reaction; Xyl, D-xylose reaction; Ino, *m*-inositol reaction; Col, reaction to colicin M; and Dun, reaction to Dundee test phage. After Selander et al. (1990a).

S. paratyphi C

The clones of *S. paratyphi* C and *S. choleraesuis* and certain clones of *S. typhisuis* (Ts 1 and Ts 2) apparently shared a common ancestor from which they evolved without modification of the serotype, except for the acquisition of the Vi capsular antigen by *S. paratyphi* C. The common ancestor presumably was invasive and already adapted to swine, as are the extant clones of *S. choleraesuis* and certain clones of *S. typhisuis*, which cause swine paratyphoid fever. If so, the evolutionary derivation of *S. paratyphi* C would have involved only a shift in host to humans. The occasional occurrence of *S. paratyphi* C in animals suggests that physiological specialization for humans is not as complete as in the case of *S. typhi* or *S. paratyphi* A. It is noteworthy that in some parts of the world humans are a not uncommon secondary host for *S. choleraesuis*, which is invasive, producing severe enteric fever with an unusually high mortality rate.

S. typhisuis apparently is polyphyletic; Ts 3 is genotypically very different from Ts 1 and Ts 2, notwithstanding the serotypic identity and physiological similarity of all three clones.

S. typhi

Seven ETs of S. *typhi*, marking two clones, Tp 1 and Tp 2, and five subclones, were distinguished. Tp 1 and Tp 2 differ in alleles at two of the 24 enzyme loci analyzed and are also distinguishable by the restriction fragment length polymorphism (RFLP) pattern of their rRNA operons. Tp 1 is the predominant clone of S. *typhi* worldwide, being represented by 82% of the total number of isolates examined, whereas Tp 2 is known only from Senegal and Togo, in Africa.

The ETs of S. *typhi* cluster apart from those of other serogroups at a genetic distance of about 0.48 (Figure 5), which means they are distinctive, on average, at about 12 of the 24 enzyme loci assayed. Because no close relative of S. *typhi* has been identified (Reeves et al., 1989; Selander et al., 1990b), there is no basis for speculation regarding the host range and pathogenicity of the ancestral population from which clones of this serovar evolved. The marked distinction in chromosomal genotypes of the clones of S. *typhi* from those of other serovars indicates that their phylogenetic lineage is old, but both their distinctive phenotypic characters and close adaptation to humans could be fairly recent developments. Indeed, the clones of S. *typhi* may have arisen so recently that there has not been sufficient time for the mutational or recombinational generation of any large amount of genotypic diversity in enzyme genes or other genes that have moderate to slow evolutionary rates. However, there is the alternative possibility that the relatively low level of genotypic diversity among strains of S. *typhi* is a consequence of a recent episode of periodic selection (Levin, 1981) affecting populations on a global scale.

Most populations of S. *typhi* are monomorphic for the d allele at the H1 flagellin locus and do not express an H2 locus. [Indeed, we have recently determined that an H2 gene is not present in the genome (unpublished data).] But the Indonesian population is polymorphic for two H1 alleles (H1:d and H1:j) and also for the expression of a Z_{66} antigen encoded by the H2 locus (Frankel et al., 1989a). Partly for this reason, it has been proposed that S. *typhi* originally was diphasic and first became adapted to humans in Indonesia, with a monophasic H1:d clone subsequently spreading throughout the world (Frankel et al., 1989a). However, an equally plausible case for an African origin of this serovar can be made on the basis of the unique co-occurrence in Senegal and Togo of the two major clones, Tp 1 and Tp 2.

In summary, through population genetic analysis of chromosomal genotypes we are beginning to construct an evolutionary framework that will be invaluable in determining the precise molecular genetic events that occurred in the course of the evolution of these important human pathogens. The human-adapted clones probably are recently evolved, and their origins are varied. One group of human-adapted clones (S. *paratyphi* C) evolved from clones already causing enteric fever in swine, and others were derived from host-generalists causing gastroenteritis.

HOST RANGE–GENETIC DIVERSITY CORRELATION

A notable feature of variation emerging from population genetic studies of *Salmonella* is a strong tendency for clones of the host-adapted serovars to be fewer in number and genotypically less diverse than those of serovars that are pathogenic for a variety of host species (Table 6). Population genetics provides two very different explanations for the observed relationship between host range and genotypic diversity, one relating to effective population size and the other to ecological niche breadth.

Effective population size model

Under the assumptions of the neutral mutation theory of molecular evolution (Kimura, 1983), the amount of allelic variation at a locus in a finite population at equilibrium between the generation of selectively neutral mutations and their loss through random genetic drift is a direct function of the long-term effective size of the population. For bacteria with a clonal population structure, effective population size may more closely correspond to the number of extant colonies than to the actual size of the total standing crop of cells (Maruyama and Kimura, 1980). For nonequilibrium populations, the evolutionary effective size is roughly the harmonic mean of the effective size of the population over all generations since its origin. Consequently, if new populations arise from one or a small number of cell lineages, younger populations are expected to be genetically less variable than older ones.

Serovars of *Salmonella* with strains capable of infecting a variety of different host species may, other things being equal, be expected to maintain larger effective population sizes and, consequently, to carry more genetic variation than serovars that are limited in distribution to humans or single species of animals. It is also probable that the clones of many or all of the host-adapted serovars have arisen more recently than have those of the common, broad-host-range serovars.

Ecological niche breadth model

The basis for this model is the premise that much or all of the allelic diversity at enzyme and other structural gene loci is adaptive and is maintained by one or more types of balancing selection (Nevo et al., 1984; Kimura, 1983; Nei, 1987). For *Salmonella*, the rationale is that the relatively narrow range of ecological conditions encountered by strains of a host-adapted serovar selects for a corresponding limited amount of genetic diversity. But in the case of the ubiquitous serovars, clones of many different genotypes may find niches to which they are especially adapted, or there may be selection for "general purpose genotypes" that are moderately well adapted to a wide range of ecological conditions provided by a variety of host species.

Because of the formidable problems of estimating evolutionary effective population size and niche breadth, it will be extremely difficult to test either hypothesis. However, it should be noted that the apparent selective neutrality or near neutrality of electromorph alleles at bacterial enzyme loci, as demonstrated experimentally and statistically for *E. coli* (Whittam et al., 1983; Hartl and Dykhuizen, 1985; Hartl et al., 1986), is compatible with the interpretation that effective population size is the major determinant of amount of protein polymorphism carried by populations. It is also relevant that repeated efforts to demonstrate ecological correlates of amount of genetic variation in populations of higher organisms have yielded little success (Schnell and Selander, 1981; Nei and Graur, 1984).

DISTRIBUTION AND EXPRESSION OF Vi CAPSULAR ANTIGEN

The polysaccharide Vi capsular antigen occurs regularly in strains of *S. typhi*, not uncommonly in isolates of *S. paratyphi C*, and rarely in those of *S. dublin*; it has also been identified in a few strains of *Citrobacter freundii*, a relative of *Salmonella*.

The structure of the Vi antigen is determined by genes of the chromosomal *viaB* region. To ascertain whether the failure of certain strains of the three *Salmonella* serovars to express this antigen is a result of gene regulation or an absence of the *viaB* genes, an 8.6-kb *Eco*RI fragment of the *viaB* region of the *Citrobacter freundii* chromosome was used as a probe in dot-blot experiments (Rubin et al., 1985; Selander et al., 1990b).

The results of probing of 10 phenotypically Vi-negative isolates of *S. typhi* demonstrated that the occasional absence of the antigen in isolates of this serovar may reflect either an absence (three isolates) or lack of expression of the *viaB* genes (seven isolates). In the case of *S. paratyphi C*, the frequent absence of Vi antigen expression in laboratory cultures of strains is, with rare exception (1 of 35 phenotypically Vi-negative isolates probed), not attributable to an absence of the structural genes (see Snellings et al., 1977). Recent work by Daniels et al. (1989) indicates that most strains of *S. paratyphi C* actually express the Vi antigen in the host but that expression is rapidly lost when isolates are grown on laboratory media.

Because of the occurence of the Vi antigen in four distantly related phylogenetic lineages, the inference is that several horizontal transfer events involving the *viaB* region have occurred. The presence of the *viaB* region in isolates of all clones (ETs) of *S. typhi* and *S. paratyphi C* suggests that it has been present in these lineages for long periods of time. If so, the instability of expression of the Vi antigen in strains of the latter serovar cannot be attributed to a recent acquisition of the genes, but the molecular mechanism regulating expression remains to be determined. In the case of *S. dublin*, however, the *viaB* region may have been acquired rather recently, for it is confined to a single minority subclone of the predominant, globally distributed clone (Du 1), which, moreover, has a limited distribution, occurring only in Europe (Selander, unpublished data). All isolates of this subclone have the *viaB* genes, and expression is stable. Expression of the antigen in *C. freundii* is highly variable as a result of the presence of an invertible insertion sequence in the *viaB* region (Snellings et al., 1981; Ou et al., 1988).

EVOLUTION OF FLAGELLIN GENES

Comparative nucleotide sequencing of the phase 1 (H1) flagellin gene in isolates of clones for which population genetic analysis has provided a phylogenetic framework is beginning to elucidate the evolutionary processes by which new antigenic forms of flagellins arise and new serotypic combinations (serovars) are formed.

Sequence invariance of H1 genes within serovars

The H1 gene is approximately 1.5 kb in length, with highly conserved ends that flank a so-called hypervariable central region in which the coding sequence for the antigenic factors is located (Joys, 1976; Wei and Joys, 1985; Frankel et al., 1989b) (Figure 8). This structure makes the central region a good candidate for direct amplification by the polymerase chain reaction (PCR) technique.

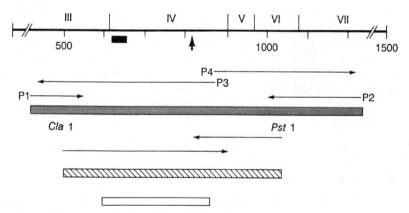

FIGURE 8. Schematic representation of the central part of the H1:i flagellin gene of *S. typhimurium* and the sequencing strategy for strain RKS284, representing ET Tm 1. The upper part of the diagram indicates the length of the entire coding region of the H1:i gene sequenced by Joys (1985) and identifies regions III to VII and the coordinates used by Wei and Joys (1985). Region IV is the segment of the gene that contains the sequence of the major epitope of the i antigen (filled box) and shows less than 33% amino acid homology with other sequenced *Salmonella* H1 flagellin genes. The several boxes in the lower part of the diagram indicate regions sequenced in a comparative study. After Smith and Selander (1990).

It was reported a few years ago that the H1 flagellins of two substrains of the laboratory strain *S. typhimurium* LT2, which has the H1:i antigen, differed in several amino acids (Joys et al., 1974). This finding led to the notion that the sequence of the central region of the H1 gene is only weakly constrained by natural selection, with the consequence that new serovars rapidly evolve one from another by random mutational drift (Wei and Joys, 1985). In other words, this region appeared to be mutating and evolving at an alarming rate! But it turns out that one of the genes sequenced was a chimera apparently formed earlier in the course of a transduction experiment. We sequenced the central region of the H1:i gene in six ETs of *S. typhimurium* and in LT2 and found that the sequences are identical—not even a third-position change was detected (Smith and Selander, 1990). Similarly, three ETs of *S. heidelberg*, which has the H1:r antigen, were identical in nucleotide sequence (Smith et al., 1990). In each case, the strains were genomically divergent, differing from one another at 1-4 of 24 enzyme loci. These findings clearly do not support the thesis that the rate of evolution of the H1 flagellin gene by point mutation is unusually high; although the central part of the gene is hypervariable among serovars, there is no evidence that it is hypermutable.

New antigenic types of flagellins may arise by intragenic recombination

Throughout most of the world, *S. typhi* is monophasic and has the H1:d allele; but in Indonesia strains carry either the H1:d allele or an H1:j allele. Frankel et al. (1989a) have shown that the H1:j arises from H1:d by intragenic recombination involving the pairing of a directly repeated 11 bp sequence that leads to the deletion of a 261-bp segment in the central antigenically determinant part of the gene (Figure 9). *S. muenchen* also has the H1:d allele, and in this serovar spontaneous mutations changing H1:d to H1:j have been observed in the laboratory.

What processes are responsible for the generation of new serotypes?

If gradual mutational change of the flagellins in diverging clonal cell lineages is a primary process by which new serovars evolve, there should be a positive correlation among strains between overall level of genomic divergence and degree of differentiation in sequence of the H1 gene. Is there such a correlation? To answer this question, we sequenced the central part of the H1 gene in strains of six serovars that have the i, r, or d H1 antigen, and for which we have estimated chromosomal genomic divergence by MLEE analysis (Figure 10).

S. typhimurium (i) and *S. heidelberg* (r). These two serovars differ only in the antigenic character of their H1 flagellins (O and H2 antigens are shared) (Table

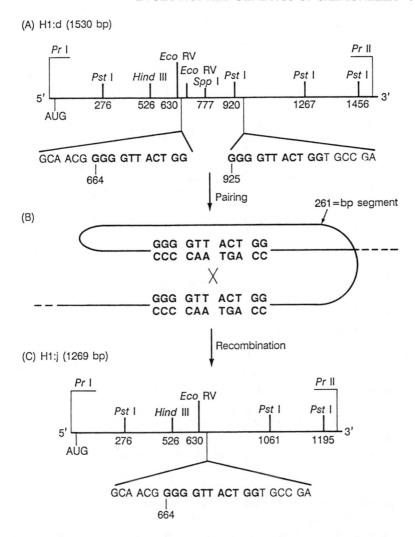

FIGURE 9. Derivation of the H1:j allele from the H1:d allele in *Salmonella typhi* by intragenic recombination involving deletion of a 261-bp segment of the H1 gene. (A) Restriction map of the H1:d allele, showing a directly repeated 11-bp sequence (bold letters). (B) Pairing of the 11 bp sequences leads to recombinational loss of a 261-bp segment. (C) Restriction map of the H1:j allele. After Frankel et al. (1989a).

2). They are closely related, with a genetic distance of 0.20 (Figure 10), which is only slightly greater than the genetic distances between ETs within either of these serovars. They are also similar biochemically (i.e., in biotype) and in pathogenicity and host range. Somewhat surprisingly, then, we found a 19% nucleotide sequence difference in the central part of their H1 genes, after

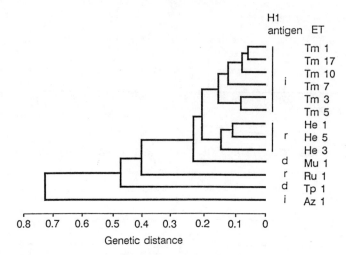

FIGURE 10. Genetic relationships among the strains of several *Salmonella* serovars for which the H1 flagellin gene was sequenced. Strains are designated by their ET numbers, as follows: *S. typhimurium*, Tm 1, 3, 5, 7, 10, and 17; *S. heidelberg*, He 1, 3, and 5; *S. muenchen*, Mu 1; *S. rubislaw*, Ru 1; *S. typhi*, Tp 1; and "Arizona" in subspecies IIIb, Az 1. The antigenic designations of the H1 flagellins of the ETs are shown. After Smith et al. (1990).

optimal alignment. Because the H1 sequence is invariant within each serovar, in neither serovar is this gene evolving fast enough by mutation to have generated this degree of difference. (Incidentally, the H2 gene is identical in sequence in strains of *S. typhimurium* and *S. heidelberg*, which is further evidence of their close phylogenetic relationship.) Thus, it is necessary to invoke a recombination event involving, say, the horizontal transfer of the H1:r allele into a strain of *S. typhimurium* (H1:i), thereby creating the *S. heidelberg* serotype.

***S. typhimurium* (i) and *S. heidelberg* (r) compared with *S. muenchen* (d).** Strains of all three of these serovars are rather closely related (genetic distance, 0.22); but in each comparison of the H1 sequences, optimal alignment yielded a 50% difference. So again, the evolution of H1:d and either H1:i or H1:r from a gene in a recent common ancestral clone would seem to be ruled out, and horizontal transfer and recombination are strongly implicated.

Similar antigens in serovars with highly divergent genomes. The H1:r alleles of *S. heidelberg* and *S. rubislaw* are only 1.6% different in sequence (Table 7), yet the genetic distance between strains of these serovars is 0.41 (Figure 10). And there is only a 1.3% difference in sequence between the H1:d alleles of *S. muenchen* and *S. typhi*, strains of which show a genetic distance of 0.50. Even more striking are the results of a comparison of the H1:i alleles of *S. typhimurium* (in subspecies I) and a strain of a serovar in subspecies IIIb.

TABLE 7. Nucleotide differences in a 577-bp segment of the central region (474 to 1050 bp) of the H1:r alleles of *Salmonella rubislaw* and *Salmonella heidelberg*

Position[a]	S. rubislaw		S. heidelberg	
	Base	Amino acid	Base	Amino acid
525	G	Val	T	Val
564	T	Val	C	Val
747	T	Ala	C	Ala
900	T	Ala	C	Ala
904	C	His	A	Thr
905	A		C	
906	C		A	
907	A	Arg	G	Ala
908	G		C	

Source: Smith et al. (1990).
[a]Based on the nucleotide sequence of the H1:r allele of *S. rubislaw* reported by Wei and Joys (1985).

Although these strains are about as far apart as any two strains of *Salmonella* can be (genetic distance, 0.72) and serotypically have only the H1:i factor in common (Table 2), there was only a 2.2% difference in the sequence of the central region of their H1 genes.

In conclusion, the nonconcordance we have found strongly implicates horizontal transfer of DNA sequences as a major mechanism generating new serovars in *Salmonella*. We have yet to find convincing evidence of convergence of H1 alleles. In our material, the sequences of H1 genes coding for the same phenotype (antigen) in clones of very different chromosomal genomes are so similar that convergence is unlikely. This is not to say convergence does not occur—we simply have not yet found it.

EVOLUTION OF SOMATIC POLYSACCHARIDE ANTIGENS

The genes encoding the biosynthetic enzymes for the O antigen of the somatic lipopolysaccharide have been mapped to the chromosomal *rfb* cluster (Brahmbhatt et al., 1988), a detailed restriction map has been constructed, and the genes conferring antigenic specificity in *S. typhimurium*, *S. typhi*, and *S. paratyphi* A have been sequenced and compared (Wyk et al., 1989; Verma and Reeves, 1989). Antigenic specificity depends on the particular dideoxyhexose that is added as a side branch to a common mannosyl-rhamnosyl-galactosyl

TABLE 8. Basis of somatic (O) antigenic specificity in three *Salmonella* serovars

Serovar	Antigen	Group[a]	Sugar	Enzyme	Gene
S. typhimurium	O4	B	Abequose	Abequose synthase	rfbB
S. typhi	O9	D	Tyvelose	Paratose synthase and tyvelose-2-epimerase	rfbS and rfbE
S. paratyphi A	O2	A	Paratose	Paratose synthase	rfbS and cryptic rfbE

Source: Verma and Reeves (1989).
[a]A classification of serovars of subspecies I based on the predominant O antigen or antigens (see Le Minor, 1984; Ewing, 1986).

backbone. In *S. typhimurium*, abequose, which specifies antigen O4 (group B), is added by abequose synthase. But in *S. paratyphi* A, antigen O2 (group A), and *S. typhi*, antigen O9 (group D), abequose is replaced by paratose and tyvelose, respectively (Table 8).

S. typhi differs from *S. typhimurium* in two ways. Abequose synthase (*rfbB* gene) is replaced by paratose synthase (*rfbS* gene), and there is an additional enzyme—tyvelose-2-epimerase, encoded by the *rfbE* gene—that converts paratose to tyvelose. The *rfbE* gene is also present in *S. paratyphi* A and is, in fact, identical in sequence to that of *S. typhi*, with one small but critical difference: a frameshift mutation involving the loss of one thymine residue has converted the fourth codon to a stop codon. Because the gene in *S. paratyphi* A is cryptic (silenced), paratose is not converted to tyvelose.

Abequose synthase and paratose synthase evolved from a common ancestor, but they are very divergent, showing only intermittent sequence similarity in the amino-terminal half and only moderate homology in the carboxy-terminal half. Their G + C content is 0.32, which is much lower than the normal value of 0.51 for *Salmonella* and suggests that these enzymes evolved in a low G + C content organism and were relatively recently transferred to *Salmonella*. Moreover, the fact that the G + C content is lower than in any of the Enterobacteriaceae, species of which range in G + C content from 0.38 to 0.60, would seem to exclude all salmonellae and closely related genera as donor sources.

RECOMBINATION IN BACTERIAL EVOLUTION

Mechanisms and significance of horizontal gene transfer in *Salmonella*

The evidence reviewed above strongly implicates horizontal transfer and recombination of the H1 flagellin gene and the genes encoding certain enzymes mediating the synthesis of the O polysaccharide as a major mechanism generating new serovars of *Salmonella* and also demonstrates that allelic variation in the H1 flagellin gene may arise through intragenic recombination.

The most likely method of transfer of flagellin genes is phage-mediated transduction. In the laboratory, flagellin genes can easily be transduced between strains of different serovars, as long ago demonstrated by Zinder and Lederberg (1952). Additionally, gene exchange between strains may be effected by the conjugational transfer of plasmids. In work now in progress, we have found that an H1:d flagellin gene is present on a very large plasmid carried by several "triphasic" strains of *Salmonella* (strains exhibiting three antigenic types of flagella), which also have "normal" H1 and H2 chromosomal flagellin genes (Ewing, 1986). (Surprisingly, this plasmid-borne H1-type flagellin gene is associated with an H2-type promoter.) On the basis of the notion that the course of evolution in *Salmonella* has involved a progressive reduction in the number of flagellin antigen types, the triphasic state was once believed to be ancestral and the biphasic condition derived (Edwards et al., 1962). However, there never was a rational basis for this concept, and the discovery that the third flagellin gene of the triphasic serovars is plasmid-borne suggests a simpler interpretation.

In the triphasic serovar *S. saliantis*, selection against the d antigen (imposed when a strain is grown in medium containing anti-d serum) leads to the irreversible loss of the ability to express this antigen; the strain becomes permanently diphasic for H1:e,h and H1:e,n,z_{15}, and now has the cell-surface antigen profile of the serovar *S. sandiego* (Table 2). Because the sequences of these plasmid-borne H1:d flagellin genes are identical or nearly so in strains of several triphasic serovars that have been shown by MLEE analysis to be distantly related, we conclude that this plasmid is a vehicle for the transfer of flagellin genes.

Evidence of horizontal transfer and recombination of chromosomal DNA sequences is also accumulating for *E. coli* and other species of bacteria (DuBose et al., 1988; Milkman and Stoltzfus, 1988; Plos et al., 1989). A particularly good example is provided by sequence analysis of the *cap* region of the chromosome of *Haemophilus influenzae*, which has clearly established DNA transfer as the mechanism responsible for the occurrence of the type b capsular polysaccharide in two widely divergent phylogenetic divisions, the clones of which differ in degree of virulence (Kroll and Moxon, 1990; Musser et al., 1990).

Types of chromosomal recombination

The H1 flagellin gene of *Salmonella* and the b-specific *cap* region sequence of *H. influenzae* are examples of substitutive recombination in which a gene or part of a gene of one strain is replaced by an homologous sequence from another strain. [In the terminology of molecular genetics, this type of event is designated as homologous recombination; see Porter (1988) for detailed discussion.] Homologous recombination tends to randomize the associations of alleles in the chromosomal genome, and thereby directly reduces linkage disequilibrium in a population. Consequently, it has important implications for the genetic structure and evolution of populations and should, therefore, be carefully distinguished from additive recombination (called nonhomologous recombination) in which a DNA sequence is added to the chromosomal genome of a strain without replacing an homologous sequence (Porter, 1988; Levin, 1988). Examples are provided by insertion sequences, certain toxin genes, and, in the case of *Salmonella* and *Citrobacter*, the *viaB* region controlling the synthesis of the Vi capsular antigen. Nonhomologous recombination does not directly affect linkage disequilibrium among other chromosomal genes, but, of course, it can be a source of genetic variation among strains, and may, therefore, have important evolutionary consequences (Riley and Sanderson, 1990).

Failure to consider the low frequency with which chromosomal recombination occurs in natural populations and to distinguish between the two types of recombination and between chromosomal and plasmid genomes has engendered a widespread notion that gene exchange is rampant among bacteria in general, even to the extent that there is only a single gene pool and a single species (Sonea and Panisset, 1983). But the reality is that genetic isolation of species is a major feature of bacterial evolution (Campbell, 1990; Mahan et al., 1990). Even such relatively closely related bacteria as *E. coli* and *S. typhimurium* (Riley and Krawiec, 1987; Ochman and Wilson, 1987) differ in the insertion sequences present in their genomes, harbor entirely different plasmids, and show well-defined differences in their restriction–modification systems (Riley and Sanderson, 1990).

Although an important role for the horizontal transfer and homologous recombination of flagellin genes in the generation of new serovars of *Salmonella* is now established, the full evolutionary significance of the exchange of DNA sequences within populations of *Salmonella* is as yet unclear. The occurrence of very extensive polymorphism in the flagellins and lipopolysaccharides and the presence of a complex molecular mechanism mediating flagellin phase variation clearly suggest that the ability to present altered cell-surface structures to the environment is selectively advantageous. Because of this advantage, the transfer and chromosomal integration of genes encoding and mediating expression of cell-surface molecules may more often be followed by an increase in frequency of the recombinant strain than in the case of most other genes. Even rare events may be important for the adaptation and evolution of populations (Hartl and Dykhuizen, 1984; Levin, 1988; Selander and Musser, 1990).

Inasmuch as the evidence from MLEE analysis demonstrates that the population structure of the salmonellae is basically clonal, it is apparent that the frequency of horizontal transfer and homologous recombination of genes in general is not high enough to randomize the chromosomal genome. A clonal population structure is characteristic of many other species of bacteria (Selander and Musser, 1990; Selander and Smith, 1990) and has recently been reported for a variety of parasitic protozoans as well (Tibayrenc et al., 1990).

Clearly, the role of the horizontal transfer and recombination of DNA sequences in the evolution of the salmonellae, as well as other bacteria, will continue to be a major area of research through the application of molecular techniques within a population genetic framework.

EVOLUTIONARY AND STRUCTURAL CONSTRAINTS IN THE ALKALINE PHOSPHATASE OF *ESCHERICHIA COLI*

Robert F. DuBose and Daniel L. Hartl

Evolutionary genetics is undergoing a major change in focus. For the past 50 years the field has been dominated by issues of genetic variation—its origin, prevalence, adaptive significance, and ultimate fate in natural populations. Preoccupation with genetic polymorphism was a natural response to a fundamental limitation of classical genetics: without genetic variation, no Mendelian analysis is possible. Genetic variation is also important in its own right because variation that is not selectively neutral is the basis of Darwinian natural selection. Preoccupation with genetic variation also narrowed the focus of the field in that theoretical and experimental attention was devoted almost exclusively to genetic polymorphism.

The advent of molecular techniques in the past 20 years has made it possible to study genes or parts of genes that are not variable. These methods overcome many previous technical limitations and make it possible to study the mechanisms of selective constraint. For example, the very extensive sequence databases that have accumulated—presently estimated at 65 million base pairs of DNA—have stimulated development of new methods of phylogenetic reconstruction and hypothesis testing, as well as many new questions about the detection and analysis of paralogous sequences (sequences homologous because of gene duplication). For the first time, it has become possible to carry out deliberate experimental manipulations at the molecular level to test evolutionary hypotheses about protein structure and function. These approaches are rapidly being incorporated into evolutionary studies, and, as the empirical basis of the field shifts, so too must the theoretical viewpoint become broadened beyond its former exclusive preoccupation with polymorphism.

In this chapter we demonstrate one way in which the comparison of DNA sequences and experimental manipulation of genes can be combined with more traditional studies of intraspecific genetic variation to obtain a more complete understanding of evolution and evolutionary constraints affecting one model enzyme—the alkaline phosphatase of *Escherichia coli*. Some of the conclusions apply narrowly to alkaline phosphatase, others bear on very general principles of protein evolution, and still others are posed as hypotheses whose generality must be evaluated in other systems.

This chapter is divided into three parts. The first focuses on intraspecific genetic variation in alkaline phosphatase and the statistical inferences that can be drawn from the sampling distributions of amino acid polymorphisms versus synonymous nucleotide polymorphisms. The second part deals with experimental manipulation of the amino acid sequence of alkaline phosphatase from the standpoint of single amino acid changes and replacements of segments of functional domains. The third part concerns inferences that can be drawn from the variation in alkaline phosphatase among related bacterial species.

The reasons for choosing the alkaline phosphatase of *E. coli* include convenience of experimental manipulation and a substantial body of information obtained by many other investigators (Torriani-Gorini et al., 1987). This enzyme has been well characterized both enzymatically and genetically and is encoded by the structural gene *phoA*, located at 9 minutes on the standard genetic map (see Figure 1, and Bachmann, 1987).

GENETIC POLYMORPHISMS IN CODING SEQUENCES

The sequence of *phoA* includes 1413 nucleotides encoding a 471 amino acid polypeptide. Complete sequences of the coding region were obtained from eight natural isolates of *E. coli* (DuBose et al., 1988). The origin of the strains is given in Table 1, along with the sample configurations represented by the nucleotide sequences. Among the eight sequences, there were 10 amino acid polymorphisms. The sample configuration {8,0} means monomorphism in the sample; {7,1} means polymorphism with seven strains sharing a consensus amino acid; {6,2} denotes six strains sharing an amino acid and two sharing a different amino acid; and so on. The sample configurations for amino acids are compared with those of nucleotides at twofold and fourfold degenerate sites in the third positions of codons. Twofold degenerate sites will accept any pyrimidine or any purine, and fourfold sites will accept any nucleotide without altering the amino acid specified (Li et al., 1985). We exclude all third-position sites that are nondegenerate and those such as leucine and arginine that are ambiguous.

The analysis of the sample configurations proceeds as in Sawyer et al. (1987), and the details are given in Hartl and Sawyer (1990). One conclusion is

(A)

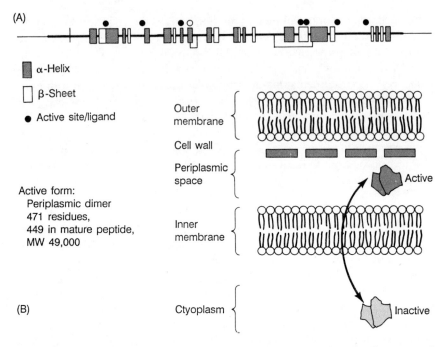

α-Helix

β-Sheet

● Active site/ligand

Outer membrane

Cell wall

Periplasmic space

Active form:
 Periplasmic dimer
 471 residues,
 449 in mature peptide,
 MW 49,000

Inner membrane

Active

(B)

Ctyoplasm

Inactive

FIGURE 1. (A) Schematic of structure of *E. coli* alkaline phosphatase. The structural elements are represented in the line at the top of the figure. The heavy central bar represents the *phoA* coding region and the short vertical line demarcates the cleavage site in the signal sequence for export into the periplasm. Regions of tertiary structure are shown as boxes; open boxes represent β-sheets, stippled boxes represent α-helices. The residues critical to the active site are represented by filled circles above the structure. The open circle represents the residue Arg-166, whose role is described in the text. Disulfide bridges are shown as connecting lines beneath. (B) Localization of alkaline phosphatase; the dimers are active in the periplasm (the space between the cell wall and the inner cell membrane), but are inactive in the cytoplasm.

immediate: because 13% of the twofold and fourfold degenerate sites, but only 0.7% of the combined first and second position sites, are polymorphic there are far too few amino acid polymorphisms. This finding implies that most amino acid replacements are harmful.

  A more refined analysis takes advantage of the degeneracy of the genetic code. We compare the sampling configurations of amino acid polymorphisms with those of synonymous polymorphisms in the same gene. The synonymous polymorphisms provide a null distribution of sites subjected to minimal selective constraints against which the polymorphic amino acid sites can be compared. The assumption that mutation, recombination, random genetic drift, migration, and other relevant aspects of population structure affect the amino acid and synonymous polymorphisms in the same way seems justified, because synony-

TABLE 1. Sample configurations among *phoA* sequences[a]

Configuration	Amino acid sites	Twofold degenerate silent sites	Fourfold degenerate silent sites
8,0	1403	150	177
7,1	6	8	14
6,2	2	5	10
5,3	1	2	7
4,4	1	0	4
Total silent sites		165	212
Polymorphic silent sites		15	35

[a]Strains of *E. coli* originally isolated from a variety of sources by Milkman (1973) and analyzed by Milkman and Crawford (1983). Sources were human (strain designation RM39A, also ECOR4 in Ochman and Selander 1984), gorilla (RM70B=ECOR70), Celebese ape (RM45E=ECOR69 and RM2021=ECOR65), leopard (RM191F=ECOR16), pig (RM201C=ECOR45), goat (RM217T=ECOR67), and giraffe (RM224H=ECOR68).

mous and nonsynonymous sites are interspersed within the same gene. The major difference is that the nonsynonymous sites may be subjected to selection pressures substantially greater than those impinging on the synonymous sites.

The population model for synonymous sites is one in which haploid individuals die at a constant rate and are replaced with possibly mutated offspring of other individuals. Under the assumption that all nucleotide substitutions are selectively neutral, independent, and equiprobable, the stationary population frequencies have a Dirichlet distribution with a parameter α measuring the combined effects of mutation, recombination, and random genetic drift (Sawyer et al., 1987).

Using the sample configurations of the 377 synonymous sites in Table 1, we estimated α by the method of maximum likelihood as $\alpha = 0.026$, with a 95% confidence interval of 0.019–0.034. The sample configurations fit a neutral distribution with $p = 0.76$ in a χ^2 test with 6 degrees of freedom. This nonsignificant result implies that the neutral distribution with the estimated value of α may be used as a null hypothesis for analysis of amino acid polymorphisms.

The next step is to use the estimated distribution to make inferences about the amino acid polymorphisms. This can be approached in two ways. First, it is a problem in hypothesis testing: can the hypothesis that all amino acid polymorphisms are selectively neutral be rejected? This hypothesis may seem redundant in view of the conclusion just stated that most amino acid replacements are harmful. In this context, however, we are inquiring about amino acid positions that are polymorphic, not about all possible amino acid replacements. That is, we are making the distinction between mutation creating a new variant and the fixation of such variants within populations. Although only a minority of possible amino acid replacements may be selectively neutral, those that are neutral are more likely to become polymorphic, and the null hypothesis is that all amino acids that are polymorphic in the sample are selectively neutral. Specifically, suppose π is the fraction of amino acid polymorphisms in the sample that are selectively neutral. Then a confidence interval for π can be estimated from the sampling configurations (Sawyer et al., 1987; Hartl and Sawyer, 1990). In the case of *phoA*, this approach is of limited utility because the 95% confidence interval for π is 0–1.39, that is, the sample configurations provide no significant evidence for or against selective neutrality of amino acid polymorphisms. In contrast, a similar analysis of nucleotide sequences in the *gnd* gene for 6-phosphogluconate dehydrogenase in the same strains indicated that at least 50% of the amino acid polymorphisms in this gene are subject to selection (Sawyer et al., 1987).

The second approach to analyzing the amino acid polymorphisms is to estimate the average selection coefficient among all amino acid replacements that would be required to generate the sample configurations in Table 1. To obtain an estimate of the average magnitude of selection, we assume that there is one favored amino acid at each site in the polypeptide and that all other possibilities are equally deleterious. With this assumption, the population frequency x of the correct nucleotide in the first or second positions in a codon has the stationary distribution

$$Ce^{\sigma x}x^{\alpha-1}(1 - x)^{3\alpha-1}$$

where σ is a measure of selection; $1 - x$ is the combined frequency of the three incorrect nucleotides; α is the parameter estimated above from the configurations of synonymous sites representing the effects of mutation, recombination, and genetic drift; and C is a normalization constant. For the *phoA* data, we set $\alpha = 0.0260$ and obtain the maximum likelihood estimate $r = 43.1$, with a 95% confidence interval of 24.3–77. The conventional selection coefficient is given by $s = \mu\sigma/(3\alpha)$, where μ is the nucleotide mutation rate. Assuming $\mu = 0.5$–2×10^{10} per nucleotide per generation (Drake, 1969 and personal communication), then the average selection coefficient against random amino acid replacements in *phoA* is $s = 2.8 \times 10^{-8}$–1.1×10^{-7}. The estimated selection coefficient is remarkably small considering that many individual replacements (for example, critical residues in the active site) have drastic effects on enzyme function. It nevertheless implies that many other amino acid replacements are selectively neutral or only slightly deleterious. Indeed, as shown in the next section, *phoA* is actually quite tolerant of single amino acid replacements.

EXPERIMENTAL MANIPULATION OF THE AMINO ACID SEQUENCE

Experimental manipulation of the amino acid sequence of alkaline phosphatase was motivated by new concepts in enzyme evolution. The point of departure is the concept of modular evolution, or genes-in-pieces (Gilbert, 1978). This model postulates that novel polypeptides do not evolve one residue at a time, but rather by the assembly of larger preexisting units (or domains). The theory contrasts with that of protein evolution by single residue replacement, which is well supported for the refinement of preexisting enzyme structure, but which stretches beyond plausibility the chance that the de novo random assembly of amino acids could lead to useful functions in any but the smallest polypeptides (Hartl, 1989).

According to the genes-in-pieces model, the units that survive today as introns originally served as the boundaries between the assembled smaller subunits. There is considerable evidence supporting this idea, for example, the observations that the positions of introns in hemoglobin and lysozyme almost exactly mark the boundaries between amino acid residues that are spatially clustered in the tertiary structures (Gō, 1981, 1983, 1985; Artymuik et al., 1981; Jung et al., 1980; Craik et al., 1982), and the finding of paralogous peptide segments that are shared among the low-density lipoprotein (LDL) receptor and other proteins (Südhof et al., 1985a,b).

From the standpoint of direct experimental manipulation, the genes-in-pieces concept is difficult to test because it is quite general and the delineation of large folding domains is often rather vague. However, a recent postulate— the principle of local functionality (Brenner, 1988)—is fundamental to the genes-in-pieces hypothesis, and is specific enough to make experimentally testable predictions. The principle states that the folding of small segments of a protein structure is mediated largely by local interactions of its amino acids. According to Brenner, these small autonomously folding segments were the evolutionary building blocks of the large folding domains specified by many of the exons existing today.

This refinement of the genes-in-pieces model is appealing both because it has theoretical support (Presta and Rose, 1988; Richardson and Richardson, 1988) and because it makes specific predictions that are amenable to direct experimental testing. Since the model concerns primarily small peptide segments, it is possible to test the theory directly by recreating the sorts of events that may have resulted in the assembly of novel proteins—namely, the replacement of structural units within a protein with similar structural units from other sources.

If the principle of local functionality does not hold or has only limited applicability, however, the modular theory of protein evolution would be undermined. It would be difficult to defend the position that large, complex domains can be assembled and continue to function in the face of observational evidence implying that smaller, less complex units cannot.

Single amino acid replacements

With the principle of local functionality at issue, we have used oligonucleotide site-directed mutagenesis (Zoller and Smith, 1983) to make specific, multiple changes in short DNA sequences. Alkaline phosphatase is nearly ideal for such experiments because it is a dimer of two identical 449 amino acid polypeptides and is located in the periplasmic space between the cell membrane and the cell wall, which makes it easy to isolate. Moreover, its three-dimensional structure has been determined by X-ray crystallography to a resolution of 2.8 Å units (Sowadski et al., 1985; and see Figure 1).

The method of oligonucleotide site-directed mutagenesis generally proceeds as follows (Figure 2). A desired sequence is first cloned into a single-stranded DNA vector. An oligonucleotide primer complementary to the sequence but containing certain mismatches (the desired mutations) is synthesized artificially and annealed to the target template. The remaining single-stranded circle is replicated by DNA polymerase and free nucleotides. *E. coli* cells are then transformed with the resulting heteroduplex molecule, and the mutant sequences derived from the DNA strand synthesized in vitro are identified by DNA hybridization and recovered. In this manner, it is possible to create point mutations, deletions, insertions, or multiple amino acid replacements in the target gene (Ghosh et al., 1986; Mark et al., 1987; Pine and Huang, 1987; Zoller and Smith, 1983).

Initially, we introduced a series of point mutations at various locations along the length of the *phoA* gene. Although the ultimate goal was to replace larger regions, point mutants were chosen first to investigate the range of activity differences. These replacements were also necessary to indicate whether the effects of larger replacements were likely to be due to one or a few critical residues. The initial replacements included a presumably conservative substitution in an α-helix (see Table 2, entry L196V), a potentially harmful charge change in a β-sheet (T203K), and a presumed knockout mutation in one of the metal ion-binding ligands (H331P). These three replacements were constructed to determine whether their predicted effects would be observed, and Table 2 shows that all three mutants had the predicted effects on activity.

The next pair of mutations (R166K and R166G) altered an arginine residue believed to stabilize the active site. Two replacements were made: R166K changed the arginine to lysine [shortening the side chain by one methylene (—CH$_2$—) group]; and R166G changed the arginine to glycine (removing the side chain entirely). The prediction was that if the active site is as cramped as the crystallography indicates, shortening the side chain on the arginine might weaken its stabilizing effect on the negatively charged hydroxyl group of the active-site serine, and removing the side chain completely should eliminate the stabilizing effect. In terms of kinetics, this means that as the side chain becomes shorter, the enzyme should be unable to capture phosphate molecules as easily (the K_m will increase), but should have a faster rate of turnover (the K_{cat} will also increase). This is precisely what happens. Moreover, in all mutants the ratio of the K_{cat} to the K_m is approximately the same, which indicates that the lessening of scavenging ability is about equal to the increase in the turnover rate. This implies that although these mutants would be selected against under

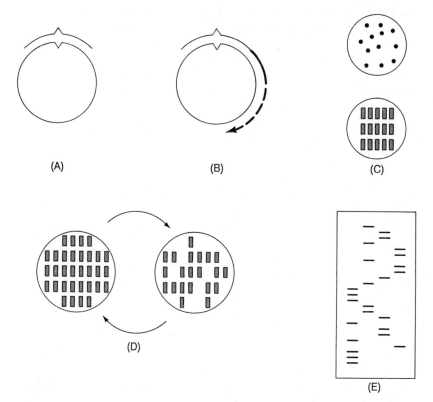

FIGURE 2. General strategy for oligonucleotide site-directed mutagenesis. (A) An oligonucleotide primer containing one or more base-pair mismatches but otherwise complementary to the region of interest is annealed to a single-stranded circular DNA molecule. (B) DNA polymerase is used to extend the annealed primer, producing a double stranded circle. (C) The DNA is transformed into a competent *E. coli* host, resulting in plaques on a lawn of cells (top panel). Individual plaques are picked and subcloned to a fresh plate for screening (bottom panel). (D) Presumptive mutants are screened by performing colony hybridizations from the subcloned plaques, using the original mutagenic oligonucleotide radiolabeled with γ-ATP as probe. The filters are washed repeatedly at increasing temperatures until only the mutant colonies remain. The washing conditions can be adjusted so that a single base pair mismatch eliminates the signal. (E) Mutants are verified by DNA sequencing (see Zoller and Smith, 1983, for more detail).

conditions of phosphate starvation (because of reduced scavenging ability), the mutations might be beneficial in phosphate-rich medium because of their enzyme's higher turnover rate (Dykhuizen and Hartl, 1983). Other studies of the catalytic mechanism of alkaline phosphatase, focusing on the role of Arg-166, have independently created the same mutations and reached similar

TABLE 2. Activities of amino acid replacements in alkaline phosphatase[a]

Enzyme	K_m	K_{cat}	K_{cat}/K_m
Wild type	36.23	103.14	2.847
Q82P	26.59	47.85	1.800
S102T	0	0	0
S102T + A104R	0	0	0
R166K	31.25	91.07	2.914
R166G	70.50	111.41	1.580
C168V	77.05	2.23	0.029
L196V	22.83	64.25	2.814
A200P	36.79	72.78	1.978
T203K	45.93	12.49	0.272
A260P	92.09	3.36	0.036
H331P	0	0	0
N334D	42.65	90.23	2.116
C336M	0	0	0
C336S	0	0	0
A348P	26.99	82.75	3.066
A396P	46.33	87.71	1.893
2c (A61G)	33.16	49.10	1.481
2ac (A61G + A65G)	16.86	20.71	1.228
4c (T107S + A108G)	28.45	38.89	1.367
7c (T174S)	42.41	115.86	2.732

[a]Replacements are designated using single letter amino acid abbreviations and position in the polypeptide chain. For example, Q82P is a replacement of glutamine at position 82 with proline. Wild type is defined as the enzyme present in E. *coli* strain K12.

conclusions (Butler-Ransohoff et al., 1987; DuBose et al., 1987; Chaidaroglou et al., 1988).

The other point mutations in Table 2 include a series of proline replacements and two changes (C168V and C336S in Table 2) that each remove one of the two disulfide bridges in alkaline phosphatase. These mutations, the locations of which are shown in Figure 3, serve as a baseline for other studies and can also be used to study the effects of multiple spatially distant mutations.

Some of the effects of the proline substitutions were quite unexpected. Of the six proline mutations, some chosen deliberately to reside in well-defined structures such as an α-helix (A348P) or a β-sheet (Q82P), only two resulted in drastically reduced activity (Table 2). One (H331P) eliminated a metal-ion binding site, and so loss of activity is expected; but the other (A260P) is located

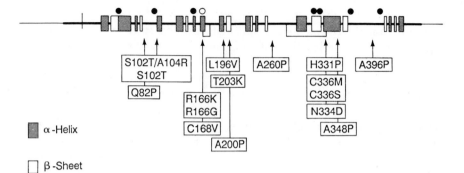

FIGURE 3. Locations of single amino acid replacements introduced into alkaline phosphatase. The structural regions are diagrammed as in Figure 1. The individual point mutations are shown in boxes below the diagram, where the first letter represents the wild-type residue, the number the position of the amino acid in the chain, and the last letter the amino acid present in the mutant. For example, R166K means that wild-type Arg-166 has been replaced with lysine. Multiple mutations replacing the same residue (e.g., R166) are enclosed in a single box.

in a region of random coil, where one might expect the least constraint in terms of amino acid replacement. The loss of activity would suggest that "random coils" are anything but random. It is also worth noting that the proline substitution that lies in the heart of an α-helix (A348P) actually possesses activity slightly greater than wild-type, despite the fact that from a physical standpoint one would expect the structure in that region to be somewhat distorted, and thus the activity lower.

In sum, the results presented in Table 2 show that alkaline phosphatase activity is insensitive to amino acid replacements at many positions. It is true that some replacements at certain key sites are nonfunctional, but most replacements exhibit some activity. This does not imply that these mutations would be selectively neutral in nature, as the analysis presented in the previous section demonstrated that most amino acid replacements in alkaline phosphatase are harmful. On the other hand, the enzyme still functions, even if poorly, which implies that mutant enzymes with single amino acid replacements would have enough activity to serve as starting material for improvement by natural selection.

Multiple amino acid replacements and the principle of local functionality

With the point mutations serving as background, we turned specifically to the testing of a prediction of the principle of local functionality. The underlying rationale was that, if the theory is correct, one should be able to substitute a region of structure in one protein with another similar structure from a different source without complete loss of function. Focusing first on α-helices, we constructed three sets of mutations using the helices designated 2, 4, and 7 in Figure 4. These three helices were chosen because Chou–Fasman analysis of predicted secondary structure (Chou and Fasman, 1978) indicated that appropriate compensating frameshifts flanking the region of interest could result in a structure that still possessed helical potential. To further increase helical potential, proline residues were eliminated in the frameshifted region by first introducing appropriate point mutations in the wild-type helix. These are the mutations listed in Table 2 as 2c, 2ac, 4c, and 7c, each of which still retained enzyme activity. None of the double frameshift mutants possessed activity. However, a problem in interpreting this result is that the predicted helical structures of the double frameshift regions, as inferred from the Chou–Fasman analysis, are not independently verifiable. For this reason, we turned to amino acid sequences known to be helical in their native state.

To make substitutions known to be helical in alkaline phosphatase or in another protein, multiple amino acid replacements were created in three steps. First, by oligonucleotide site-directed mutagenesis, the DNA sequence coding

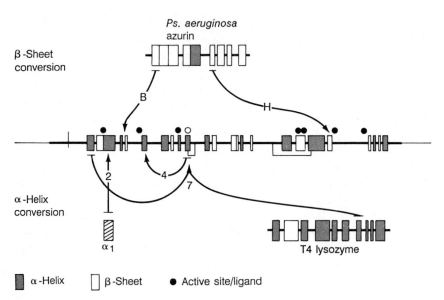

FIGURE 4. Replacements of local structural regions in alkaline phosphatase. The three α-helices (2, 4, and 7) and the two β-sheets (B and H) are indicated. The base of each arrow indicates the origin of the donor element and the arrowhead the target element in the enzyme. Helix 7 was replaced with helix 1 from alkaline phosphatase and helix 6 from bacteriophage T4 lysozyme. Alkaline phosphatase helix 7 also served as donor for a replacement of helix 4.

for residues in the amino-terminal (A) and carboxyl-terminal (B) halves of an α-helix were replaced with the corresponding sequence encoding another helix. The sequence encoding carboxyl-terminal residues was then replaced in a molecule already containing the amino-terminal part, yielding a new protein in which both segments of the structure were replaced (AB; see Figure 5). With this three-step strategy, we could study the effects of helical replacement as well as the interaction of substitutions within helices. The results imply considerable interaction between the two parts of the replacements.

The same three helices used in the frameshift experiments—2, 4, and 7—

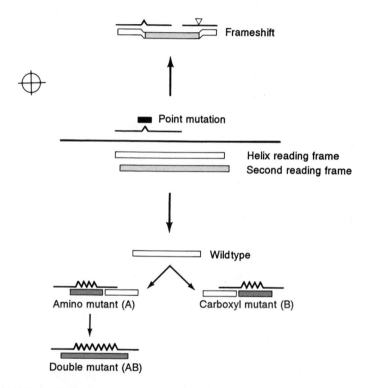

FIGURE 5. Mutagenesis strategy for local structural replacements. A portion of the wild-type *phoA* DNA is shown as the thick black line at the center of the figure, with wild-type and frameshifted reading frames indicated below. Frameshift mutations (upper panel) were created by introducing compensating single-base-pair insertions and deletions at the 5′ and 3′ ends of the DNA region of interest. These mutations result in a short run of amino acid replacements translated in an altered reading frame. Structural replacements (lower panel) were created in two stages. First, oligonucleotide site-directed mutagenesis was used to replace either the amino (A) or carboxyl (B) residues in the region of interest. Second, the carboxyl residues were introduced into a structure already containing the changed amino residues, resulting in a double mutant (AB) with all the amino acids in an entire α-helix or β-sheet element replaced.

were targeted for replacement. These represent an α-helix sensitive to amino acid replacement (helix 4, adjacent to the active site), an α-helix insensitive to replacement (helix 7, exposed to solvent on the edge of the active pocket), and one that interacts with other structures more distal to it (helix 2, in the vicinity of overlap of the alkaline phosphatase monomers; see Sowadski et al., 1985). The donor helices were chosen from alkaline phosphatase itself, from bacteriophage T4 lysozyme (Matthews and Remington, 1974), and from an artificially designed helix termed "α1" (Eisenberg et al., 1986; Ho et al., 1987; Regan and DeGrado, 1988) (see Figure 4).

In each case of helix replacement, a hierarchy was observed in which the rank order of the activities followed the general pattern: $(7 > 1) > (7 > T4) > (4 > 7) > (2 > \alpha1)$, and $B > A$ (the activity of AB ranged from closer to A in some mutants to closer to B in others). The activities are given in Table 3.

The data show two salient patterns. First, mutations in the amino-half of the helix (A) invariably result in less active enzymes than do mutations in the carboxyl-half (B). One explanation for this effect is that improper folding at the initiation of a helical run perturbs overall structure more than improper folding at the end of a helical run. Regardless of the physical cause, comparisons of DNA sequences of *E. coli* and *Serratia marcescens* (discussed below) reveal an excess of amino acid replacements in the carboxyl halves of α-helices, suggesting

TABLE 3. Activities of α-helix replacements[a]

Enzyme	K_m	K_{cat}	K_{cat}/K_m
$7 \rightarrow 1.A$	14.14	8.32	0.588
$7 \rightarrow 1.B$	22.28	90.14	4.046
$7 \rightarrow 1.AB$	21.29	73.57	3.456
$4 \rightarrow 7.A$	29.09	1.25	0.043
$4 \rightarrow 7.B$	9.42	26.13	2.774
$4 \rightarrow 7.AB$	26.21	0.42	0.016
$7 \rightarrow T46.A$	40.96	2.65	0.065
$7 \rightarrow T46.B$	18.59	65.24	3.509
$7 \rightarrow T46.AB$	42.01	8.90	0.212
$2 \rightarrow \alpha1.A$	≈ 0	≈ 0	≈ 0
$2 \rightarrow \alpha1.B$	≈ 0	≈ 0	≈ 0
$2 \rightarrow \alpha1.AB$	≈ 0	≈ 0	≈ 0
Wild type	36.23	103.14	2.85

[a]In each case the target α-helix in alkaline phosphatase is given to the left of the arrow and the source of the replacement α-helix at the right. A denotes the amino half of the helix, B the carboxyl half, and AB the complete replacement. Wild type is defined as the enzyme present in *E. coli* strain K12.

that this pattern is preserved over evolutionary time. Second, there is a non-additive interaction between the residues that comprise individual structural units. In each case, the activity of the double mutant differs from the combination of the two individual single mutants. In addition, the K_m and K_{cat} are affected differently. The harmful effect of multiple substitutions in a helix is epistatic in that the double mutant shows the K_m of the least active single mutant. The K_{cat} does not follow such a simple pattern; in some cases, the double mutant has a turnover rate similar to the amino-terminus single mutant, whereas in others it is more similar to the carboxyl-terminus mutant.

Replacement mutations were also created for two β-sheets in alkaline phosphatase (sheets B and H in Figure 4), with *Pseudomonas aeruginosa* azurin as the donor template molecule (Canters, 1987; Norris et al., 1983). The sequence analysis discussed below indicates there are significantly fewer polymorphisms in β-sheets than expected at random, and thus the effect of β-sheet replacement might be expected to be quite severe. This was the observed result: of the six β-sheet replacements (A, B, AB for sheets B and H), only one had enzyme activity—the carboxyl half of *phoA* sheet B → azurin sheet 1.

In summary, by site-directed mutagenesis we have been able to test predictions about the course of protein evolution at the molecular level. We have found that small structural elements (α-helices and β-sheets) can be replaced without loss of enzymatic function. For helical structures in alkaline phosphatase, it appears that replacement of one helical-forming unit with another unrelated helix does not necessarily eliminate activity. Not all replacements work equally well, which provides ample scope for divergence and specialization of function as chimeric proteins evolve. But β-sheets in alkaline phosphatase appear to be much more highly constrained, an observation that is supported by the comparative sequence data discussed below.

COMPARISONS AMONG SPECIES OF ENTERIC BACTERIA REVEAL SELECTIVE CONSTRAINTS

To complement our experimental studies, we have examined the patterns of amino acid replacement in alkaline phosphatase by cloning and sequencing the gene from *Serratia marcescens*, a close relative of *E. coli*. Previous work had shown that *E. coli* and *Serratia* hybrid dimers are functional (Bhatti, 1973, 1975; Bhatti and Done, 1973), and that the regulatory network in *E. coli* is capable of regulating the *Serratia* gene (Levinthal et al., 1962; Signer et al., 1961). As might be expected, the *Serratia* sequence shows evidence of selective constraints on both the DNA and the protein sequences. At the DNA level, there are many more synonymous substitutions than those causing amino acid replacements, with the number of synonymous changes at near saturation level, which is expected, given the estimated divergence time of the two bacterial species (Ochman and Wilson, 1987; Wilson et al., 1987; Woese, 1987). The synonymous polymorphisms are scattered at random throughout the gene. How-

□ α-Helix   ☐ β-Sheet

FIGURE 6. Spatial distribution of amino acid differences between the alkaline phosphatase of *E. coli* and *Serratia marcescens*, aligned with respect to structural elements. Vertical lines denote residues that differ between *E. coli* and *S. marcescens*. Structural elements are shown below the line, with stippled boxes representing α-helices and open boxes β-sheets. The signal sequence is not included. The distribution of differences along the polypeptide is significantly nonrandom as judged using tests of Feller (1957) and Stephens (1985).

ever, there is a striking nonrandom pattern in the spatial distribution of differences in first and second positions of codon. In particular, there are long stretches of amino acids with no substitutions ($p < 0.0001$ from computer simulation studies). Relative to the three-dimensional structure, the regions devoid of replacements correspond to β-sheets (Figure 6).

The relatively low frequencies of replacements in the β-sheets agree with the results from site-directed mutagenesis, in which only one of the six replacement mutations in the β-sheets did not eliminate activity. The deficiency in the number of replacements within the β-sheets presumably occurs because amino acid changes in these regions are harmful to enzyme function and deleterious to the organism. Consistent with this view, the point mutations listed in Table 2 that are near β-sheets have the lowest enzyme activity.

The histogram in Figure 7 shows the distribution of replacements across the helical regions of *S. marcescens* alkaline phosphatase compared with that of *E. coli*. Even for replacements within α-helices, there is significant deviation from spatial randomness, indicating another level of selective constraint. In particular, when the α-helices are aligned at their amino and carboxyl ends and the replacements are tallied with respect to position within the helix, a significant excess of polymorphisms occurs at the carboxyl ends of the helices ($p < 0.025$). The large bar in the center (M) represents all the middle residues pooled together (because the helices are not all the same length). The numbers in the boxes below the histogram are the results of χ^2 tests performed on the three sections of helix (amino terminus, middle, and carboxyl terminus), as well as the first two combined versus the carboxyl end of the structure. The results of both comparisons are statistically significant. These results agree with those obtained from the site-directed mutagenesis experiments, in which the activity of the carboxyl-terminus mutants was invariably greater than that of the amino-terminus mutants.

One further level of unequal selective constraint in alkaline phosphatase evolution should be mentioned. The 5′ end of the *Serratia* DNA sequence, extending from the start codon to a few residues beyond the signal sequence, is

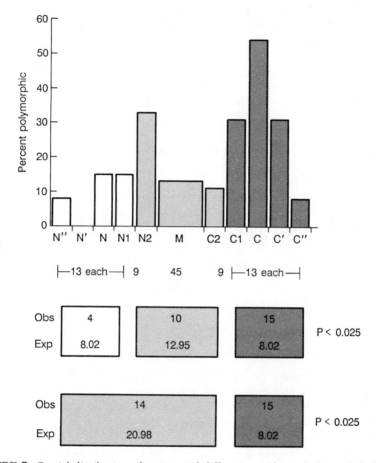

FIGURE 7. Spatial distribution of amino acid differences within α-helices of alkaline phosphatase in *E. coli* and *Serratia marcescens*. All 13 α-helices were aligned with respect to four residues around the amino (N) end and four around the carboxyl (C) end, leaving gaps in the middle (M) where necessary. The number of residues that differed in each position was tallied. Numbers below the histogram represent the total number of α-helices with residues in each position. N″ and N′ define the two residues upstream of the amino end of the α-helix, and C′ and C″ define the two residues downstream of the carboxyl end. Results of χ² tests are shown below the histogram, comparing the amino, middle, and carboxyl regions with each other and the amino and middle regions combined versus the carboxyl region. Expected numbers of differences are proportional to the total number of residues in each class.

not recognizably homologous with the *phoA* of *E. coli* under the same alignment stringency as used in the remainder of the sequence. Many gaps must be introduced to align the sequences, and even then the alignment is poor. The protein sequences show somewhat greater similarity. The common characteristic of the two regions is that they both encode signal peptides for transport of the enzyme across the cell membrane. Thus, if the *E. coli* and *Serratia* alkaline phosphatase signal sequences have descended from a common ancestor, this region has evolved at a much higher rate than the rest of the protein, as seems to be true of signal sequences in general (von Heijne, 1985).

Alkaline phosphatase may be unusual in having fewer replacements in β-sheet structure than in α-helices. We examined a series of bacterial protein sequences (azurin, amylase, LDH, subtilisin, and thioredoxin) in the National Biomedical Research Foundation (NBRF) database and found that while there were regions of putative selective constraint, they did not always map to β-sheets, as in alkaline phosphatase. It may be that the β-sheets of alkaline phosphatase are in some manner more constrained than those of the other enzymes, but the constraints are fully consistent with the lack of activity of our multiple amino acid replacements in these regions. If alkaline phosphatase is exceptional in this regard, it should be possible to replace β-sheet structures in other proteins, as the principle of local functionality implies.

CONCLUSIONS

Statistical comparison of the sample configurations of synonymous and nonsynonymous nucleotide substitutions in *phoA* indicates that most amino acid replacements in alkaline phosphatase are harmful. From the observed distribution of polymorphisms among natural isolates, the average selection coefficient against an amino acid replacement at polymorphic sites is estimated to be in the range 2.8×10^{-8}–1.1×10^{-7}. At the same time, the sample configurations are compatible with selective neutrality, implying that the amino acid replacements are selectively neutral or only very slightly deleterious.

By combining experimental manipulation of the amino acid sequence of alkaline phosphatase with comparison of amino acid sequences between species we have gained insights regarding the functional and selective constraints that operate at several levels of structure of this enzyme. Sequence comparisons alone do not always translate into an experimental framework, and experimental manipulation carried out to reveal enzyme mechanisms does not usually yield evolutionary generalizations. By combining the approaches, one is able to infer evolutionary patterns and test specific predictions experimentally. The feedback between these approaches yields a clearer understanding of the forces that shape protein evolution.

The experimental replacement of oligonucleotide site-directed mutagenesis for replacing single amino acid residues has indicated that alkaline phosphatase can accommodate many single amino acid replacements without loss of function. The larger structural units chosen for replacement were α-helices and β-sheets. This choice was motivated by the principle of local functionality, which is well defined and lends itself to direct experimental testing. The principle is relevant to the origin of novel enzyme functions: since the length of most extant polypeptides precludes the possibility that they were created by random assembly of amino acids, a reasonable alternative is that they were originally assembled from ancestral smaller preexisting units capable of relatively independent folding into specific conformations. The smaller units themselves might have originated residue-by-residue, as the combinatorial odds against them are less unfavorable. Brenner (1988) has suggested that the smaller units correspond to local structures such as α-helices, β-strands, or centers catalyzing certain reactions, such as active-site serines. According to the principle of local functionality, amino acids in close physical proximity in folded proteins delineate the folding domains. Such local folding units are much smaller than those envisioned in the genes-in-pieces model, which often coincide with entire exons. If such local folding units provide building blocks for protein assembly, it should be possible to replace existing units with similar structures from other sources. From an evolutionary standpoint, it is not necessary that the replacements possess wild-type levels of activity, but only that some function be preserved so that natural selection may improve function by means of successive single amino acid replacements (Hartl, 1989). Although limited in scope, the replacements of α-helices with different helical structures that we have carried out support the principle of local functionality.

Comparisons of DNA sequences among species aid in understanding the evolution of protein structure because regions of conserved amino acid sequence reflect structural constraints on the molecule. In bacterial alkaline phosphatase, the conserved regions map primarily to β-sheets. In six experimental replacements of β-sheets, only one was tolerated, a result again indicating stronger constraints on β-sheets than on α-helices. Similarly, within the α-helices of alkaline phosphatase, there may be stronger constraints at the amino end than at the carboxyl end. Some constraints may be quite general across structural features, while others are unique to particular proteins. For example, comparisons of the sequences of other crystallized proteins among species have failed to demonstrate a conservation of β-sheets similar to that in alkaline phosphatase.

Our experiments are far from exhaustive. In work in progress, we are attempting to select for greater enzyme activity in the multiple amino acid replacement mutants in which activity is greatly reduced. From the standpoint of enzyme structure, the interesting question is whether mutations that enhance activity are found within the local folding unit disrupted by the mutations, or whether they are located elsewhere in the molecule. These experiments mimic early evolution in that a local folding unit that is not optimal for activity is refined via single amino acid replacements.

Methods similar to ours can be used more generally to test the genes-in-pieces hypothesis through domain sharing. Since it is possible to use oligonucleotide site-directed mutagenesis to introduce restriction sites at strategic places in any DNA sequence, larger regions of protein structure encompassing entire domains can be replaced.

In the more general context of the change in focus that evolutionary genetics is undergoing, the incorporation of molecular biology into evolutionary studies has provided a valuable set of tools with which to explore the mechanisms of molecular evolution and the nature of selective constraints. By combining the insights gained from more traditional studies of genetic variation within and among species with those obtained from direct experimental manipulation, we are able to test theories previously out of reach and thus achieve a clearer understanding of the forces that shape the course of evolution at the molecular level.

CHAPTER 4

ORTHOLOGOUS AND PARALOGOUS DIVERGENCE, RETICULATE EVOLUTION, AND LATERAL GENE TRANSFER IN BACTERIAL *trp* GENES

Irving P. Crawford and Roger Milkman

Sound bacterial phylogeny, as determined by the molecular analysis of certain highly conserved genes (such as 16 S rDNA), has now emerged from the uncertain background of a taxonomy based mainly on metabolic capabilities and a few structural features. Even before the establishment of essentially universal phylogenetic criteria, it was known that microorganisms vary broadly in chromosomal arrangement, regulatory mechanisms, and coding sequences of genes with related functions. The tryptophan synthetic pathway genes (Figure 1) are as well studied as any; although the biochemical reactions of this pathway are constant, a variety of gene arrangements, gene fusions, and regulatory mechanisms has been discovered over the past 30 years of study of bacteria and fungi (Crawford, 1989).

Now this variation can be superimposed on a phylogenetic tree. Complete congruence between the bacterial phylogeny and the evolution of the *trp* gene complex could imply an evolutionary indivisibility of the bacterial genome, with no exchange possible between species. Complete discordance between the two might indicate a massive interchange during evolution, at least of those parts of the chromosome not directly involved in replication, transcription, and

FIGURE 1. Tryptophan biosynthetic pathway, together with the branch to p-aminobenzoic acid (folate pathway). Chorismic acid is the last intermediate common to the synthesis of aromatic amino acids and vitamins. The reactions are catalyzed, singly or in combination, by the polypeptides (e.g., TrpE) coded by corresponding genes (e.g., trpE). AS, anthranilate synthase; PABS, p-aminobenzoate synthase. Modified from Crawford, 1989.

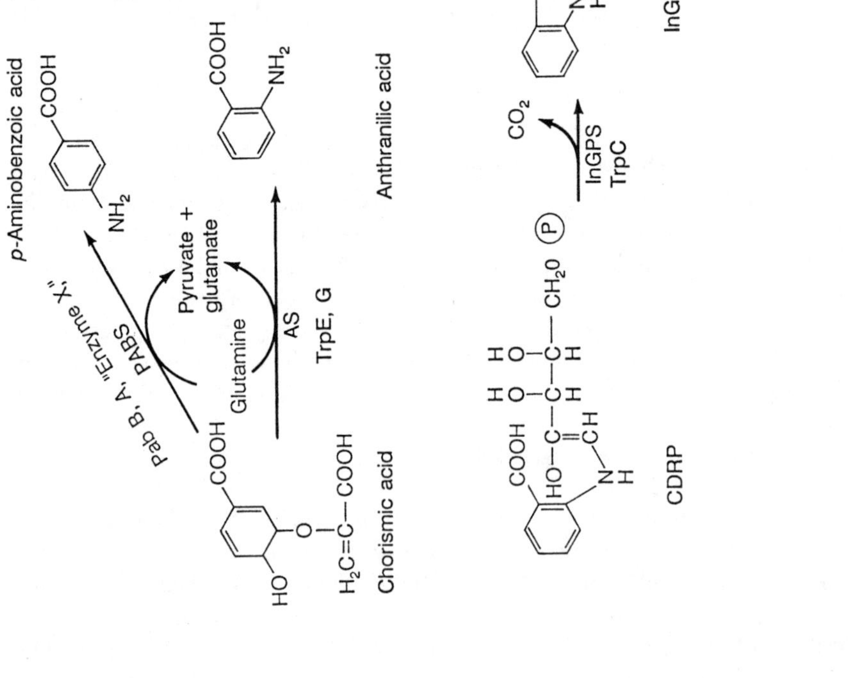

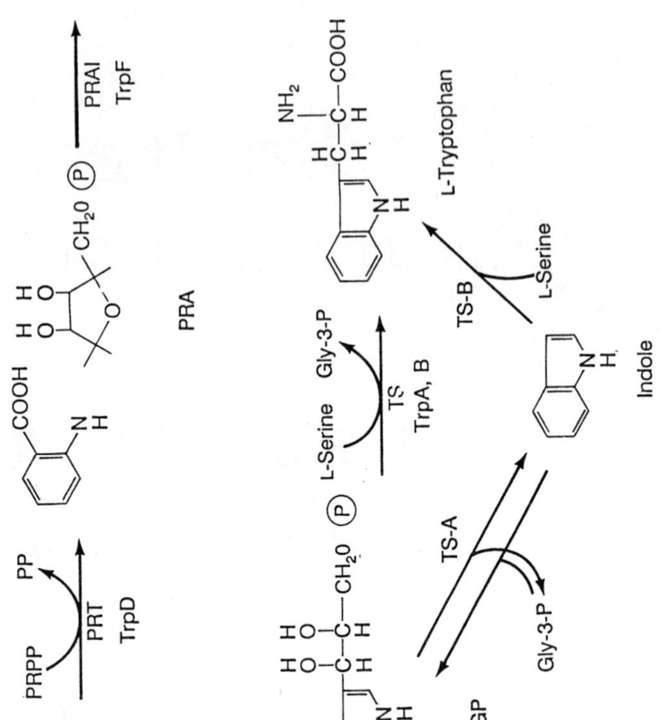

translation. Alternatively, such lack of association could be explained by the operation of discordant adaptive pressures over vast periods of time.

But in fact, congruence is prevalent, though not complete. Although the organization of the *trp* genes is generally similar among closely related bacteria and varies along phylogenetic divisions, at least one striking exception (to be described) suggests lateral gene transfer between bacterial "phyla."

A comparison of *trp* sequences in 28 organisms (1 archaebacterium, 18 eubacteria, 8 fungi, and 1 plant) shows that the seven polypeptides used in tryptophan synthesis are highly conserved in catalytic properties, as well as in secondary and tertiary structure. Amino acid sequence is slow to change, although some polypeptides vary a great deal more than others. Regulatory mechanisms are less well conserved than gene structure; they do reflect phylogeny (for example, fluorescent pseudomonads have a characteristic system of regulation), but they are occasionally lost and rebuilt on a different pattern. There is also great variation in the chromosomal distribution of the *trp* genes, and it is clear that proteins can be fused in a variety of combinations without loss of function. Thus an enzyme may be characterized by its amino acid sequence (the direct determination of secondary and tertiary structure, species-by-species, is not practical at present); catalytic properties; the mechanisms, both coordinate and otherwise, regulating synthesis and function; by any fusions with other proteins; and by the position on the chromosome of the gene that codes it. As we shall see, these properties do not always coincide completely, and the amino acid sequence is evidently the best indicator of phylogenetic position.

Homologous characters, including genes, must share a common ancestor. Macromolecular homology has been divided by Fitch and Margoliash (1970) into **orthology** and **paralogy**. The common ancestor of two orthologous characters precedes cladogenesis; the common ancestor of two paralogous characters precedes gene duplication (although the term could be extended to include serial homology at higher levels). We shall speak of orthologous divergence and paralogous divergence in referring to the divergence of orthologous and paralogous characters, respectively.

One or indeed several corresponding steps in two different pathways may be catalyzed by paralogous enzymes. There are occasional strong signs of **paralogous** transfers: three instances will be described in which the replacement of an enzyme in a pathway by a paralogous enzyme catalyzing a similar reaction in a different pathway has been inferred.

Figure 2 is a reminder of the overall scheme of bacterial phylogeny, as revealed by 16 S RNA sequences. In Table 1, which lays out the composition of the two bacterial kingdoms, taxa in boldface type are those in which at least some *trp* genes have been sequenced (with or without regulatory information). These include one archaebacterium (*Methanococcus voltae*), two spirochetes, one high-GC and several low-GC Gram-positive eubacteria, and numerous purple bacteria, mostly in the γ subdivision, with one in the β subdivision and two in the α. There is still much need for survey work using methods that

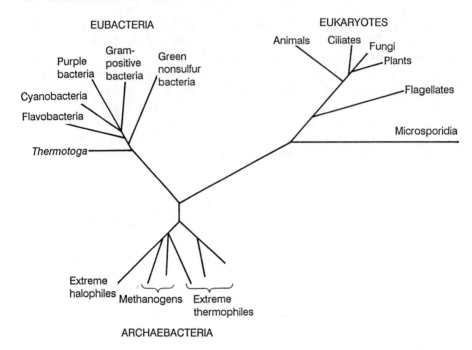

FIGURE 2. The two bacterial phylogenies, taken from the universal phylogenetic tree determined from rRNA sequence comparisons (Woese, 1987).

center on the complementation of *Escherichia coli* auxotrophs or on what we will call *transplacement* (also called "reverse genetics"). Reports on new organisms are now coming in monthly.

Table 2 lists many of the eubacterial species in which the chromosomal organization of the *trp* genes has been determined. (In most of the other species indicated in boldface type in Table 1, only one or two enzymes have been cloned.) The genes in the tryptophan pathway (Figure 1) show considerable variation in operon composition and in gene fusions. Clearly, the grouping of genes reflects their regulation (Figure 3) and leads to inferences about their roles in the ordinary conduct of metabolism in their particular organism. Gene fusions (Table 2) result in protein fusions, often reflecting a close physical association of enzyme activities that goes beyond the coordination of their synthesis, but note that the activities of TrpC and TrpF are, as far as we know, merely successive reactions.

In the fungi, as expected, there are no operons, but several novel gene fusions are seen, namely, (G)C(F), (G)C, and (A)B. The well-known enteric bacterial *trp* operon (all genes contiguous, two pairs fused, regulated by both repression and attenuation) is seen with or without the (G)D fusion in all enteric bacteria, together with their close relative *Vibrio*. This arrangement is seen

TABLE 1. Eubacteria and archaebacteria and their subdivisions[a]

EUBACTERIA

Purple bacteria

α Subdivision
Purple nonsulfur bacteria (*Rhodobacter, Rhodopseudomonas*), **rhizobacteria**, agrobacteria, rickettsiae, *Nitrobacter, Thiobacillus* (some), *Azospirillum,* **Caulobacter**

β Subdivision
Rhodocyclus (some), *Thiobacillus* (some), *Alcaligenes, Bordetella, Spirillum, Nitrosovibrio, Neisseria*

γ Subdivision
Enterics (**Acinetobacter, Erwinia, Escherichia, Klebsiella, Salmonella, Serratia, Shigella,** *Yersinia*), **vibrios, fluorescent pseudomonads,** purple sulfur bacteria, *Legionella* (some), *Azobacter, Beggiatoa, Thiobacillus* (some), *Photobacterium, Nanthomonas*

δ Subdivision
Sulfur and sulfate reducers (*Desulfovibrio*), myxobacteria, bdellovibrios

Gram-positive eubacteria

High G + C species
Actinomyces, **Streptomyces**, *Actinoplanes, Arthrobacter, Micrococcus, Bifidobacterium, Frankia, Mycobacterium,* **Corynebacterium**

Low G + C species
Clostridium, Bacillus, *Staphylococcus, Streptococcus*, mycoplasmas, **lactic acid bacteria**

Photosynthetic species
Heliobacterium

Species with Gram-negative cell walls
Megasphaera, Sporomusa

Cyanobacteria and chloroplasts
Oscillatoria, Nostoc, Synecococcus, Prochloron, Anabaena, Anaystis, Calothrix

Spirochaetes and relatives

Spirochaete group
Spirochaeta, *Treponema, Borrelia*

Leptospira group
Leptospira, *Leptonema*

Green sulfur bacteria
Chlorobium, Chloroherpeton

Bacteroides, flavobacteria, and relatives

Bacteroides group
Bacteroides, Fusobacterium

Flavobacterium group
Flavobacterium, Cytophaga, Saprospira, Flexibacter

(Continued)

TABLE 1. (Continued)

EUBACTERIA (Continued)

Planctomyces and relatives
Planctomyces group
  Planctomyces, Pasteuria
Thermophiles
  Isocystis pallida

Chlamydiae
  Chlamydia psittaci, C. trachomatis

Radio-resistant micrococci and relatives
Deinococcus group
  Deinococcus radiodurans
Thermophiles
  Thermus aquaticus

ARCHAEBACTERIA

Extreme halophiles
  Halobacterium, Halococcus morrhuae
Methanobacter group
  Methanobacterium, Methanobrevibacter, Methanosphaera stadtmaniae, Methanothermus fervidus
Methanococcus group
  Methanococcus
"Methanosarcina" group
  Methanosarcina barkeri, Methanococcoides methylutens, Methanothrix soehngenii
Methanospirillum group
  Methanospirillum hungatei, Methanomicrobium, Methanogenium, Methanoplanus limicola
Thermoplasma group
  Thermoplasma acidophilum
Thermococcus group
  Thermococcus celer
Extreme thermophiles
  Sulfolobus, Thermoproteus tenax, Desulfurococcus mobilis, Pyrodictium occultum

---

[a]**Boldface type** indicates taxa in which at least some *trp* genes have been sequenced (with or without regulatory information).

nowhere else except, remarkably, in *Brevibacterium lactofermentum*, a high-GC Gram-positive organism. This leads to the strong inference of lateral gene transfer at some time after the loss of the pathway. The argument is based on identical regulation and (C)F fusion, as well as greater similarity in sequence to enterics than to others, notably Gram positives (Matsui et al., 1986). It may

TABLE 2. *Trp* gene fusions and gene organization in bacteria and archaebacteria

Organism	Gene organization[a]
EUBACTERIA	
Phylum: Purple bacteria	
α Subdivision	
Rhizobium meliloti and relatives	E(G) D·C F·B·A
Caulobacter crescentus	E G? D·C F·B·A
β Subdivision	
Pseudomonas acidovorans	E G·D·C F·B·A
γ Subdivision	
Escherichia coli and relatives	E·(G)D·C(F)·B·A
Serratia marescens and relatives	E·G·D·C(F)·B·A
Pseudomonas aeruginosa and relatives	E G·D·C F B·A
Acinetobacter calcoaceticus	E G·D·C F·B·A
Phylum: Gram-positive eubacteria	
Low (G + C) subdivision	
Bacillus subtilis and relatives	G E·D·C·F·B·A
Lactobacillus casei	E? G? D·C·F·B·A
High (G + C) subdivision	
Brevibacterium lactofermentum	E·G·D·C(F)·B·A
Phylum: Spirochetes	
Spirochete subdivision	
Spirochaeta aurantia	E G? D? C? F? B? A?
Leptospira subdivision	
Leptospira biflexa	E·G D? C? F? B? A?
ARCHAEBACTERIA	
Phylum: *Methanococcus*	
Methanococcus voltae	E? G? D? C? F·B·A

[a]Contiguous genes are indicated with centered dots; fused pairs are shown with one element in parentheses.

be noted in passing that, at least when the entire pathway is lost, the most likely effective replacement would be a single complete operon, such as that found in the enterics. No other likely cases of lateral transfer of *trp* genes have emerged from this survey, which is, however, far from exhaustive.

PARALOGOUS RETICULATE EVOLUTION

The term *reticulate evolution* was first used in its modern sense by Wagner (1954) to describe a pattern of evolutionary interchanges due to recurrent species

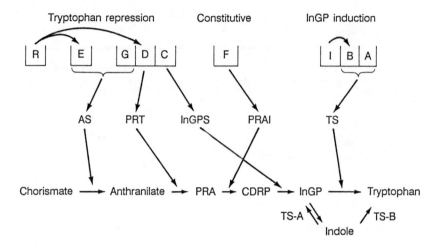

FIGURE 3. In *Pseudomonas aeruginosa* and *Pseudomonas putida* the genes responsible for tryptophan synthesis are located in four chromosomal clusters. The proposed *trpR* gene encodes a repressor acting on *trpE* and on the *trpGDC* cluster. The *trpI* gene encodes an activator inducing the *trpBA* pair. From Chang et al. (1990).

hybridization, followed by the sorting out of new combinations of morphological traits in the descendant species. The resulting reticulate pattern is different from the usual pattern of simple divergence (see also Wanntorp, 1983, and accompanying articles). We here apply the concept of reticulate evolution to paralogous evolution, the evolution of gene families by repeated gene duplication, a process clearly envisioned in the 1950s (Lewis, 1951) but examined extensively only recently. An example in the present context is the assumed duplication and divergence of a gene to make separate glutamine amidotransferases for the tryptophan and folate pathways, followed by the loss of one pathway with double duty assumed by the other. In principle, this process could be followed by a further duplication and divergence to produce separate enzymes again. This process results in a reticulate pattern in the sense that the corresponding steps in the respective pathways are catalyzed by a common enzyme originally, then two that diverged, then a common one again, then two, as Figure 4 illustrates.

So far, five of the seven elementary polypeptides in the tryptophan pathway have been crystallized, and the structures of four of these have been determined. Three (A, C, and F) of the four are TIM barrels (Plapp, 1982; Richardson, 1981; Zubay, 1985), and one of the as-yet-uncrystallized proteins (G) may also be a barrel. The barrels' substrates are all sugar–phosphate relatives.

The eightfold "α-helix———turn———β-pleated-sheet———turn—" pattern imposes substantial structural constraints on these proteins. In the group of proteins sequenced so far, the minimum amino acid identity is 27% for TrpC, 20% for TrpA, and 11% for TrpF. TrpF also contains several "cadres" of highly conserved sequences of five or six residues, with gaps at the turns.

TrpB, which is more complex than a simple TIM barrel, is the most highly

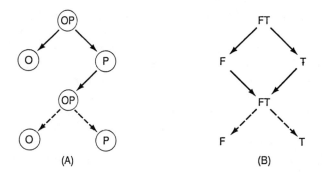

FIGURE 4. (A) An amphibolic enzyme, capable of placing an amino group either *ortho* or *para* to the carboxyl group, evolves paralogously into two enzymes with separate functions. Next, one becomes amphibolic and the other is soon lost. Finally, this amphibolic enzyme evolves once more into two enzymes with separate functions. (B) The reticulate pattern is seen in the relationship of the first steps of the folate and tryptophan pathways, respectively. Initially they are mediated by a common enzyme; next, they are separate; next, they have a common enzyme once again; and finally, they are separate once more. The last step (dashed lines) is hypothetical.

conserved of the seven polypeptides, with a minimum of 47% identity among 12 sequences ranging from yeast to enteric bacteria. In *Brevibacterium lactofermentum*, whose *trp* operon's similarity to the enterics' has already been noted, TrpB has 65% identity with *Vibrio*, 62% with *E. coli*, and less than 54% with the others. The high degree of conservation of TrpB has several likely causes. Tryptophan synthase functions as a tetramer of two α and two β polypeptides. This structure is preserved when, as in yeast, the α and β polypeptides are fused (Crawford et al., 1987). There is a meeting of α and β faces, and a meeting of two β faces: a serine/pyridoxal-5'-phosphate binding region, and an indole *tunnel* that connects the two α and β active sites in the formation of tryptophan (Miles et al., 1987).

At the beginning of the tryptophan pathway is another pair of active sites. These are concerted, if not connected. They are usually present on separate polypeptides, TrpE and TrpG. Although crystal structures are not yet available, a variety of sequences (more than 25 for TrpG) are known. These polypeptides are the subunits of the pathway's first enzyme, anthranilate synthase, which holds particular interest. The large (E) subunit can independently catalyze the formation of anthranilate from chorismate in the presence of high concentrations of NH_4^+ at rather high pH. Its tryptophan binding site implements feedback inhibition. TrpE resembles the product of the *E. coli pabB* gene, except that PabB produces *p*-aminobenzoate in the folate pathway instead of anthranilate (= *o*-aminobenzoate). Also, PabB is not feedback-inhibited.

The small (G) subunit of anthranilate synthase is a glutamine amidotransferase in that it provides an —NH_2 group by deamidating glutamine. In this function it is exactly like the product of the *E. coli pabA* gene. Clearly the elements of reticulate evolution are coming into view. The earliest reactions in the folate pathway resemble the first steps of the tryptophan pathway. Moreover, in three eubacteria (*Acinetobacter calcoaceticus, Pseudomonas acidovorans*, and *Bacillus subtilis*), one small subunit serves both the AS large subunit and the PABS large subunit. In the two purple Gram negatives (one γ, one β) this *amphibolic* gene (Kane et al., 1972; Kane, 1977) is in an operon with *trpD* and *trpC*. Presumably *pabA* has been lost in each case. In *Bacillus subtilis*, a Gram-positive bacterium, the gene is in an operon that includes the PABS genes and another gene in the folate pathway: apparently *trpG* has been lost in *Bacillus* and its function taken over by the gene for the folate pathway's glutamine amidotransferase.* This conclusion is supported by the sequence. Finally, Brian Nichols and coworkers compared the AS and PABS large subunit sequences in a few enteric species (Kaplan et al., 1985; Goncharoff and Nichols, 1988). In each of the several comparisons made, the two sequences appear to be homologous, and the same is true for the small subunits.

Ordinarily, then, anthranilate synthase produces anthranilate from chorismate and glutamine. After amination and the removal of the enolpyruvyl group, the product aromatizes spontaneously to form anthranilate. *Para*-aminobenzoate synthase, the dimer formed from the products of *pabB* and *pabA*, mediates a similar reaction, except that a third enzyme is required to complete the job. This enzyme, called Enzyme X, was partially purified and characterized, by Nichols and coworkers (Nichols et al., 1989), who previously demonstrated that an extract now known to contain both PABS and Enzyme X does not convert isochorismate to *p*-aminobenzoate. Thus, the folate pathway does not branch off with an isomerization of chorismate into a form whose amination simply leads to the formation of the *para*, rather than the *ortho*, product. In the enteric bacteria the large subunits of AS and PABS are clearly homologous; and so are the small subunits. The third PABS gene (coding Enzyme X) has not yet been cloned and sequenced.

So far, so good: straightforward scenarios whose basic patterns are fairly easy to analyze. But a further degree of complexity, involving three pathways, takes a bit more unraveling and therefore requires a little more experimental documentation. As we shall see, this case, plus the possible lateral gene transfer event, indicates that during evolution *trp* genes occasionally have not only transposed and fused, but have been deleted, leaving the cell in subsequent times of need to improvise or import other genes. Clearly, this set of events is not thought to have included an abrupt removal of available tryptophan. Rather, following the loss of *trp* genes during a period of constant abundance, tryptophan availability may be concluded to have declined to a suboptimal level. The supply was adequate for survival and growth, but cells lacking *trp* genes were at a slight or occasional disadvantage with respect to those with the genes,

---

*In *B. subtilis*, the gene orthologous to *E. coli*'s *pabB* is called *pabA*, and the amphibolic gene referred to here, presumably orthologous to *E. coli*'s *pabA*, is called *gat* for glutamine amidotransferase (Zeigler and Dean, 1987).

perhaps in terms of growth rate. At that point, improvisation or import would have conferred a benefit.

THE PHENAZINE PATHWAY

The story of the third pathway developed from the study of duplicate anthranilate synthase genes in fluorescent pseudomonads, in particular, *Pseudomonas aeruginosa* and *P. putida*. These species belong to the γ subgroup of the purple phylum. Surprisingly, no *trpG* (small subunit) mutants have been isolated until recently, although anthranilate synthase has been purified from *P. putida* and the amino acid sequence of the small subunit of glutamine amidotransferase has been determined (Kawamura et al., 1978). To learn the reason for the biochemists' inability to find *trpG* mutants, Crawford's laboratory began to study a clone that complemented AS⁻ *E. coli* mutants. This clone was derived from an R′ plasmid containing about 18 kb of *P. aeruginosa* chromosomal DNA (Hedges et al., 1977). This plasmid, which has a broad host range, complemented *trpE* and only *trpE* mutants of *E. coli*. The plasmid, therefore, produces anthranilate synthase or an enzyme with anthranilate synthase function. The DNA subcloned from the active portion showed *trpE*-like and *trpG*-like reading frames, overlapping by 23 bp. The simplest interpretation was that these are the *trpE* and *trpG* genes of *P. aeruginosa*. Yet the deduced amino acid sequence of the putative TrpG was less than 40% similar with Zalkin's TrpG sequence from *P. putida*. In addition, although the anthranilate synthase produced by *Pseudomonas* cells is feedback inhibited by tryptophan, the "AS" enzyme produced by Hedges' plasmid is not. More detailed evidence will follow, but the evolutionary and functional relationships of AS and PABS suggest the existence of a third pathway. In fact, there is a third pathway in *P. aeruginosa*, probably involving anthranilate and known to lead to the formation of a phenazine pigment called *pyocyanin* (not *pyocin*, a destructive agent derived from phage tails but also found in *Pseudomonas*; it will enter our story later). Figure 5 illustrates pyocyanin, the blue-green phenazine pigment of *P. aeruginosa*, together with its hypothetical synthetic pathway. Although it is known that two molecules each of chorismate and glutamine are used, the formation of anthranilate is hypothetical. Indeed, externally supplied anthranilate does not get into pyocyanin. We have tentatively renamed the genes cloned by Hedges *phnA* (large subunit) and *phnB* (small subunit). The assignment of the genes cloned by Hedges to the phenazine pathway is supported decisively by the following information.

Shinomiya and Ina (1989) identified *P. aeruginosa trpE* on some R′ plasmids that complement *trpE* mutants in *P. aeruginosa* and also cotransduce with *trpD* and *trpC*. [These genes in *P. putida* are known to be close from mapping studies (Gunsalus et al., 1968).] Restriction maps of these *trpE* clones do not resemble those of the subcloned *phn* large subunit gene, derived from Hedges' 18-kb plasmid. Supportive evidence of a third pathway came from a series of

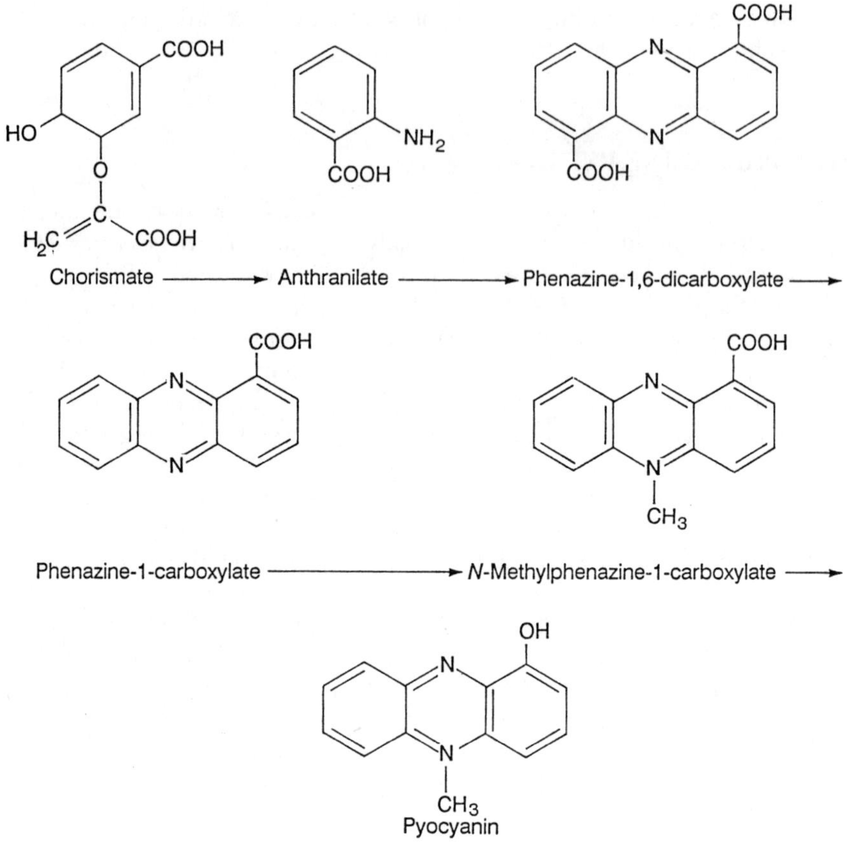

FIGURE 5. Hypothetical scheme for pyocyanin synthesis. The anthranilate molecule has been turned over (cf. Figure 1) to make clear how two anthranilates might join to form phenazine-1,6-dicarboxylate. After Essar et al. (1990).

transplacement experiments. Transplacement experiments comprise the following steps:

1. A gene is cloned into a plasmid of a type devised by Puhler and coworkers that can replicate in one host species but not in the species under investigation (Simon et al., 1983); however, genes cloned on the Puhler plasmid can be expressed in both species.
2. The cloned gene is inactivated by the insertion of an antibiotic resistance cassette.
3. The resulting plasmid is amplified, purified, and then transformed into the bacterium under investigation.
4. The antibiotic whose resistance is conferred by the cassette is added promptly. Resistant colonies can arise only through the incorporation of the Puhler plasmid into the bacterial chromosome; this is generally achieved by ho-

mologous recombination; heterodiploids resolve promptly; in such cases, the selected resistant strain is found to have lost the function of the cloned gene.

In transplacement experiments involving *phn* genes (Essar et al., 1990c), the resulting haploid *P. aeruginosa* strains did not require tryptophan, produced normal anthranilate synthase, and were defective in pyocyanin production.

Essar et al. (1990a,b) then identified and localized the early *trp* genes (*E*, *G*, *D*, and *C*) in both *Pseudomonas* species in a series of steps summarized in Figure 6. They first used a derivative of Hedges' R′ plasmid to clone *phnA* and *phnB* in *P. aeruginosa* to confirm their identity. The cloned small subunit gene (*phnB*) provided a probe with which to recognize *trpG* in *P. aeruginosa* chromosomal DNA digests by low stringency Southern hybridization. The *P. aeruginosa trpG* was now cloned and yielded a probe, which in turn hybridized to *P. putida trpG*, again under low stringency conditions. Note also that in both cases a walk led to the adjacent *trpD* and *trpC*. Next, a slightly longer walk from *trpG* in the opposite direction led to the *P. putida trpE*. And finally, the *P. putida trpE* was cloned, providing a probe that, again under low stringency, hybridized to *P. aeruginosa*'s *trpE*, which was then cloned.

These experiments revealed the following information regarding structure. First, in both species *trpG* is part of a three-gene (GDC) operon, and *trpE* is solitary. And second, in *P. putida* there is a 2.3-kb open reading frame on the opposite strand between *trpG* and *trpE*; but in *P. aeruginosa*, a 25-kb segment containing *pyocin R2* genes intervenes. Recalling that pyocin is a phage tail component, this finding suggests previous occupancy by a complete prophage (Shinomiya et al., 1983; Shinomiya and Ina, 1989).

Returning to transplacement experiments, inactivations of *Pseudomonas*

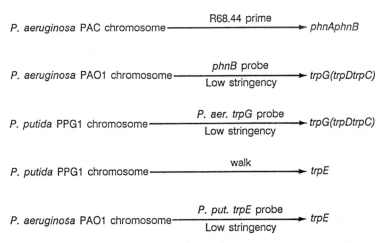

FIGURE 6. Outline of the cloning of the *Pseudomonas aeruginosa* and *Pseudomonas putida* anthranilate synthase genes. Each line indicates the DNA source (left), the source of the probe used in Southern hybridizations, and any departure from normal stringency (center), and the designation of the gene(s) recovered. From Essar et al. (1990).

trpE and *trpG* provided mutants with the expected tryptophan requirement and the concomitant absence of anthranilate synthase in extracts, hence the presence of normal *phnA* and *phnB* did not complement the TrpEG defect. This proved that TrpEG is required in the tryptophan pathway. One further detail should be added: the transplaced *trpG* mutants would grow in normal minimal medium, but this medium contains enough NH_4^+ to permit the large subunit (made by the normal *trpE*) to produce anthranilate without the collaboration of the small subunit made by *trpG*. A decrease in the NH_4^+ level was accompanied by a failure to grow. This experiment explains why *trpG* mutants had never appeared in earlier mutant collections.

To complete the transplacement story, the inactivation of either the *phnA* or *phnB* gene results in defective pyocyanin synthesis. The phenazine pathway may thus have anthranilate as an intermediate, as diagrammed in Figure 5; but if so, why is it inaccessible to the subsequent enzymes of the tryptophan pathway? This question will be addressed shortly.

PHYLOGENY OF THE *trp*, *pab*, AND *phn* GENES

To measure the relatedness of various anthranilate synthase genes, Doolittle's (1986) normalized alignment score (NAS) method is used. Here the question is not simply one of homology (all the comparisons under discussion indicate homology), but the degree of the similarity on which the conclusion of homology is based. The greater this similarity, the closer the relationship (and thus the more recent the common ancestry). An examination of Table 3 and Figure 7 shows that *P. aeruginosa* and *P. putida* TrpE proteins appear to be homologous to *E. coli*'s PABS large subunit, PabB, by the criterion similarity for 500 residues, for which a score of 160 indicates that common ancestry is probable, and 280 is the threshold of essential certainty. By these criteria, they are also homologous to *E. coli* TrpE. More dramatically (Table 4), *P. aeruginosa* and *P. putida* TrpG proteins appear to be far more similar to *E. coli* PabA than

TABLE 3. Normalized alignment scores (NAS) for anthranilate synthase genes from various organisms

Organism	Organism		
	P. putida trpE	*E. coli trpE*	*E. coli pabB*
P. aeruginosa trpE	862	257	269
P. putida trpE	—	255	265
P. aeruginosa phnA	244	456	184

$$NAS = \frac{\text{All identities} + \text{Cys identities} - 2.5 \times \text{gaps}}{\text{Average number of residues}} \times 1000$$

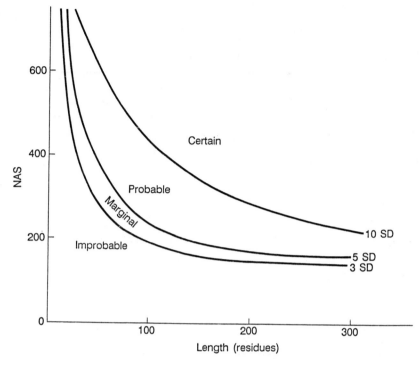

FIGURE 7. Significance of NAS values as a function of sequence length. After Doolittle, 1986.

to *E. coli* TrpG, although they are homologous to both. Note that TrpG and PabA have only about 200 residues, so that the scores must be higher to qualify for homology. Also, *P. aeruginosa* PhnA shares homology with all the TrpE polypeptides and with *E. coli* PabB, but the similarity to *E. coli* PabB is considerably lower. A similar pattern is seen in the homology of PhnB with *E. coli* TrpG and the less similar PabA. Thus, analysis of the primary structure of the two anthranilate synthase gene pairs in *P. aeruginosa* shows, surprisingly, that the *trp* genes are more closely related to the *pab* genes of *E. coli* than to its own *trp* genes. Moreover, the *phn* genes of *P. aeruginosa's phn* genes resemble the *trp* genes of *E. coli*. And, the *trp* genes of *P. putida* are also more similar to the *pab* genes of *E. coli* than to its *trp* genes. In the *P. putida* strain used in these experiments—and, of course, in *E. coli*—there are no *phn* gene pairs to compare.

Are these results also indicative of reticulate evolution? We propose the following model. Beginning with two adjacent ancestral genes that encode a large subunit and a small subunit of an AS-type enzyme, two evolutionary processes occur. First, the pair duplicates and diverges: this paralogous event leads to two related pairs of genes in the line of descent. Second, the line of

TABLE 4. Normalized alignment scores for anthranilate synthase and p-aminobenzoate synthase α and β subunits from various organisms[a]

β Subunit similarities

Organism	S.t. TrpG	S.m. TrpG	V.p. TrpG	P.a. PhnB	E.c. PabA	S.t. PabA	S.m. TrpG	B.s. TrpG	P.p. TrpG	P.a. TrpG	A.c. TrpC
E.c. TrpG	965	839	622	475	398	366	406	399	383	396	317
S.t. TrpG		834	627	470	409	372	395	384	363	381	302
S.m. TrpG			591	396	409	356	395	401	380	401	327
V.p. TrpG				415	356	350	340	369	367	379	362
P.a. PhnB					301	312	304	225	270	316	278
E.c. PabA						877	773	626	676	687	586
S.t. PabA							762	610	676	671	575
S.m. PabA								607	707	718	593
B.s. TrpG									633	585	563
P.p. TrpG										843	655
P.a. TrpG											649

α Subunit similarities

Organism	V.p. TrpE	P.a. PhnA	B.l. TrpE	R.m. TrpE	S.a. TrpE	E.c. PabB	P.p. TrpE	P.a. TrpE	B.s. TrpE	S.c. TrpE
E.c. TrpE	580	456	426	236	269	198	255	257	240	235
V.p. TrpE		431	415	208	275	209	253	260	230	220
P.a. PhnA			392	208	247	184	244	260	225	204
B.l. TrpE				198	222	208	233	237	205	192
R.m. TrpE					208	151	214	213	201	217
S.a. TrpE						213	305	305	261	251
E.c. PabB							265	269	270	217
P.p. TrpE								866	325	299
P.a. TrpE									319	286
B.s. TrpE										255

Source: From Essar et al. (1990c).
[a]Scores obtained from Crawford (1989).
Organisms: A.c., *Acinetobacter calcoaceticus*; B.l., *Brevibacterium lactofermentum*; B.s., *Bacillus subtilis*; E.c., *Escherichia coli*; P.a., *Pseudomonas aeruginosa*; P.p., *Pseudomonas putida*; R.m., *Rhizobium meliloti*; S.a., *Spirochaeta aurantia*; S.c., *Saccharomyces cerevisiae*; S.m., *Serratia marcescens*; S.t., *Salmonella typhimurium*; V.p., *Vibrio parahaemolyticus*.

descent splits: this orthologous event produces the ancestors of *Pseudomonas* on one hand and *E. coli* on the other. In the evolutionary path to *Pseudomonas*, one set becomes *trpE* and *trpG*, and the other set becomes *phnA* and *phnB*. In the evolutionary path to *E. coli* it is the second set that becomes *trpE* and *trpG*, while the first set becomes *pabB* and *pabA* (Figure 8). *Acinetobacter calcoaceticus* can be grouped with the fluorescent pseudomonads on the basis of further comparisons. These (Table 4) show that *A. calcoaceticus* is like *P. aeruginosa* and *P. putida* in that its *trpE* and *trpG* are more like the *pab* genes of *E. coli* than its *trp* genes. Also *A. calcoaceticus'* amphibolic glutamine amidotransferase is also more like the PABS small subunit of *E. coli* than the AS small subunit of *E. coli*.

These data illustrate complexities at the level of enzyme structure produced by combinations of paralogous and orthologous events. Figure 8 will become more interesting when the *pab* genes of the fluorescent pseudomonads can be added to the picture. We can also expect interesting results from an extension of these comparisons to the *phn* genes of other phenazine producers. The changes shown in Figure 8 do not yet qualify as an example of reticulate evolution, even in a paralogous sense, in that no functional divergence followed by convergence or even the trading of functions, has been demonstrated: we cannot yet say what TrpEG1 and TrpEG2 did before the divergence of the enteric and fluorescent pseudomonad groups.

To return to the accessibility question, are there actually two functional anthranilate synthase gene pairs in *Pseudomonas*, and, if there are, are they completely sequestered? The answers are yes, there are two, and they are connected by a one-way street. Phn AS cannot ordinarily substitute for Trp AS in tryptophan synthesis, but Trp AS can supply some anthranilate to the phenazine pathway. Only the double (*trp/phn*) mutant is devoid of pyocyanin.

The tryptophan pathway certainly leads through anthranilate, and the phenazine pathway probably does also. Thus, *trpE* mutants should still have functional *phnA* and *phnB*, and, therefore an enzyme capable of making anthranilate. But, if this enzyme is made, why does it not compensate for the absence of functional TrpEG enzyme? Is it possible that some form of channeling is involved—either of a spatial or temporal nature?

First of all, there is no absolute separation of the pathways, for the barrier appears to operate in one direction only: *phnAB* mutants produced by transplacement form some pyocyanin, demonstrating the contribution of TrpEG-produced anthranilate to the phenazine pathway, as depicted in Table 5. This is sufficient to explain the failure of standard mutant screens to yield pyocyanin-negative mutants which lack the anthranilate synthase of the phenazine pathway. The basis of this asymmetrical block appears to be temporal: while the tryptophan pathway operates throughout the growth cycle, pyocyanin is a secondary metabolite in that it is normally produced only in stationary phase. Thus, the *phn* pathway is probably not operating when its anthranilate could make it possible for *trpE* mutants to grow.

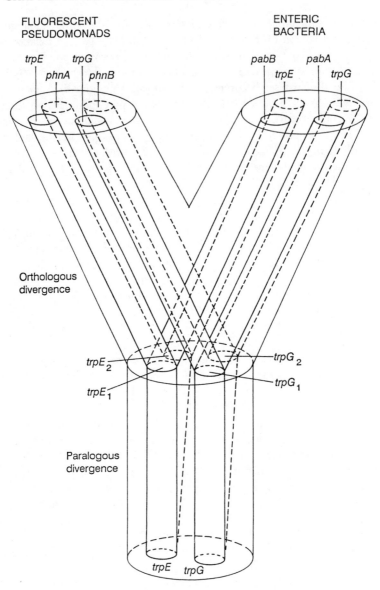

FIGURE 8. Evolution of anthranilate synthase in enterics and fluorescent pseudomonads.

Second, the barrier to the passage of anthranilate from the phenazine pathway into tryptophan production is not general. Channeling can explain the appearance of *trpE* mutants under the conditions of the screen. Under these conditions, TrpE⁻ mutants (and TrpG⁻ mutants in low NH_4^+ medium) are

TABLE 5. Pyocyanin production by *P. aeruginosa* PA01 and its derived mutants[a]

Strain	Pyocyanin production[b] (μg/ml)	Percent wild-type production
PA01 (wild type)	6.30	100
PADE *trpE1*	5.44	86
PADE *phnA47*	2.13	34
PADE *phnA48*	1.45	23
PADE *phnA47trpE1*	0.02	0.3

Source: From Chang et al. (1989)
[a]Cells are grown 16–20 hours in 5 ml of complex broth at 37°C with aeration.
[b]Pyocyanin concentrations are reported as micrograms pyocyanin produced per milliliter of culture.

not leaky. But there is no barrier to the utilization of anthranilate from the phenazine pathway in some other circumstances in which *P. aeruginosa* is cultured. For example, Essar et al. (1990c) found in *P. aeruginosa* that deleted TrpE mutants, as well as inactivated ones, could "revert" to prototrophy, whereas similar mutants of our *P. putida* strain (which has no phenazine pathway) could not. These *P. aeruginosa* revertants had much higher levels of *phnAB* mRNA than normal, and they produced it in log phase as well as stationary phase. Clearly, this observation suggests a mechanism for the reversion: a change in the regulation of the *phnAB* genes, permitting them to contribute anthranilate in the time of need.

Perhaps this and similar cases, together with the recent dramatic realization that bacterial life is not described sufficiently by log growth in culture (Cairns et al., 1988; Hall, 1990), will help bring greater depth to the study of bacterial evolution.

CHAPTER 5

MOLECULAR EVOLUTION OF HUMAN IMMUNODEFICIENCY VIRUSES AND RELATED RETROVIRUSES

Shozo Yokoyama

In June of 1981, the Centers for Disease Control recognized the first five cases of a fatal disease that came to be known as the acquired immunodeficiency syndrome (AIDS). All of these victims died of pneumonia caused by the protozoan *Pneumocystis carinii*. In addition to the unusual lung infections, many AIDS patients have other infections and the purplish lesions of Kaposi's sarcoma (Goedert and Blattner, 1985).

The causative agent of AIDS, originally referred to as lymphadenopathy-associated virus (LAV) (Barre-Sinoussi et al., 1983), was identified remarkably quickly. Other names for the virus that causes AIDS have included human T-cell lymphotropic virus type III (HTLV-III) (Popovic et al., 1984), AIDS-associated retrovirus (ARV) (Levy et al., 1984), and, more recently, human immunodeficiency virus (HIV) (Coffin et al., 1986). It has been estimated that more than 250,000 cases of AIDS have already occurred, that between 5 and 10 million people worldwide are infected with the HIV, and that within the next 5 years 1 million new AIDS cases may occur (Mann et al., 1988). AIDS has become an acute public health problem throughout the world.

Information on AIDS comes from a variety of disciplines, including medicine, epidemiology, molecular biology, cell biology, and immunology. Through vigorous research in these fields, the patterns of HIV infection have been clarified, a blood test has been formulated, and the virus's targets in the body have been established (Gallo and Montagnier, 1988; however, see Duesberg,

1989). With extensive molecular analyses, nucleotide sequences of HIV and related viruses are accumulating very rapidly, and comparisons have revealed extensive genetic variation.

How was the genetic variability of HIVs generated and how are these viruses maintained in human populations? When did HIV and AIDS first appear? What is the evolutionary relationship between HIV and other viruses? How quickly will the number of AIDS cases increase in the future? To answer these and other questions, evolutionary genetic analyses are indispensable. In the future, therefore, evolutionists are expected to play an important role in AIDS research. In this chapter, I will review the results of molecular evolutionary studies of HIV and related viruses. Before going into detail, however, the general features of AIDS and HIV will be briefly discussed.

EPIDEMIOLOGY OF AIDS

HIV infections began spreading before AIDS was even recognized and before the HIV virus was isolated. Blood stored as early as 1959 in Zaire has been found to contain antibodies against HIV (Mann et al., 1988; see also Saxinger et al., 1985).

Although some people seem to experience an early, brief, mononucleosis-like illness with fever, malaise, and, possibly, a skin rash between 2 weeks and 3 months after infection with HIV, a person may remain symptom-free for years. The incubation time between infection and development of AIDS is estimated to range from 4 to 11 years (Medley et al., 1987; Costagliola et al., 1989). The interval between diagnosis and death varies greatly, but about 50% of patients die within a year and half, and 80% die within 3 years of diagnosis (Mann et al., 1988).

From analyses of AIDS and seroprevalence data, three epidemiological patterns of AIDS have been recognized (Piot et al., 1988; Mann et al., 1988). Pattern 1 is typical of industrialized countries, such as the United States and many Western European countries, where most patients are either homosexual or bisexual males, or intravenous drug users, and the male-to-female ratio of reported AIDS cases ranges from 10:1 to 15:1. Pattern 2 is observed in some areas of central, eastern, and southern Africa and, increasingly, in some Latin American countries, where HIV infects many heterosexuals. The ratio of infected males to females is almost 1:1 and perinatal transmission is common. Pattern 3 prevails in areas such as eastern Europe, North Africa, the Middle East, and Asia. In these regions, AIDS generally occurs in people who have traveled to pattern 1 or 2 areas and have had sexual contact with individuals in such areas. Some cases have been caused by imported blood or blood products.

As of June 1989, 177 countries had reported at least one AIDS case to the World Health Organization (Table 1). Worldwide, thousands of AIDS cases are reported every year, about 60% of which are from the United States.

TABLE 1. AIDS cases reported to WHO (June 2, 1989)

Continent	Number of cases	Number of countries
Africa	24,686	52
Americas	108,830	44
Asia	369	37
Europe	21,855	30
Oceania	1,451	14
Total	157,191	177

RETROVIRUSES

The HIV contains an RNA genome and RNA-dependent DNA polymerase enzymatic activity, as do all members of the Retroviridae virus family. Retroviruses are divided into three subfamilies—oncovirus, lentivirus, and spumavirus—depending upon morphology and biological properties (Teich, 1982). The oncoviruses are characterized by their oncogenic potential. In contrast, the lentiviruses cause slow progressive inflammatory disease, whereas the spumaviruses induce persistent infection without any clinical disease.

The genome

The genome within the retrovirus particle is a complex of two identical chains of RNA, making retroviruses diploid. During the synthesis of the viral DNA, sequences near the ends of viral RNA (U3 and U5) are duplicated to generate long terminal repeats (LTRs), several hundred base pairs in length, at the ends of proviral DNA. Between these regulatory regions are coding sequences for the major structural proteins of the virus: (1) the *gag* gene (about 1.5 kb in size) encodes the core proteins; (2) the *pol* gene (about 3 kb) encodes protease, reverse transcriptase, and integrase; and (3) the *env* gene (about 2.6 kb) encodes the envelope glycoproteins.

Many oncoviruses contain genes (oncogenes) whose activities are responsible for both the initiation and maintenance of neoplastic transformation. The retroviral oncogenes (v-*onc*) originated from cellular homologues (c-*onc*) by transduction, and cellular and viral oncogenes have very similar DNA sequences (for a review, see Bishop, 1983). Retroviruses are often replication-defective, having lost some of their genome in the transduction process. Accordingly, they depend on the presence of related helper viruses to provide those missing components when they are replicating.

Life cycle

The life cycle of the retrovirus begins when the particle binds to the outside of a eukaryotic cell and injects its core. DNA polymerase first makes a single-stranded DNA copy of the viral RNA, an associated enzyme ribonuclease destroys the original RNA, and the polymerase synthesizes a second DNA copy using the first one as a template. These two enzymes together are known as reverse transcriptase. The double-stranded DNA migrates to the cell nucleus, and a third viral enzyme, integrase, then splices the viral genome into the host cell's DNA. The proviral DNA is duplicated, together with the cell's own genes, whenever the cell divides. These exogenous retroviruses are ordinarily benign in the germline.

The second half of the viral life cycle takes place only sporadically. It begins when nucleotide sequences in the LTRs direct enzymes belonging to the host cell to copy the DNA of the integrated virus into RNA. Some of the RNA provides the genetic material for a new generation of viruses, whereas other RNA strands serve as the mRNAs that guide the cellular machinery in producing the structural proteins and enzymes of the new viruses. The completed virion encloses itself in a patch of host-cell membrane as it buds from the cell. This envelope carries the final structural element of the retrovirus, the envelope protein. These endogenous retroviruses have varying degrees of infectivity and pathogenicity.

For more details on these topics, readers may consult Varmus (1988) and Haseltine and Wong-Staal (1988).

PHYLOGENY OF HIV AND RELATED VIRUSES

Initially, LAV$_{BRU}$ was isolated early in 1983 from a Frenchman with a lymph-adenopathy (Barre-Sinoussi et al., 1983) and several HTLV-III isolates were recovered from blood specimens obtained in late 1982 and early 1983 from patients on the east coast of the United States (Popovic et al., 1984). ARV2 was isolated from a Californian about a year later (Levy et al., 1984). From these isolates, complete clones LAV$_{BRU}$ (Wain-Hobson et al., 1985), ARV2 (Sanchez-Pescador et al., 1985), and BH5, BH10 (Ratner et al., 1985), and pv.22 (Muesing et al., 1985) (HTLV-III) were sequenced.

Immediately after the isolation and characterization of these HIV strains, Chiu et al. (1985) compared the amino acid sequences of the highly conserved reverse transcriptase of an HIV strain (BH10) with those of lentiviruses and oncoviruses. From this comparison, they concluded that HIV is more closely related to the lentivirus subfamily than to the oncovirus subfamily (see also Sonigo et al., 1985; Gonda et al., 1985).

The proportion of different amino acids for the *gag* proteins was 0.6–0.8% (for LAV$_{BRU}$ vs. HTLV-III), 3.4% (for LAV$_{BRU}$ vs. ARV2), and 3.8–4.0% (for HTLV-III vs. ARV). These data show that LAV$_{BRU}$ and HTLV-III strains are most closely related and that their common ancestor diverged from the ancestor of ARV2. On the assumption that the rates of amino acid substitution for the *gag* protein of HIVs are the same as those for the Moloney murine leukemia virus (MMLV) and Moloney murine sarcoma virus (MMSV), i.e., 1.24×10^{-3} per site per year, it was concluded that ARV2 diverged from the others after 1968 and that the LAV$_{BRU}$ and HTLV-III strains diverged after 1977 (Yokoyama and Gojobori, 1987). These estimates were consistent with the observation that LAV$_{BRU}$ was isolated from a Frenchman who last visited New York in 1979 (Barre-Sinoussi et al., 1983).

Following Chiu et al. (1985) and Sonigo et al. (1985), Yokoyama et al. (1987) compared the amino acid sequences of the highly conserved reverse transcriptase (RT), endonuclease, and integrase (EN) regions of different HIV, lentivirus, and oncovirus strains. In this analysis, the new HIV strains LAV$_{ELI}$ and LAV$_{MAL}$, both isolated from Zaire (Alizon et al., 1986), were included. Assuming again that the rate of amino acid substitution in these regions is 1.24×10^{-3} per site per year, we concluded that the common ancestor of the HIVs isolated in the United States and France diverged from that of the Central African strains within the last 40–50 years.

These results were consistent with the suggestion made at that time that AIDS emerged as a new disease in recent decades (Kanki et al., 1985). However, this notion had to be modified quickly when a new HIV strain was characterized in 1987.

The new strain HIV type 2 (HIV2 or HIV2$_{ROD}$) was isolated from a West African AIDS patient and completely sequenced (Guyader et al., 1987). The HIV2 and the previously characterized HIV strains (now called HIV type 1 or HIV1) showed such a large number of nucleotide differences they could not be considered strains of the same viruses (Guyader et al., 1987); apparently the two groups had been evolving independently for some time.

To study the phylogenetic relationships among the HIV1 strains, HIV2, lentiviruses, and oncoviruses, the DNA sequences of the RT and EN regions of these viruses were compared (Yokoyama et al., 1988). The lentiviruses used were equine infectious anemia virus (EIAV) and visna lentivirus (VISNA), and oncoviruses included the human T-cell leukemia virus type I (HTLV-I), bovine leukemia virus (BLV), Rous sarcoma virus (RSV), and Moloney murine leukemia virus (MMLV).

A representative evolutionary tree (Figure 1) was obtained by the neighbor-joining method (Saitou and Nei, 1987), based on the total numbers of nucleotide substitutions in the RT and EN regions (Yokoyama et al., 1988). From Figure 1, it is clear that (1) the seven HIV1 strains form one cluster and their common ancestor diverged from the ancestor of HIV2, (2) the common ancestor of the HIV1 and HIV2 strains diverged from the ancestor of the lentiviruses, and (3) the common ancestor of the HIV and lentivirus groups and that of the oncovirus group diverged even earlier. This topology is consistent with the results obtained by Xiong and Eickbush (1988b) and by Doolittle et al. (1989).

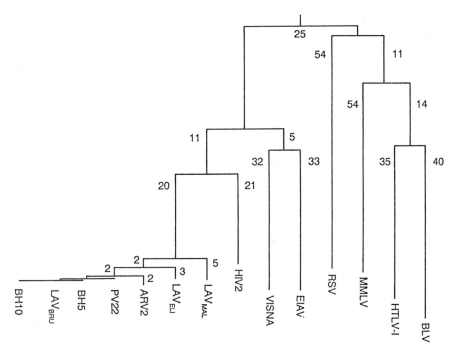

FIGURE 1. Phylogenetic tree for the different strains of HIV1, HIV2, lentivirus, and oncovirus inferred from an analysis of nucleotide sequences of the RT and EN regions. The numbers next to the different branches are branch lengths measured by percent nucleotide substitutions per site. After Yokoyama et al. (1988).

If we accept the hypothesis that the HIV1 strains diverged within the last 40–50 years, then the branch lengths (Figure 1) indicate that HIV1 and HIV2 diverged about 250 years ago, a time inconsistent with the hypothesis of a recent origin of AIDS.

Soon after the characterization of HIV2, the simian immunodeficiency viruses STLV-III$_{AGM}$ (Franchini et al., 1987; Hirsch et al., 1987), which were supposedly isolated from an African green monkey, and SIV$_{MAC}$ (or SIV$_{MAC-142}$; Chakrabarti et al., 1987), isolated from a macaque, were sequenced. The former strains were later shown to be contaminated by a SIV$_{MAC-251}$ strain in laboratories (Kestler et al., 1988). It should be noted that AIDS researchers have paid special attention to African green monkeys, because when these primates are infected with simian immunodeficiency virus (now known as SIV$_{AGM}$), they do not seem to develop simian AIDS (SAIDS) (Kanki et al., 1985). Fortunately, a genuine SIV$_{AGM}$ was completely sequenced about a year later (Fukasawa et al., 1988).

Before the DNA sequences of SIV$_{AGM}$ were published, Smith et al. (1988) compared the nucleotide sequences of the *env* genes of HIV1, HIV2, and

SIV$_{MAC}$ strains to estimate their phylogenetic relationship. They showed that HIV2 and SIV$_{MAC}$ form one cluster and HIV1 strains form another cluster. The same tree topology was obtained with the nucleotide sequences of the RT and EN regions (Yokoyama, 1988). The numbers of nonsynonymous substitutions have been evaluated in these regions for pairwise comparisons of representative viruses (Table 2), where VISNA was taken as the outgroup. A phylogenetic tree for these viruses, constructed by the neighbor-joining method based on these values (Figure 2), reveals that the ancestor of SIV$_{AGM}$ diverged from the common ancestor of the HIV1 group (see also Sharp and Li, 1988; Doolittle, 1989). Within the HIV2 group, STLV-III$_{AGM}$ and SIV$_{MAC}$ have most recently diverged, and their common ancestor diverged from the ancestor of HIV2.

The relationship of SIV$_{AGM}$ to the HIV1 group may not be as close as Figure 2 indicates. When the same method is applied to different regions of the viral genomes, somewhat different tree topologies can be obtained. For example, when the nucleotide sequences are compared either at the LTR or at the *pol* gene, SIV$_{AGM}$ clusters with the HIV1 group, whereas it clusters with the HIV2 group when either the *gag* gene or the *env* gene is compared (T. Gojobori, personal communication). At present, therefore, it seems most reasonable to conclude that HIV1, HIV2, and SIV$_{AGM}$ diverged from each other about the same time, as originally noted by Fukasawa et al. (1988).

More recently, a feline immunodeficiency virus (FIV; Talbott et al., 1989) and an SIV$_{SM}$ (Hirsch et al., 1989) isolated from an African primate, the sooty mangabey, have been completely characterized. A phylogenetic tree for these and other retroviruses is shown in Figure 3, which was constructed by the neighbor-joining method, based on the numbers of nonsynonymous substitutions in the RT and EN regions. Figure 3 reveals important evolutionary details within the HIV2 group. Clearly, SIV$_{SM}$ is most closely related to SIV$_{MAC}$, as

TABLE 2. The numbers of nonsynonymous substitutions per site for pairwise comparisons of HIV genes at the RT and EN regions ($\times 10^{-3}$)[a]

	LAV$_{MAL}$	LAV$_{ELI}$	HIV2	STLV-III$_{AGM}$	SIV$_{MAC}$	SIV$_{AGM}$	VISNA
BH10	28	21	287	311	292	283	549
LAV$_{MAL}$		23	280	306	288	285	547
LAV$_{ELI}$			286	311	289	283	539
HIV2				93	65	298	542
STLV-III$_{AGM}$					45	336	564
SIV$_{MAC}$						299	532
SIV$_{AGM}$							595

[a]The number of nonsynonymous sites is 926.

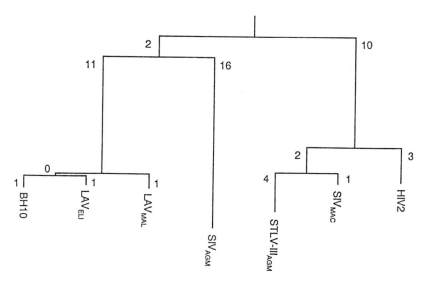

FIGURE 2. Phylogenetic tree for different HIV1, HIV2, SIV, and VISNA (the outgroup) strains inferred from an analysis of nonsynonymous changes at the RT and EN regions. The numbers next to the different branches are branch lengths in percent nonsynonymous substitutions per site.

noted by Hirsch et al. (1989). SIV_{SM}-infected sooty mangabeys inhabit West Africa, but SIV_{MAC} has not been found in Asian macaques in the wild. Therefore, Hirsch et al. (1989) concluded that the macaques have most probably been exposed to SIV_{SM} in captivity. Interestingly, like African green monkeys, sooty mangabeys infected with SIV_{SM} do not seem to develop SAIDS. Thus, Hirsch et al. further argued that SIV from a West African sooty mangabey, or a closely related species, infected a human and then evolved as HIV2 in West Africa, where HIV2 infection of humans is endemic.

Figure 3 also shows that FIV diverged before the radiation of the HIV1, HIV2, and SIV_{GM} groups (see also Talbott et al., 1989).

EVOLUTIONARY RATES OF RETROVIRUSES

To understand the mechanisms of maintenance of genetic diversity within and among retroviruses, it is important to determine how rapidly their genomes are changing. For this purpose, evolutionary rates will be discussed by considering oncogenes, HIVs, and SIVs separately.

Oncogenes

Before the evolution of retroviral genomes was studied, it was known that the evolutionary rates of RNA genomes were more than a million times greater

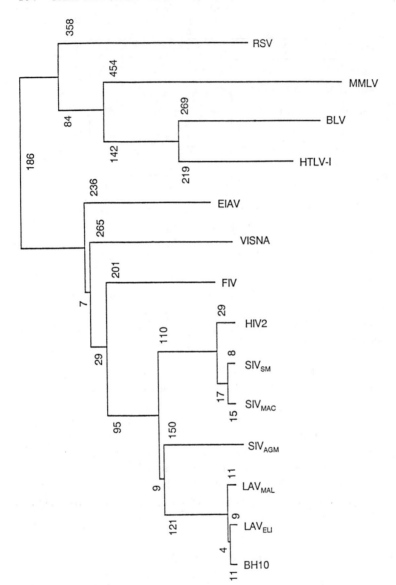

FIGURE 3. Phylogenetic tree for different strains of HIV, SIV, FIV, lentivirus, and oncovirus inferred from an analysis of nonsynonymous changes at the RT and EN regions. The numbers next to the different branches are branch lengths ($\times 10^{-3}$).

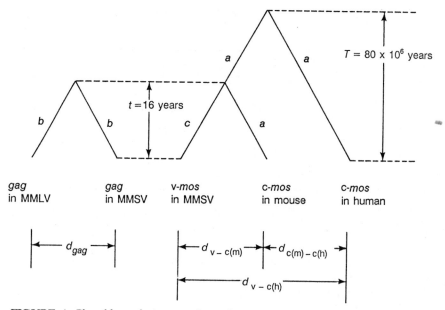

FIGURE 4. Plausible evolutionary relationship among mouse c-*mos*, human c-*mos*, v-*mos*, and *gag* genes of MMLV and MMSV. After Gojobori and Yokoyama (1985).

than those of DNA genomes (Holland et al., 1982). The hemagglutinin (Air, 1981) and neuraminidase (Krystal et al., 1983; Martinez et al., 1983) genes in the influenza A virus were shown to be typical examples of RNA genome evolution. (For more recent studies on this topic, see Hayashida et al., 1985; Saitou and Nei, 1986.)

Gojobori and Yokoyama (1985) developed a method of computing the rates of nucleotide substitution for the retroviral (v-*onc*) and cellular (c-*onc*) oncogenes and the retrovirus genome simultaneously. Figure 4 shows a probable evolutionary pattern for the *gag* gene of MMLV, the *gag* and v-*mos* genes of MMSV, and the c-*mos* genes of mouse and human. In Figure 4, d_{gag}, $d_{v-c(m)}$, $d_{c(m)-c(h)}$, and $d_{v-c(h)}$ are the numbers of nucleotide substitutions per site between the *gag* genes of MMLV and MMSV, between v-*mos* and mouse c-*mos*, between mouse c-*mos* and human c-*mos*, and between v-*mos* and human c-*mos*, respectively. Parameters a, b, and c are the rates of nucleotide substitution per site per year in c-*mos*, *gag*, and v-*mos* genes, respectively.

The divergence time (T) between mouse c-*mos* and human c-*mos* is taken as 80 million years (Myr); and the time since transduction of mouse c-*mos* to the MMLV genome (t) is about 16 years, because MMSV was obtained from a sarcoma induced in BALB/c mice after passage of high doses of MMLV in 1966 (Moloney 1966), and v-*mos* was cloned in 1981. The complete nucleotide sequences of c-*mos* of human and mouse, as well as the genomes of MMLV and MMSV containing v-*mos* and *gag* genes were also obtained in 1981. (For

TABLE 3. The rates of nucleotide substitutions per site per year for c-*mos*, v-*mos*, and *gag* genes

Codon position	c-*mos* ($\times 10^{-9}$)	*gag* ($\times 10^{-3}$)	v-*mos* ($\times 10^{-3}$)
First	1.21 ± 0.16	0.59 ± 0.19	1.31 ± 0.50
Second	0.91 ± 0.14	0.72 ± 0.22	0.56 ± 0.31
Third	3.32 ± 0.34	0.59 ± 0.19	2.06 ± 0.63
All	1.71 ± 0.12	0.63 ± 0.13	1.31 ± 0.31

[a]Modified from Gojobori and Yokoyama (1985).

the source of these data, see Gojobori and Yokoyama, 1985). From these nucleotide sequences and divergence times, *a*, *b*, and *c* were estimated from the following relationships:

$$a = d_{c(m)-c(h)}/2t$$
$$b = d_{gag}/2t \quad\quad\quad (1)$$
$$c = [d_{v-c(h)} - d_{c(m)-c(h)}]/t + a$$

(Gojobori and Yokoyama, 1985).

The evolutionary rates for c-*mos* (*a*), *gag* (*b*), and v-*mos* (*c*) are shown in Table 3, where we can see that the genes in the viral genome evolve 4–8 $\times 10^5$ times faster than the cellular oncogenes in the mammalian genome. This extremely high rate of evolutionary change of the v-*mos* and *gag* genes may be expected because RNA viruses have a high mutation rate due to a lack of proofreading enzymes that assure fidelity of DNA replication (Holland et al., 1982).

The same method has been applied to a total of nine sets of viral oncogenes and two corresponding cellular homologues (Gojobori and Yokoyama, 1987). The rates of synonymous and of nonsynonymous substitutions for cellular oncogenes are $(3.16 \pm 0.50) \times 10^{-9}$ and $(0.37 \pm 0.05) \times 10^{-9}$, respectively, and the corresponding values for viral oncogenes were $(1.09 \pm 0.63) \times 10^{-3}$ and $(0.48 \pm 0.27) \times 10^{-3}$. Thus, for cellular and most viral oncogenes, the rate of synonymous substitution is higher than that of nonsynonymous substitution, suggesting that amino acid substitutions in both cellular and viral oncogenes are functionally constrained.

In these analyses, the time between the transduction experiment and virus isolation is taken as the divergence time between v-*onc* and the corresponding c-*onc* genes. In practice, the divergence time may be overestimated, however, because the viruses isolated may have been stored frozen for a period of time.

TABLE 4. Rates of nucleotide substitution per site per year in the HIV genome[a]

Gene	Synonymous sites		Nonsynonymous sites	
	$(\times 10^{-3})$	R^b	$(\times 10^{-3})$	R^b
gag	9.7	1.05	1.7	0.33
pol	11.0	1.19	1.6	0.31
env	9.2	1.0	5.1	1.0

[a]Modified from Li et al. (1988).
[b]Ratio of the substitution rate for each gene to that for the env gene.

Human immunodeficiency viruses

To evaluate the rate of nucleotide substitution in HIVs, Hahn et al. (1986) sequenced viruses isolated sequentially over a 1- or 2-year period from HIV-infected individuals. Their intention was to observe directly the pattern of mutation accumulation through time. They found that the HIVs within an individual were genetically heterogeneous to start with and evolved in parallel fashion. This was an important finding, but it was not helpful in directly estimating mutation rates. From these independently diverged HIV strains, they estimated the rates of nucleotide substitution to be 3.7×10^{-4} to 1.8×10^{-3} for the gag gene and 3.2×10^{-3} to 1.6×10^{-2} for the env gene.

Assuming that the rates of amino acid substitution in the gag proteins of HIV, MMLV, and MMSV are the same and using the nucleotide sequences of LAV$_{BRU}$, BH5, BH10, pv.22, and ARV2, Yokoyama and Gojobori (1987) estimated the rate of nucleotide substitution to be on the order of 10^{-3} per site per year, with the rate in the second half of the genome being twice that in the first half. More recently, Li et al. (1988) proposed a method to estimate the rate of nucleotide substitution in HIV more directly. They demonstrated that the rate of nonsynonymous substitution per site per year is lowest in the pol gene, which is 1.6×10^{-3}, and that the average rate of synonymous substitution is about 10^{-2} per site per year (Table 4). Because the proportion of nonsynonymous sites is 70–75%, the rate of nucleotide substitution for all sites is on the order of 10^{-3} per site per year. Table 4 also shows that the evolutionary rate for the env gene is much higher than those of the gag and pol genes.

Simian immunodeficiency viruses

As mentioned above, STLV-III$_{AGM}$ isolates were derived from cell cultures infected with a macaque isolate SIV$_{MAC-251}$ (Franchini et al., 1987; Hirsch et al., 1987). This circumstance provides an opportunity to measure the rate of nucleotide substitution in the env gene of STLV-III$_{AGM}$ directly. This rate can then be used to extrapolate the evolutionary rates at different regions of the SIV genome. Unlike the method of Li et al. (1988), the present method does not require the estimation of divergence times of different strains.

Franchini et al. (1987) sequenced the entire genome, whereas Hirsch et al. (1987) sequenced only the *env* gene. Furthermore, the entire genome of SIV_{MAC} (or $SIV_{MAC-142}$; Chakrabarti et al., 1987) and a part of the *env* gene of $SIV_{MAC-251}$ (Kestler et al., 1988) were sequenced. Let $d_{251-AGM}$, $d_{251-MAC}$, and $d_{MAC-AGM}$ be the total numbers of nucleotide substitutions per site between $SIV_{MAC-251}$ and $STLV-III_{AGM}$, between $SIV_{MAC-251}$ and SIV_{MAC}, and between SIV_{MAC} and $STLV-III_{AGM}$, respectively. Furthermore, let K be the number of nucleotide substitutions per site in $STLV-III_{AGM}$ after it diverged from $SIV_{MAC-251}$. Since $SIV_{MAC-251}$ and $STLV-III_{AGM}$ are most closely related among the three,

$$K = (d_{251-AGM} + d_{MAC-AGM} - d_{251-MAC})/2 \qquad (2)$$

The co-culture of $STLV-III_{AGM}$ and HUT-78 cells seems to have been contaminated by $SIV_{MAC-251}$ early in 1984 (see Kestler et al., 1988). Because the two $STLV-III_{AGM}$ strains were cloned in 1986, it may be appropriate to consider that the DNA sequences of $STLV-III_{AGM}$ reflect about 1.5 years of mutation accumulation. Thus, the rate (k) of nucleotide substitution per site per year in the *env* gene of $STLV-III_{AGM}$ is simply given by $K/1.5$.

Differences in the cloned DNA sequence may also reflect heterogeneity originally present in the virus stock in HUT-78 cells. We have seen such an example in a study by Hahn et al. (1986). Thus, the 1.5 years may underestimate the actual divergence time of the cloned DNAs and the estimate k may be regarded as the maximum evolutionary rate.

After aligning more than 1 kb of the *env* genes, the numbers of synonymous and nonsynonymous substitutions were evaluated for each pair by the method of Miyata and Yasunaga (1980) (Table 5). With these estimates, the average

TABLE 5. Average number of nucleotide substitutions per site[a]

Source of STLV-III$_{AGM}$	$d_{251-AGM}$	$d_{MAC-AGM}$	$d_{251-MAC}$	k ($\times 10^{-3}$)
Synonymous changes				
Franchini et al. (1987)	0.014	0.089	0.084	6.3
Hirsch et al. (1987)	0.018	0.084	0.084	6.0
Average				6.2
Nonsynonymous changes				
Franchini et al. (1987)	0.004	0.025	0.023	2.0
Hirsch et al. (1987)	0.008	0.029	0.023	4.7
Average				3.3

[a]The numbers of synonymous and nonsynonymous sites were 218 and 801, respectively.

rate of synonymous substitution at th 5)te *env* gene for the two STLV-III$_{AGM}$ data is 6.2×10^{-3} per site per year, whereas that of nonsynonymous substitution is 3.3×10^{-3} per site per year (Table 5). These estimates are slightly smaller than the corresponding estimates 9.2×10^{-3} and 5.1×10^{-3} obtained by Li et al. (1988).

From the result obtained by Li et al. (1988), these two *k* values can be converted to those for the other parts of the SIV genome by use of the ratio of the substitution rate for each gene to that for the *env* gene. For synonymous sites, the ratios are 1.05 and 1.19 between the *gag* and *env* genes and between the *pol* and *env* genes, respectively, whereas the corresponding values for the nonsynonymous sites are 0.33 and 0.31 (Table 4). Thus, the rates of synonymous substitution for *gag* and *pol* genes are 6.5×10^{-3} and 7.4×10^{-3} per site per year, respectively, and the corresponding values for nonsynonymous substitution are 1.1×10^{-3} and 1.0×10^{-3} per site per year (Table 6).

Because the RT and EN regions have been used extensively in molecular evolutionary studies of retroviruses, it is also important to determine the evolutionary rates in these regions. The ratio of the number of synonymous substitutions in the RT and EN regions and that for the entire *pol* gene is 1.39 when HIV2 and STLV-III$_{AGM}$ (Franchini et al., 1987) are compared, whereas it is 1.50 when HIV2 and SIV$_{MAC}$ are compared, with an average value of 1.44. Similarly, the ratio of the number of nonsynonymous substitutions at the RT and EN regions and the number in the entire *pol* gene is 0.79 when HIV2 and STLV-III$_{AGM}$ are compared and 0.72 when HIV2 and SIV$_{MAC}$ are compared, with an average value of 0.76. Thus, the rates of synonymous and nonsynonymous substitutions at the RT and EN regions are 10.7×10^{-3} per site per year ($= 1.44 \times 7.4 \times 10^{-3}$) and 0.76×10^{-3} per site per year ($= 0.76 \times 1.0 \times 10^{-3}$), respectively (Table 6). The latter rate is about twice the rate of nonsynonymous substitution for oncogenes, i.e., 0.48×10^{-3}, but slightly smaller than the estimate 0.96×10^{-3} obtained by Sharp and Li (1988).

TABLE 6. Rates of nucleotide substitution per site per year in the SIV genome

Gene	$k (\times 10^{-3})$	
	Synonymous sites	Nonsynonymous sites
gag	6.5	1.1
pol	7.4	1.0
RT and EN	10.7	0.8
env	6.2	3.3

DIVERGENCE TIME OF HIV1, HIV2, SIV$_{AGM}$, AND FIV GROUPS

The rate of synonymous substitution appears to be more uniform than that of nonsynonymous substitution within a genome (Tables 4 and 6). However, this uniformity may not necessarily hold when the RT and EN regions of different strains are compared (Yokoyama et al., 1988). Furthermore, because of their rapid evolutionary rate, the synonymous changes are not useful in studying molecular evolution of different members of retrovirus subfamilies. Because of the conservative nature and rather uniform rate of nucleotide substitution, the number of nonsynonymous substitutions in the RT and EN regions has been useful in molecular evolutionary studies of retroviruses, as we have already seen.

To estimate the divergence times among HIV1, HIV2, and SIV$_{AGM}$ strains, we shall assume that the rate of nonsynonymous substitution in the RT and EN regions of HIV and SIV is 0.5×10^{-3} per site per year. Then, from the branch lengths in Figure 3, the divergence time between HIV1 and HIV2 groups is estimated to be 290 $\pm$ 28, whereas that between HIV1 and SIV$_{AGM}$ is 284 $\pm$ 27. Similarly, HIV2 and SIV$_{MAC}$/SIV$_{SM}$ seem to have diverged about 58 $\pm$ 11 years ago. Because of the large standard errors associated, it is possible that HIV1, HIV2, and SIV$_{AGM}$ groups diverged about the same time, 200–350 years ago (see also Yokoyama et al., 1988; Yokoyama, 1988). The divergence time between SIV$_{SM}$ and SIV$_{MAC}$ is 22 $\pm$ 7 years, an estimate that agrees with the suggestion that the two strains diverged in the late 1960s or early 1970s (Hirsch et al., 1989).

The divergence between FIV and other immunodeficiency viruses occurred 440 $\pm$ 35 years ago, whereas the common ancestor of HIV, SIV, and FIV diverged from the ancestor of VISNA 514 $\pm$ 39 years ago. The latter value is much larger than the divergence time (300 years) obtained by Sharp and Li (1988). Thus, there is some discrepancy in these estimates, but the divergence time between the HIV, SIV, and FIV strains and lentiviruses is about twice that among the HIV1, HIV2, and SIV$_{AGM}$ groups.

The estimates for the divergence times among HIV1, HIV2, and SIV$_{AGM}$ are controversial. Smith et al. (1988) estimated that the HIV1 and HIV2 groups diverged 37 years ago, which is an order of magnitude smaller than the present estimate. On the other hand, Fukasawa et al. (1988) implicitly suggested that HIV1 and SIV$_{AGM}$ diverged millions of years ago. More recently, assuming that the rate of nonsynonymous substitution was 0.96×10^{-3} per site per year in the RT and EN regions, Sharp and Li (1988) estimated the divergence time of the HIV1, HIV2, and SIV$_{AGM}$ to be about 150 years. This estimate is somewhat smaller than those obtained by Yokoyama et al. (1988) and Yokoyama (1988).

How can we explain the difference in the estimates obtained by Smith et al. (1888) on the one hand and by Yokoyama et al. (1988) and Sharp and Li (1988) on the other? It should be noted that Smith et al. (1988) considered only the numbers of nucleotide substitution at the third position of codons in the *env* gene, whereas the numbers of nonsynonymous substitutions in the most conserved region of the retrovirus genome were used by Yokoyama et al. (1988) and Sharp and Li (1988). Thus, it is likely that the divergence time between HIV and SIV was underestimated because of multiple substitutions at rapidly evolving nucleotide sites.

CONCLUSIONS

From the comparative studies of nucleotide and amino acid sequences of HIV and related retroviruses, we have learned several basic evolutionary features of HIV and AIDS: (1) HIV, SIV, and FIV strains belong to the lentivirus subfamily rather than the oncovirus subfamily; (2) the retrovirus genome is evolving more than a million times faster than the DNA genome; (3) HIV1, HIV2, and SIV$_{AGM}$ are equally distantly related and apparently diverged from one another more than 200 years ago; and (4) HIV and, presumably, AIDS have existed much longer than previously thought. As more nucleotide sequences of the HIV-related viruses are determined, these points will be refined further.

Although Hirsch et al. (1989) clarified the evolutionary relationships within the HIV2 group, they were unable to determine the origin of HIV1, because the divergence among the ancestral strains of HIV1, HIV2, and SIV$_{AGM}$ occurred much earlier than that of HIV2 and SIV$_{MAC}$ (or SIV$_{SM}$) (Figure 2). Thus, to understand the evolution of the three viral groups, additional sequence data of other viruses, possibly isolated from as yet unsampled primates, are needed (see also Doolittle, 1989).

Molecular and evolutionary analyses have not resolved the problem of whether nonpathogenic viruses were derived from pathogenic viruses or vice versa. Once this relationship is clarified, it may be possible to locate nucleotide and amino acid substitutions that are related to pathogenicity. Such detailed evolutionary analyses may provide information crucial to understanding of the biological functions of HIV and the prevention and control of AIDS.

It is important to consider the differences in pathogenicity of HIV1 and HIV2 groups. People infected with HIV1 strains have a greater chance of developing AIDS than those infected with HIV2 strains (Essex and Kanki, 1988). The mechanisms underlying the variation in pathogenicity remain to be clarified. As already noted, SIV-infected African green monkeys and sooty mangabeys do not exhibit the symptoms of SAIDS. However, human cells infected with SIV$_{AGM}$ develop signs of cytopathic effects. Similarly, when SIV$_{SM}$ is injected into a macaque, the animal develops SAIDS (Doolittle, 1989). Elucidation of the mechanisms of immunity against SAIDS in these African primates will provide important information about the pathogenicity of AIDS. Finally, comparative studies of SIV$_{AGM}$ and SIV$_{SM}$ may provide useful information on specific nucleotide and amino acid substitutions that are associated with pathogenicity. These lines of evolutionary research may provide important clues for the cure of AIDS.

CHAPTER 6

EVOLUTION AND POPULATION GENETICS OF ORGANELLE GENES: MECHANISMS AND MODELS

C. William Birky, Jr.

Genes in chloroplasts and mitochondria were first identified by two features that distinguished them from nuclear genes—maternal inheritance and vegetative segregation. Additional peculiarities and complexities were quickly discovered when organelle heredity in *Chlamydomonas* and yeast was studied with powerful cytological and molecular tools. For reviews of organelle genetics, see Gillham (1978), Kirk and Tilney-Bassett (1978), Harris (1989), and Birky (1978, 1983a,b, 1988). The obvious analogy between the many organelle genomes in a cell and a population of organisms led to the application of population genetic theory to the study of the cellular phenomena. Further analysis revealed the new phenomena of intracellular selection and random drift, and led to more general explanations of maternal inheritance and vegetative segregation. I will briefly review these and other features of organelle genomes and their inheritance and show how they have been incorporated into models and theory for organelle population genetics and evolution.

Population genetic theory cannot provide complete explanations for the complex phenomena of gene diversity and sequence evolution, but it guides experimental and theoretical studies by identifying potential mechanisms, showing which ones are likely to be important, and indicating which parameters need to be measured and to what accuracy. In this chapter some preliminary answers to these questions will be given, and some of the many significant gaps in the available theory will be identified. Several unpublished arguments that have not been rigorously proved are included as plausible conjectures. Theoretical studies have focused on sequence diversity, but many of their findings will also apply to the diversity and evolution of genome structure.

UNIQUE FEATURES OF ORGANELLE GENOMES AND THEIR INHERITANCE

Mitochondria and chloroplasts each have on the order of 100 genes. In contrast to nuclear genes, organelle genes encode only proteins needed for the essential functions of photosynthesis and respiration and for parts of the transcription and translation apparatus needed to express those genes. "Junk" sequences, such as pseudogenes, have been found only in plant mitochondrial DNA. Mitochondria, chloroplasts, and nuclei use different enzymes for replication and repair and are exposed to different kinds and levels of natural mutagens. Consequently, these three genomes may be subject to different degrees of selection and have different mutation rates.

The entire organelle genome is usually a single DNA molecule, and each gene is normally present in a single copy per genome, or molecule. Organelle genomes are always present in many copies per cell, usually on the order of 10^2–10^4. Thus a cell contains a population of genomes; and each genome has one copy of each gene, although there are some exceptions. Chloroplast genomes in most plants and some algae, and mitochondrial genomes in some fungi, have a large inverted repeat that carries the ribosomal RNA genes and some other genes. Because mutations that occur on one copy rapidly spread to the other, presumably by gene conversion, these duplicated genes effectively behave as a single copy. Another exception is the mitochondrial genome of plants, in which the full-sized molecule regularly undergoes intramolecular crossing-over to produce smaller circles with partial genomes; the consequences for evolution are not well known.

The number of organelle genomes (the intracellular population size) varies with the physiological state of the cell and among differentiated cell types. In animals it is several orders of magnitude larger for eggs than for sperm; there are no data for gametes in plants. The mitochondrial DNA (mtDNA) or chloroplast DNA (cpDNA) molecules are variously packaged into one or many mitochondria or chloroplasts, depending on the organism and, again, on the physiological activity and differentiated state of the cell. Mitochondria regularly fuse and share mtDNA molecules in many organisms, notably yeast, but not in others, such as *Paramecium*. Chloroplasts fuse during mating in *Chlamydomonas*, but plastid fusion seems to be extremely rare in the angiosperm species where it has been looked for genetically. The packaging of organelle genomes probably influences the rate of vegetative segregation and that of random changes in organelle gene frequency within cells.

In a mitotic cell cycle, each nuclear chromosome is replicated once and each daughter cell receives one copy. The cell thus exerts stringent control over the replication of nuclear genes and their partitioning among daughter cells; the same is true for the cell genome of prokaryotes and for many plasmids. In contrast, the replication and partitioning of organelles and organelle genomes are usually under relaxed control, and therefore can be viewed as stochastic phenomena (Birky, 1983b); the evidence for this is largely indirect but compelling.

The replication of organelle genomes is random in the sense that some molecules replicate more than others by chance, without regard to their genotype. This may happen because (1) DNA polymerases are unable to distinguish between unreplicated and replicated molecules; (2) there is some random degradation of DNA with compensating replication; (3) cells must sometimes replicate a partial set of molecules as they differentiate or to compensate for receiving too few or too many genomes at cell division; (4) there is some degree of randomness in the choice of entire organelles for replication; or (5) replication of some mtDNA molecules is inhibited by unrepaired lesions (Stephen R. Palumbi and Allan C. Wilson, personal communication). A consequence of random replication is that the allele frequencies of organelle genes undergo stochastic changes precisely analogous to random drift in populations of organisms. This **intracellular random drift** was first predicted to occur as a consequence of repeated rounds of random pairing and gene conversion among organelle DNA molecules. It was later demonstrated in experiments with mitochondrial genes in yeast and chloroplast genes in *Chlamydomonas*. Genetic analyses showed that drift is largely due to random replication, although gene conversion may also be involved.

Organelle genomes segregate during mitotic (vegetative) as well as meiotic cell divisions. The classic explanation of this vegetative segregation is that organelles are partitioned between the daughter cells at random with respect to genotype. This leads to a gradual sorting out of organelles until all the progeny of a heteroplasmic cell are homoplasmic. More recently it has become clear that vegetative segregation is due to a combination of random partitioning and random replication.

As a result of these phenomena, gene frequencies in a clone or an entire organism undergo random drift. Variance in cell replication also contributes to the random drift of organelle gene frequencies in yeast colonies and probably also in animal tissues that turn over rapidly, e.g. blood cells, and in plants with stochastic meristems (Klekowski, 1988). It is convenient to use the term random drift to refer to the combined effects of random partitioning, random replication, gene conversion, and turnover of cells and genomes. In fact all of these mechanisms result in the loss of heteroplasmicity and transform genetic variation within cells to variation between cells; and in combination they can completely fix or eliminate alleles from an organism.

These unique phenomena of organelle genetics come together in the problem of the origin of mutant cells. Initially, a mutation produces a single mutant allele to create a heteroplasmic cell, which is unlikely to be phenotypically mutant. The mutation may be either lost or expressed in a homoplasmic mutant cell, depending on the combined effects of intracellular random drift, intra- and intercellular selection, and random partitioning.

The inheritance of organelle genomes is often characterized as maternal, biparental, or paternal. But these terms are misleading because crosses often produce more than one kind of offspring, and microorganisms without separate sexes also show uniparental inheritance. The mechanisms of uniparental inheritance are of two types.

1. *Deterministic mechanisms cause inheritance to be biased to varying degrees in favor of one parental sex or genotype.* In animals, inheritance is biased in favor of the maternal genomes in the large egg. The paternal input can be reduced or totally eliminated in animals or plants at any stage: by producing sperm or pollen without mitochondria, preventing sperm mitochondria from entering the egg, degrading them in the egg, or perhaps by preferentially partitioning them into extraembryonic tissue. In microorganisms without dimorphic sexes, inheritance may be uniparental by genotype. In *Chlamydomonas reinhardtii*, cpDNA is preferentially transmitted from the mt$^+$ parent due to degradation of molecules from the mt$^-$ parent in the zygote; but mtDNA is inherited from the mt$^-$ parent.
2. *Stochastic mechanisms fix alleles or entire genomes from one or the other parent by chance.* Fixation may be due to intracellular random drift (random replication, gene conversion) or to random partitioning of organelles between embryonic and extraembryonic tissue.

Most patterns of inheritance can be explained in terms of a two-step process in which directional mechanisms establish the input frequencies of genomes from the two parents (often strongly biased), followed by stochastic mechanisms that cause extensive variation in the final genome frequencies among the progeny. Examples are shown in Figure 1, as frequency distributions of offspring with various frequencies of organelle gene alleles from the two parents. Exclusively maternal inheritance is seen in the mitochondria of many fungi, animals, angiosperms, and some gymnosperms, and in the chloroplasts of about two-thirds of all angiosperm species (UM). Paternal inheritance is characteristic of cpDNA in all gymnosperms examined to date and of mtDNA in some gymnosperms (UP). Biparental inheritance of chloroplasts is shown for *Oenothera*, where most zygotes transmit only maternal alleles but some transmit both (UBa); the geranium, which may show three different patterns depending on the nuclear genotype (UBa,b,c); and alfalfa, where there are primarily paternal and biparental zygotes (UBb) (Smith, 1989). For microorganisms with mating types instead of sexes, one must speak of uniparental inheritance instead of maternal or paternal inheritance. Yeast mitochondria may show almost exclusively bi-parental zygotes, or a mixture of uniparentals and biparentals with a bias toward either parent (UBe,f); *Chlamydomonas reinhardtii* has mainly uniparental inheritance from the mt$^+$ parent, with a few biparental zygotes (UBa), except in

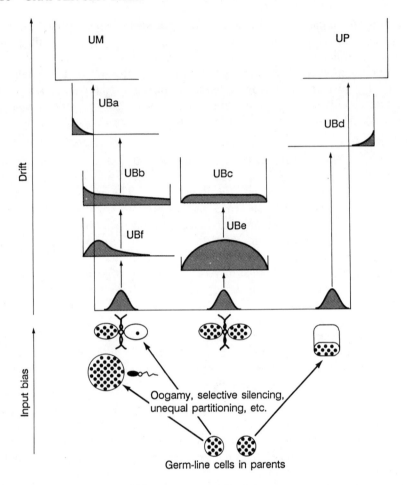

FIGURE 1. Patterns of organelle gene inheritance shown as frequency distributions, and their explanation in terms of input bias and random drift of gene frequencies within cells and clones or individuals. Each horizontal line represents gene frequencies from 0 (left end of line) to 1 (right end). The long line shows various input frequency distributions in zygotes with no bias (center distribution) or bias in favor of one or the other parent due to directional mechanisms. The shorter lines are frequency distributions of gene frequencies in the mature clones or individuals.

the case of the mitotically dividing zygotes, which show patterns like those of yeast.

A general consequence of uniparental inheritance and within-generation drift is that heteroplasmic cells and organisms are rare. However, there are important exceptions. In animals, heteroplasmy of mitochondria is due to high mutation rates for length changes in the D-loop (Boyce et al., 1989); in plant chloroplasts, the mechanisms causing heteroplasmy are unknown, although high levels of biparental transmission (Johnson and Palmer, 1989), or slow rates of vegetative segregation (Vaughn, 1981), have been proposed.

Because organelle genomes are housed in self-reproducing organelles and replicated under relaxed control, they are potentially able to replicate at different rates depending on their genotype. There is good evidence for **intracellular selection** among the mitochondrial genomes of yeast and *Paramecium* and the chloroplast genomes of *Chlamydomonas*. Selection has been shown to favor wild-type cells over cells with mutant white plastids in variegated plants, which I will call **intercellular selection**.

Organelle genes recombine only within an individual organelle in some cases (plant chloroplasts), and possibly never in some (animal mitochondria). In yeast and plant mitochondria and *Chlamydomonas* chloroplasts, organelle genomes recombine during vegetative cell cycles as well as gametogenesis; and they can recombine repeatedly and indiscriminantly with respect to partners, in striking contrast to nuclear genes. Gene conversion may be an effective mechanism for maintaining the homogeneity of genomes within a cell or organelle, but recombination between genomes from different individuals is limited by uniparental inheritance and random drift.

NEUTRAL THEORY FOR ORGANELLE GENOMES

This section considers single-locus neutral theory, examining the effects of random drift, mutation, and migration on genetic diversity and rates of substitution. When there is no recombination, the theory can be applied to entire organelle genomes. The analysis of genetic diversity using the classic Wright-Fisher model, modified for organelles, will be followed by results obtained from the application of coalescence theory.

A Wright-Fisher model

The classical Wright-Fisher model for nuclear genes incorporates the finite size of populations by sampling N_e genes from one generation to make the next generation. The model can be extended to organelle genomes by inserting a second level of sampling between cell generations and allowing for input bias (Figure 2). Random drift of gene frequencies within cells and organisms is modeled by defining an effective number of genomes per cell, n_e, which is generally much smaller than the real number of genomes n. At each cell division, each daughter cell gets a binomial random sample of n_e genomes from the parent cell. This process is repeated c times, where c is the number of cell generations from egg to adult. Between sexual generations, the creation of an input bias by deterministic mechanisms is described by the input parameter β, which is the proportion of genomes in a zygote that are derived from a sperm (and not degraded or otherwise excluded from the embryo), while $\alpha = 1 - \beta$ is the proportion of genomes contributed by the egg. When organelles are

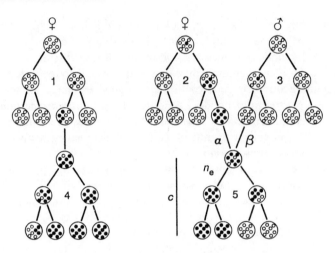

FIGURE 2. Wright-Fisher model for organelle genes. Five zygotes undergo $c = 2$ cell generations to produce clones or adult organisms with four cells. Solid and open circles within the cells represent organelle genomes of two different genotypes. In pedigree 1, a mutant genotype increases in frequency by drift in one cell lineage and is transmitted to the next generation (pedigree 5), where one heteroplasmic cell remains. The male gamete that gave rise to zygote 4 contributes no organelle genomes and is not shown. A mutant allele appears and is eliminated by drift in pedigree 3, while another mutation in pedigree 2 is fixed and transmitted. Zygote 5 comes from a cross in which $\beta = 2/8 = 0.25$, but as a result of drift within the individual, the final frequency of the paternal allele has risen to $2/4 = 0.50$.

inherited strictly maternally, $\beta = 0$. This model substitutes the mathematically tractable binomial sampling process and empirically measurable parameters for the complex and poorly understood details of vegetative segregation and uniparental inheritance. The above description applies to animals and plants with separate sexes, but it is readily adaptable to monoecious or hermaphroditic species, or to protists and fungi.

The parameter used to measure genetic diversity is the probability that two genes are identical by descent (hereafter called the identity, J), or the probability that they are different (the diversity, $K = 1 - J$). For organelle genes there are at least three levels of population structure with corresponding diversity measures: populations of genes within a cell (diversity K_a, or K_z in a fertilized egg); populations of cells within organisms (K_b); and of course populations of organisms (K_c). K_c corresponds to the nuclear gene diversity or heterozygosity H. The components K_a, K_z, and K_b are not usually measured directly; instead, DNA is extracted from a whole organism or from large portion thereof and subjected to restriction analysis. What is being measured in this case is $K_{ab} = [K_a + K_b(C - 1)]/C$ where C is the total number of cells in the organism (Birky et al., 1989). Since C is quite large, $K_{ab} \approx K_b + K_a/C \approx K_b$, so the distinction between K_b and K_{ab} can be ignored.

Inferences about genetic diversity from the Wright-Fisher model

In the past ten years, mathematical treatments of organelle gene diversity under mutation, drift, and migration have become increasingly realistic and comprehensive (Takahata and Maruyama, 1981; Takahata, 1983b; Takahata and Palumbi, 1985; Birky et al., 1983, 1989). All treatments concern multicellular organisms with separate sexes; when organelle genes are compared to nuclear genes, the organism is assumed to be diploid. Most mathematical studies are based on the infinite alleles model, which can be applied to single genes or entire genomes. However, in the case of genomes, some diversity measures may be too close to 1 to be useful. Only small changes are required to extend the models to the nucleotide level, to other measures of diversity, and to models with a finite number of alleles or infinite sites. The following important conclusions are well established:

1. Because the mean time to fixation or loss of alleles due to intracellular random drift is short, most cells are homoplasmic unless the mutation rate is very high or there is substantial paternal contribution of organelles to the zygote. This important feature of organelle population genetics can be seen by considering the mean times to fixation or loss of mutant alleles in a cell lineage due to intracellular random drift (Chapman et al., 1982; Birky et al., 1983). The equation for mean time to fixation or loss of an allele with frequency p is $\bar{t}(p) = -2n_e[p \ln p + (1 - p) \ln(1 - p)]$. The solution of this equation for a new mutation ($p = 1/n$) shows that for any reasonable values of n_e and n, the mutation will be, on average, either fixed or lost within one organismal generation (e.g., 8 cell generations if $n = 10^3$ and $n_e = 500$, or 20.4 cell generations if $n_e = n = 10^4$). Occasionally, however, mutations will persist in the heteroplasmic state for several sexual generations, so that they can be detected and used to study the rate of segregation. If there is any paternal contribution, the mean time to fixation or loss of an allele introduced into a zygote from the male parent is obtained by setting $p = \beta$. For example, if $\beta = 0.01$ and $n = 10^3$, $\bar{t}(\beta) = 112$ cell generations and it may take more than one sexual generation to complete the sorting out of organelle alleles. Takahata and Maruyama (1981) and Chapman et al. (1982) found the simple relationship that $\hat{K}_a \approx 0$ when $\beta < c/n_e$. This condition holds only when the mutation rate is not unusually high ($\mu > 10^{-5}$), as it appears to be for some length variants in animal mtDNA; Birky et al. (1989) and Clark (1988) showed that in such cases, heteroplasmicity ($\hat{K}_a > 0$) may be common even if inheritance is strictly maternal.

2. Much of the population genetic theory for nuclear genes can be adapted for use with organelles, simply by using different effective numbers of genes. Under the infinite alleles models, the inbreeding effective population number of organelle genes, which takes into account the different contributions of males and females to the zygote, is given by

$$N_{eo} = \frac{N_m N_f}{\alpha^2 N_m + \beta^2 N_f} \tag{1}$$

where N_f and N_m are the number of females and males, respectively (Takahata and Maruyama, 1981; Birky et al., 1983). N_{eo} can be substituted for $2N_e$, the effective number of nuclear genes in a population, provided $K_a << K_c$. When $\beta = 0$, N_{eo} reduces to N_f. This is as expected, because in that case only females transmit organelle genes, and they are counted as haploid rather than diploid if $K_a \approx 0$. N_{eo} and N_f have been widely applied as effective gene numbers. For example, the expected equilibrium diversity in a panmictic population is $\hat{K}_c \approx 2N_{eo}u/(2N_{eo}u + 1)$, compared to $\hat{H} \approx 4N_eu/(4N_eu + 1)$ for nuclear genes (where u is the mutation rate per sexual generation).

The joint effects of mutation, migration, and drift in a subdivided population were also treated by Takahata and Maruyama (1981), Birky et al. (1983), Takahata and Palumbi (1983), and Birky et al. (1989). An important difference between nuclear and organelle genes in subdivided populations is that, in the latter case, the migration rates of males and females must be considered separately because they make different contributions to the gene pools of their new colony (local population, subpopulation, deme). This can be done by defining an effective migration rate (Takahata, 1983b):

$$m_e = \alpha m_f + \beta m_m \tag{2}$$

For a finite island model of population structure, Birky et al. (1989) obtained equations for the equilibrium diversities within populations ($\hat{K}_c$), between colonies ($\hat{K}_d$), and in the population as a whole ($\hat{K}_t$). They found that G_{ST} is a useful measure of population subdivision because it goes to equilibrium very rapidly for organelles as it does for nuclear genes, and they obtained its equilibrium expectation. These equations are the same as the corresponding nuclear equations, with N_{eo} and m_e substituted for $2N_e$ and $2m$.

The K diversity measures can easily be extended to any number of levels. Rand and Harrison (1989) extended them to diversity between two species of crickets and showed how one can define a set of "C statistics" to partition the total genetic diversity among the four levels. As they noted, the equations of Birky et al. (1989) are not strictly applicable to the cases where diversity is highest—the length variants in animal mtDNA—because they are based on an infinite alleles model, while there is usually a limited number of different length variants. However, Crow and Aoki (1984) showed that, under reasonable conditions, the number of alleles is unimportant.

Birky et al. (1983) defined a third effective gene number, N_λ, which is more accurate than N_{eo} under some conditions. Equilibrium equations for $\hat{K}_c$ and $\hat{K}_a$ were also obtained by Takahata and Maruyama (1981) using N_{eo} but assuming that all gametes from one individual are identical. Contrary to previous statements (Birky et al., 1983), it appears that the equations of Takahata and

Maruyama (1981), those of Takahata (1983b), and our equations using N_{eo} and N_λ, all give very different results with some sets of parameters, especially when K_a is not negligible. None of the equilibrium equations takes into account the effects of high mutation rates and at the same time $\beta > 0$. Resolving the discrepancies and deciding which equations are best will require computer simulations of the basic organelle genetics model that include all levels of diversity. In the meantime, it is safe to use N_{eo} and the approximate equilibrium equations given above under most conditions, provided $K_a \approx 0$ or $K_a << K_c$. Effective gene numbers have also been obtained for monoecious or hermaphroditic organisms (Birky et al., 1983). However, measures of effective migration rates have not been devised for plants, where there can be differential migration of pollen, fertilized eggs, and plant segments.

When is K_a significantly larger than zero? To answer this question, Birky et al. (1989) compared the approximate values of $\hat{K}_c$, $\hat{K}_t$, and $\hat{G}_{ST}$ from equations using N_{eo} to values obtained by reiteration of exact equations, for different values of $\hat{K}_a$. The approximate equations require that $K_a \leq 0.01$; K_c, K_d, and K_t must be ≥ 10 times larger than K_a; $\beta \leq 0.1$ for the approximate diversity equations; and $\beta < 0.1$ for the approximate G_{ST}. K_a must be small because when it is large, each individual has to be counted as having more than one allele, in which case N_{eo} and n_e underestimate the effective numbers of genes in populations or colonies and in migrants.

Takahata and Maruyama (1981) stated that when $\beta < c/n_e$, the paternal contribution can be ignored. This is not correct; for example, if $\beta = 0.1$, $c = 50$, $n_e = 50$, and $N_m = N_f = 1000$, the predicted equilibrium diversity in the population is $\hat{K}_c \approx 0.0024$ using N_{eo} as an effective population number that takes β into account, but it is 0.0020 if N_f is used.

3. The rate and path of approach to equilibrium are different for diversity within and between individuals and populations. K_a equilibrates much faster than K_c when transmission from the two sexes is very unequal, i.e., when the product $\alpha\beta$ is small (Takahata and Maruyama, 1981). When $\alpha\beta$ is large, both K_a and K_c equilibrate slowly. This result was extended to subdivided populations and to all diversity levels by Birky et al. (1989), who also showed that G_{ST} equilibrates faster than diversity for organelle genes, just as it does for nuclei.

4. The effective population size and migration rate of organelle genes are often, but not always, lower than those of nuclear genes in the same population. Conventional wisdom holds that organelle genes will show about fourfold less diversity than nuclear genes (with the same mutation rate) because the effective number of genes is about fourfold smaller (N_f vs. $2N_e$). But this depends strongly on the breeding sex ratio, and is true only when $N_m/N_f = 1$ (Birky et al., 1983). When there are more than seven females for every male, the effective number of genes is larger, and the expected diversity is higher, for organelle genes. This situation may not be uncommon, for example, among mammals where males have harems. The difference be-

tween organelle and nuclear gene diversities also depends on β, especially when there are more males than females.

It is also intuitively evident that organelle genes will show more subdivision than nuclear genes, because only female migrants carry organelle genes between populations, and they are only counted as carrying one allele. But again, this depends strongly on the breeding sex ratio, on β, and also on the sex ratio of migrants (Birky et al., 1989). For small β, nuclear genes will often be more subdivided if the breeding sex ratio is less than $1/7$ and females migrate approximately as much as males. The relationship of the total gene diversity (K_t) for organelle genes to that of nuclear genes is complex, because organelles often have lower total effective population sizes, which tends to reduce K_t, while they tend to be more subdivided, which tends to increase K_t.

No rigorous theory has been developed for the effects of fluctuating population size on organelle gene diversity, but clearly they will be more sensitive than nuclear genes to bottlenecks. This can be seen first by noting that the ratio of equilibrium diversities is approximately $H/K_c \approx 4(Nu + 1)/(4Nu + 1)$ if the sex ratio is 1 and the mutation rates are the same in the two genomes. If a population slowly enters a bottleneck, this ratio rises from 1 to 4 as $4Nu$ goes from $>> 1$ to $<< 1$. But in a severe bottleneck of brief duration, organelle diversity might go to zero while nuclear diversity remains substantial; in a longer bottleneck, both diversities could go to zero. Wilson et al. (1985) noted that the ratio of genetic diversities in nuclear and organelle genes might be used to make inferences about the severity and timing of bottlenecks in a population.

Results of the coalescent approach

An alternative approach to population genetic theory is to focus on genealogies of genes, i.e., on the lines of descent from existing genes back to their most recent common ancestor, the coalescent. This seems to be a natural way of looking at organelle genes, because when inheritance is strictly maternal one can trace maternal phylogenies of entire molecules, without the complications of recombination. Moreover, sequences for organelle genes are being widely used to generate phylogenetic trees. Lineages of genomes (or genes or alleles) undergo extinction due to random rift, so all of the extant genomes in a population are descendants of a single genome, which is the coalescent for the entire population. The distribution of the time (in generations) back to the coalescent is the same as that of the time to fixation (Maynard Smith, 1987). The mean time to fixation of an organelle genome is approximately $2N_{eo}$, which is $2N_f$ if the genome is maternally inherited, compared to $4N_e$ for a nuclear gene (Birky et al., 1983). This is the time to fixation in the population of organisms, of a genome that is already fixed or nearly so in an individual female. The coalescent of the mtDNA molecule itself may have been in an ancestor of this female, a few sexual generations back. Motivated in part by some confusion in the literature about the human population size at the time of the coalescent of the human mitochondrial genome, Avise et al. (1984) generated the complete distribution of probabilities of lineage survival for various demographic models and were able to show that the population size was probably fairly large.

Neigel and Avise (1986; Avise et al., 1984) noted that genealogies based on gene sequences are really genealogies for the genes, not necessarily for the species, and that it is sometimes observed that mtDNA lineages in two different species are more closely related to each other than to other lineages in their own species. Species A and B tend to be polyphyletic, i.e., each shares lineages with the other, when they have diverged fewer than N_f generations ago. Species A is paraphyletic with respect to species B if it shares lineages with B but not vice versa; this is common for divergence times from N_f to $4N_f$ generations. Finally, A and B are usually monophyletic, i.e., all lineages in A are more closely related to each other than any is to any lineage in B, and vice versa, if they have diverged more than $4N_f$ generations ago.

Avise et al. (1988) showed that the coalescence times for pairs of mtDNA haplotypes in eels, catfish, and redwing blackbirds imply mean numbers of breeding females much lower than those estimated from present-day census data, or a variance of progeny number per female that is much greater than the Poisson variance. They propose that eels and catfish, and other animals with high fecundity, have extremely high variance in survival among families, whereas blackbirds may have undergone a bottleneck due to Pleistocene glaciation.

Slatkin (1989) used the method of coalescents to estimate levels of migration, but the coalescent method developed in this paper proved to be insensitive when there is less than one migrant per generation.

Rates of neutral evolution in organelle genomes

For organelle genes, as for nuclear genes, the substitution rate at neutral sites is equal to the mutation rate. Any differences in substitution rate will be attributable to differences in the mutation rate. For sites subject to selection, the rate of base pair substitution mutation, selection, and drift is shown by the equation

$$E = MF = [uN]\,[F] = uN\,\frac{1 - e^{-2(N_{eo}/N)s}}{1 - e^{-2N_{eo}s}} \tag{3}$$

where E is the substitution rate per gene per generation, M is the total mutation rate, F is the probability of fixation, N and N_{eo} are the number and effective number of organelle genes, and s is the selection coefficient. Because the number and effective number are $2N$ and $2N_e$ for nuclear genes in diploids, the total mutation rate (M) is twofold lower for organelle genes with the same mutation rate per gene, while F is higher for detrimental and lower for advantageous mutations unless the sex ratio is extremely biased. The smaller effective population size of organelles will cause more organelle mutations to be effectively neutral.

SELECTION WITHIN CELLS AND ORGANISMS

No theory has been developed to show the effect of intracellular selection on diversity, but some general features can be seen from the following considerations. We begin by making an analogy between the population of genomes within a cell and a haploid, asexual population of organisms. We ignore the fact that the genomes are packaged into organelles and that selection may, in fact, operate on the organelles rather than on the individual genomes. Under the action of intracellular random drift and selection, each mutation will eventually be fixed in the cell in which it occurs or in one or more progeny cells, or it will be lost from the lineage. From the viewpoint of evolution or population genetics, organelle mutations that are lost in this way will be of little consequence, so we will focus on those that are fixed in a cell lineage, giving rise to a homoplasmic mutant cell. The rate of appearance of such cells is the cell mutation rate, v. This cell mutation rate is related to the genome mutation rate μ by $v = n\mu\phi$, where ϕ is the probability that a new mutant allele is fixed. The fixation probability for neutral mutations will be $\phi_n = 1/n$, so that $v_n = n\mu_n(1/n) = \mu_n$.

For the fixation probability of mutants subject to directional selection, we can use the classic equation for the fixation probability of a new mutant allele in a haploid population.

$$\phi = \frac{1 - e^{-2(n_e/n)\sigma}}{1 - e^{-2n_e\sigma}} \tag{4}$$

where σ is the intracellular selection coefficient. For advantageous and detrimental mutations, the fixation probability will be $\phi_a > 1/n$ and $\phi_d < 1/n$, respectively. Consequently, $v_a > \mu_a$, but $v_d < \mu_d$. A priori it seems almost certain that many more mutations are detrimental than are advantageous. Consequently the total cell mutation rate is actually reduced by intracellular selection: $v < \mu$. As was noted by Birky et al. (1983), this disproves the common notion that because there are many copies of an organelle genome in a cell, the total mutation rate per cell ought to be higher for organelle genes than for nuclear genes when the mutation rates per gene are similar.

The effect of intracellular selection may be very important. For example, consider an organism for which $\mu = 10^{-10}$, $n = 1000$, and $n_e = 500$. Advantageous mutations with an intracellular selective advantage $\sigma = 0.01$ will have $v \approx 10^{-9}$, a tenfold increase over the case with no intracellular selection, while detrimental mutations with $s = -0.01$ will have $v \approx 5 \times 10^{-12}$. On the other hand, if $n_e = 50$ and the other parameters are the same, then drift is so strong that selection will change μ by less than twofold. Note that the mutation rate per sexual generation is $u = vc$, so that u is affected in the same proportion as v, if we can ignore selection between cells.

We can now see what effect intracellular selection is likely to have on genetic diversity. If most of the diversity is due to mutations that are neutral or effectively neutral, there will be no effect. If a substantial part of the organelle gene diversity is due to mutations under directional selection, intracellular selection will appear as an apparent reduction in the mutation rate overall, and will reduce the overall diversity at all levels. Conceivably, if the data allow one to distinguish between sites subject to strong selection and those subject to weak or no selection, intracellular selection would increase the difference between them.

For the rate of molecular evolution, we have

$$E = MF = [uN]\,[F] = [\mu nc\phi]\,N\left[\frac{1 - e^{-2(N_f/N)s}}{1 - e^{-2N_f s}}\right]$$

$$= \left[\mu nc\,\frac{1 - e^{-2(n_e/n)\sigma}}{1 - e^{-2n_e \sigma}}\right] N \left[\frac{1 - e^{-2(N_f/N)s}}{1 - e^{-2N_f s}}\right] \tag{5}$$

(An equivalent equation was derived rigorously by Takahata, 1984.) We would thus expect intracellular selection to increase the rate of advantageous substitutions, decrease the rate of detrimental substitutions, and have no effect on neutral evolution.

It seems likely that mutations that are advantageous, neutral, or detrimental at the organismal level will tend to be advantageous, neutral, or detrimental with respect to intracellular selection. To the extent that this is true, intracellular selection will look like an enhancement of the effect of selection between organisms.

This preliminary treatment ignores a host of potential complications. First, intercellular selection is not considered. Second, there is no treatment of selection on heteroplasmic cells or individuals; it is assumed that heteroplasmic individuals have the fitness of the homoplasmic zygote from which they arose. Third, if there is paternal transmission, an allele that has already been subject to intracellular selection while being fixed in a cell lineage may now be subjected to intracellular selection again, possibly against a different allele or alleles. Fourth, not all new mutations will be fixed in the individual in which they occur. The relationship of the fitnesses of heteroplasmic cells and individuals to the gene frequencies within them is probably complex. Fifth, intracellular selection increases the variance in replication of organelle genomes and thus reduces n_e, as described in the next section. Perhaps some relatively simple theory and experimentation will show that some of these complications can be ignored. It may well turn out that if K_a and K_b are small, it will be sufficient to treat the net effect of intracellular selection as modifying the cell mutation rate as we did above, even if $\beta > 0$.

RECOMBINATION OF ORGANELLE GENOMES

Recombination is limited by uniparental inheritance and vegetative segregation

Uniparental inheritance limits recombination because it limits the opportunity for organelle genomes from different individuals to come together in the same cell; and when different genomes do come together, vegetative segregation tends to separate them again. This was verified by Takahata (1983a), who showed that the within-cell linkage disequilibrium tends to be large even if the recombination frequency per cell generation is high, and is usually larger than the population disequilibrium. The population disequilibrium for organelle genes is larger than that of nuclear genes unless there are many heteroplasmic cells in which recombination is occurring. Takahata's treatment does not include recombination that is limited to molecules in the same organelle, as in plant chloroplasts. This recombination will reduce the intracellular linkage disequilibrium but might have relatively little effect on the disequilibrium in the population as a whole.

Hitchhiking reduces genetic diversity

An important consequence of strong linkage is hitchhiking, in which an advantageous mutation (the driver) is fixed by selection and carries with it one or more linked mutations, which otherwise might not have been fixed (hitchhikers). It is well known that hitchhiking reduces genetic diversity. The most extreme form of hitchhiking, with the strongest effects on diversity, is periodic selection. This happens in genomes with no or very low recombination, as in some microorganisms and organelles. Driver mutations periodically occur and fix linked hitchhiker alleles while eliminating alternative alleles from the population. If the driver goes to fixation quickly enough, diversity in the population may be reduced temporarily to zero, after which it builds up again until the next hitchhiking event. For organelles a simple expression for the equilibrium gene diversity in a population with periodic selection is

$$\hat{H}_p \approx \frac{2Nu_n}{1 + 2Nu_n + 2Nu_n rLNs} \approx \frac{\theta}{1 + \theta (1 + rLNs)} \tag{6}$$

where N is the effective number of genes; u_n is the neutral mutation rate; u_a and s are the average mutation rate and selection coefficient, respectively, for advantageous driver mutations; $\theta = 2Nu_n$; $r = u_a/u_n$; and L is the total number of genes (or sites) in the genome (Takeo Maruyama and C. W. Birky, Jr., unpublished results). This is the same as the equilibrium diversity $\hat{H}$ for a population without periodic selection, with the addition of the factor $2Nu_nrLNs$ to the denominator. Richard Hudson (personal communication) has shown that Equation (6) can be derived from Equation (18) of Kaplan et al. (1989), who treat the more general case with variable recombination.

If one is measuring nucleotide diversity and θ is small, the effect of periodic selection may be substantial. Consider an animal mitochondrial genome with 15 kb, or $L = 1.5 \times 10^4$, $u_n = 2 \times 10^{-8}$, and a measured nucleotide diversity of $\hat{H}_p = 0.01$. Suppose further that driver mutations are rare, with $r = 10^{-7}$, and are strongly selected, with $s = 0.062$. Then Equation (6) can be solved for θ to give $\hat{u} = 0.01884$, from which we find that the diversity in the absence of periodic selection would be 0.01849, nearly twice as high as the observed diversity. Essentially the same result is obtained if the drivers are more common but less strongly selected ($r = 10^{-7}$, $s = 1.3 \times 10^{-6}$).

Biased gene conversion may affect the cell mutation rate differently in organisms with different genome copy numbers

A gene conversion event occurs when two organelle DNA molecules carrying different alleles pair and recombine, resulting in conversion of one allele to the other. Gene conversion is unbiased if it is equally likely to occur in either direction, or biased if one allele is more likely to be converted to the other. Unbiased conversion in a population of organelle genomes within a cell is a form of genetic drift, which fixes or eliminates alleles with prior probabilities equal to their frequencies (Birky and Skavaril, 1976; Ohta, 1977). Consequently, it has no effect on the ultimate fixation probability of new mutant organelle alleles, and the cell mutation rate v remains independent of the number of genome copies in the cell.

To understand the role of biased gene conversion, we can adapt the theory for nuclear genes (Walsh, 1983; and references therein) by substituting n_e for $2N_e$ and n for N. If mutant and wild-type alleles are selectively neutral, the fixation probability of a newly arisen mutant allele is obtained by replacing σ with d in Equation (4), where d is a measure of the directional pressure of biased conversion. Gene conversion thus has a net effect on the cell mutation rate if it tends to be biased in favor of mutant alleles, or against them. Walsh (1983) pointed out that the existing neutral alleles in a population may have undergone conversion several times and consequently have been selected for having a conversion advantage relative to many kinds of mutant alleles. Thus ϕ may tend to be lower than $1/n$ and the effective mutation rate will be correspondingly reduced. This conclusion also holds for selected alleles, because Equation (4) applies to alleles subject to both intracellular selection and biased conversion if $d + \sigma$ is substituted for σ. Moreover, ϕ depends on n_e and n, so that the cell mutation rate will also depend on these parameters. This may help to explain the puzzling observation of Wolfe et al. (1987) that the rate of synonymous substitution in plant chloroplast genomes is lower in the inverted repeat than in the single-copy regions. Since n is twofold higher for the inverted repeat, we expect ϕ, and hence u, to be lower for neutral mutations in the inverted repeat if gene conversion tends to be biased against mutant alleles.

How the rate of evolution at a site is affected by selection on linked and unlinked background sites

The study of the effect of background selection on substitution rates has been motivated by suggestions that hitchhiking might affect the rate of molecular evolution as well as the genetic diversity in organelle genomes. But it is misleading to think of the problem in terms of hitchhiking, because the fixation probability and substitution rate of a neutral mutation at a site (base pair, gene, etc.) are not affected by selection on mutations in the background, even if there is complete linkage. However, selection in the background reduces the substitution rate of advantageous mutations and increases the rate for detrimental mutations at a site, whether the background mutations are advantageous or detrimental (Birky and Walsh, 1988). This is a generalization of the Hill-Robertson effect, in which selection increases the variance in offspring number among individuals, thereby reducing the effective population size and increasing the effectiveness of random drift so that all rates of evolution become closer to the neutral rate (Hill and Robertson, 1966; Robertson, 1961). This Hill-Robertson effect is greater the larger the background genome. By computer simulations, J. Bruce Walsh and I found that with a sufficiently large background, advantageous or detrimental mutations at a site behave as if they were neutral, even when $|Ns| \gg 1$. Although the effect is greatly reduced by recombination, it is not abolished, even if there is no linkage. In fact, the effect of a large background that is completely unlinked to the site can be greater than the effect of a small background with complete linkage. This raises the intriguing possibility that the nuclear genome has a stronger Hill-Robertson effect on organelle genomes than the organelles have on themselves.

Takahata and Slatkin (1983) used computer simulations to show that selection is more effective in a model that includes within-generation random drift as well as recombination than in a model without either, but it is not possible to tell whether the drift, recombination, or both were responsible for reducing the Hill-Robertson effect.

NUCLEAR–CYTOPLASMIC INTERACTIONS

Modeling nuclear and cytoplasmic genomes together

Nuclear–cytoplasmic interactions in which phenotypes are determined jointly by nucleus and cytoplasm (e.g., cytoplasmic male sterility) have long intrigued biologists; in fact it is a model system for the interactions of hosts and parasites and of multigenome systems in general. The only model of these interactions that has been analyzed consists of a diploid organism with a maternally inherited organelle genome with no heteroplasmy or within-individual selection. Two alleles each at a nuclear and a cytoplasmic gene locus are subjected to varying kinds of selection but no mutation or drift.

Conditions for balanced polymorphism

Clark (1984) showed that a random mating population with viability selection acting identically on males and females cannot maintain a stable polymorphism for both nuclear and organelle genes. Gregorius and Ross (1984) found that a stable polymorphism can be maintained in a model with partial selfing provided male and female fitnesses are different. Because these models require selfing, they apply to many plants but to very few animals; the models do not permit paternal transmission; there is no drift; and only a few kinds of selection have been studied. We are far from knowing whether cytoplasmic or nuclear–cytoplasmic systems will differ from strictly nuclear systems in being able to maintain polymorphism simultaneously at many loci by selection. But the joint effects of nuclear and organelle genes on male fertility in plants are of great importance in their own right, because they are likely to be important in the evolution of mating systems (Frank, 1989).

Early speculations that cytoplasmic male sterility might initiate sympatric speciation or act as a mating barrier between species have not been supported by the analysis of models similar to the above (Watson and Caspari, 1960; Caspari et al., 1966).

Nuclear–cytoplasmic disequilibria as analytic tools

Because organelle and nuclear genes are physically unlinked, nonrandom associations between organelle and nuclear genotypes (linkage disequilibrium) must be due to historical factors, assortative mating, or epistatic interactions. Information regarding the relative contributions of these factors can be obtained from the analysis of linkage disequilibrium (Arnold et al., 1988; Asmussen et al., 1987, 1989). In hybrid zones in which females mate preferentially with males of the same species (positive assortative mating), all disequilibria between nuclear and cytoplasmic genes go to zero with random mating, or with positive assortative mating. However, different models can be distinguished by the rates and trajectories of decay of the various disequilibria. Permanent disequilibria can be created by continuous immigration into the hybrid zone. A model with both immigration and multilocus assortative mating was shown to account for the gene frequencies and linkage disequilibria seen in a hybrid zone of tree frogs.

A key extension of disequilibrium theory for plants incorporates measures of disequilibrium among a diploid nuclear locus, a mitochondrial locus, and a chloroplast locus (or any two cytoplasmically inherited genomes), with varying degrees of selfing (Schnabel and Asmussen, 1989). In this theory, the cytoplasmic loci are both inherited strictly uniparentally, either from the same parent, or from different parents as in some conifers. The resulting theory should be a powerful tool for studying population genetics and evolution in plants, including important forest trees. The treatment of three genomes, subject to different selection pressures and modes of dispersal, may make it easier to isolate and identify the effects of selection and various kinds of migration.

Nuclear–cytoplasmic incompatibility

Striking examples of nuclear–cytoplasmic incompatibility are seen in some organisms, such as *Oenothera* and *Paramecium* (reviewed in Gillham, 1978; Kirk and Tilney-Bassett, 1978), but not in others (Clark, 1985; Clark and Lyckegaard, 1988). Nuclear–cytoplasmic incompatibility can inhibit the flow of organelles between species that hybridize, provided that the incompatibility involves nuclear genes that cause selection against hybrids between the species (Takahata, 1985).

Discordant behavior of organelles and nuclei in hybrid zones

Several investigators have found differences in the behavior of nuclear and mitochondrial genes in hybrid zones, in which mitochondrial genomes of one species have spread across the hybrid zone to a greater extent than have nuclear genes (references in Vanlerberghe et al., 1988). Several explanations have been proposed for this differential introgression.

One hypothesis begins with the assumption that species barriers arise due to selection on nuclear genes, because nuclear genes control many of the traits known to be involved in species isolation, such as habitat and mating preference. It follows that any given nuclear gene is likely to be physically linked to, and in linkage disequilibrium with, a gene which is subject to strong selection, whereas mitochondrial genes will be show linkage disequilibrium with the selected nuclear genes due only to epistatic interactions or historical factors. Consequently, mitochondrial genes will move faster (Barton and Jones, 1983).

Takahata and Slatkin (1984) considered a model with two cytoplasmic alleles and a single nuclear locus with two alleles that differentiate two species. Only females migrate and hybrid males are sterile. The transfer of the nuclear gene alleles between species is inhibited by viability selection against heterozygotes. They showed that this selection on the nuclear genes would not inhibit the movement of organelle genes unless the selection was very strong. Takahata (1985) examined a model with paternal transmission and finite population size to show that organelle genes introgress preferentially from larger populations to smaller ones.

It should be noted that discrepancies in mitochondrial and nuclear gene frequencies may arise by founder effects during migration (Wilson et al., 1985) or by subsequent random drift (Vanlerberghe et al., 1988), without the intervention of any directional effect such as selection. Also, if the two species have been isolated for less than $4N_f$ generations, nuclear and organelle genes may have discordant phylogenies because of differences in shared ancestral polymorphisms, which can give the illusion of differential introgression (Neigel and Avise, 1986).

SOME PARAMETERS ARE MORE IMPORTANT THAN OTHERS BUT MOST ARE NOT KNOWN WITH SUFFICIENT ACCURACY

What parameters are most important and how accurately must they be measured? Answering this question is one of the most important contributions that theory can make to the study of the causes of genetic diversity. Clues can be obtained by solving the approximate equations for the equilibrium diversities and G_{ST}, using N_λ as the effective population size. Some preliminary efforts, of which examples are shown in Figure 3, lead to the following conclusion:

1. The effect of $\beta = 0$ vs. $\beta = 0.01$ is small unless the mutation rate is very high ($u \geq 10^{-3}$), so it is usually not important to measure values of β less than about 0.1. This is fortunate because in most of the studies that use RFLP markers, the number of progeny examined has been too small to detect minor paternal contributions if most zygotes are uniparental. Such

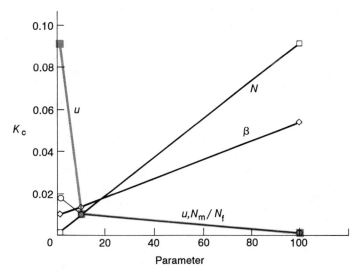

FIGURE 3. Effects of varying parameters on diversity, K_c. The parameters are population size N (open squares), sex ratio N_m/N_f (x's), mutation rate u (solid squares), and β (diamonds).

contributions can be amplified by repeated backcrosses to the paternal parent (Lansman et al., 1983; Gyllensten et al., 1985). Unfortunately, the upper limits of β calculated from these and many other experiments are too high because they ignored all sources of random drift.

2. N and u are almost always by far the most important parameters to measure accurately, because changes in these parameters usually produce changes of the same order of magnitude in measures of gene diversity. Little attention has been given to measuring u for organelle genes and there are fewer data for this parameter than for any other. The best estimates are obtained indirectly from evolutionary studies, by equating the synonymous substitution rate to the mutation rate (Sharp and Li, 1989; Wolfe et al., 1987). This method depends on accurate estimates of generation times and species divergence times, which are not available for most eukaryotes. It would be preferable to have a direct method for measuring u; in the future it may be practical to do this by sequencing many different copies of the same organelle gene from a single individual or clone. This has already been attempted with humans (Monnat et al., 1985; R. J. Monnat and L. A. Loeb, personal communication), but the resulting estimate of u is unreliable because the calculation did not take drift into account, and because it is possible that the individual whose mtDNA was cloned came from a heteroplasmic egg.

3. It is much less important to know the sex ratio if females are in the majority, but the effects of sex ratio are important when there is at least a tenfold excess of males (or whichever sex contributes fewer genes to the zygote).

4. For neutral theory it may not be very important to measure c and n_e accurately, so long as we are confident that they are of the magnitudes $c \geq 10$ and $n_e \approx 10-10^3$. On the other hand, because n_e and c determine the rate of intracellular drift and thus, indirectly, modify the effect of intracellular selection, it may be extremely important to measure them accurately if intracellular selection is important. Also they will be needed to interpret most experimental measures of the mutation rate.

The effective number of genomes per cell has been estimated for yeast mitochondria and plant and algal plastids, from the number of cell divisions required for heteroplasmic cells to produce homoplasmic progeny (reviewed by Birky et al., 1983). Another approach used for insects is to calculate n_e from the increase in variance in the frequency of an allele over several sexual generations, or between the first and last offspring of a female (Rand and Harrison, 1986; Solignac et al., 1987). Unfortunately these methods require accurate estimates of c. The number of cell generations in a sexual generation is known accurately for only a few organisms, of which only *Drosophila* is widely used to study organelle evolution (Solignac et al., 1987). For other animals, estimates of c for germ cells or a somatic tissue are usually based on estimates of the number of cells ψ, from which the number of cell generations is calculated from $c = \log \psi / \log 2$. But this assumes that the

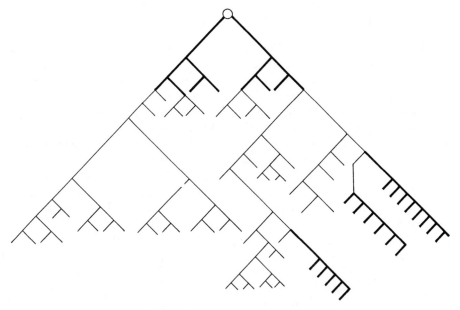

FIGURE 4. Mixed patterns of cell division in a multicellular organism. The pedigree begins with a purely clonal pattern (bold lines, top) and includes three stem lines (bold lines, bottom right). The remainder could be viewed as a clonal pattern but with some variance in offspring number per cell.

pattern of cell division is clonal, in which each cell divides each cell generation. The cells might have been produced by stem lines, in which case $c = \psi$. Also it is unlikely that all the cells divide the same number of times. Any variation in the number of cell divisions per cell will cause estimates of c made as described above to be too low for the average cell, and will decrease n_e. These patterns are shown in Figure 4. Of course, c may be different for germ cells and different somatic cells, and is usually higher by some unknown amount in males than in females. Sex differences could contribute to different evolutionary rates between genomes with different values of β.

5. Finally, if we want a deep understanding of the behavior of organelle genes, we must be able to estimate the strength of selection at all levels: within cells and between cells as well as between organisms. Better theory is required to assist in the design of the necessary experiments.

Given the difficulty of getting accurate information about c, μ, and σ, population geneticists may be content to know and use the apparent mutation rate per sexual generation, u, and not worry much about the factors that control it.

SUMMARY

The genetics and evolution of organelles are both daunting and fascinating because they are so rich in detail. One of the special complexities of organelle genetics—random drift within cells and organisms — can be treated by a model widely used for nuclear genes: binomial sampling of an effective number of genes to make the next cellular generation. Binomial sampling combined with sexually biased transmission provides an effective model for uniparental inheritance. These measurable parameters in turn can be used to develop inbreeding effective population sizes and migration rates that can often be substituted directly into existing theory for neutral alleles. A start has been made on the problems of selection within organisms, recombination between and within organelle genomes, and some of the many kinds of interactions between organelle and nuclear genes.

The resulting theory has explained why organelle gene diversity is often high in a population of organisms but virtually zero within individual organisms, and has also explained the exceptions where it is high within organisms. The theory offers explanations for differences in evolutionary rates of organelle genes between animals and plants, and variation among genomes within taxa. It is being used to study gene flow within and between species, and the evolution of mating patterns in plants. It supports the use of organelle genomes in studying biogeography and phylogenetics. Theory has also inspired the analysis of general evolutionary questions such as the effects of selection at linked and unlinked loci on diversity and evolutionary rates. For many of these applications, theory and experiment are moving together hand in hand, as they should.

But much work remains to be done. Recombination and selection theory are not yet developed to the point where they can direct the next step of experimentation by showing what parameters must be measured or how the experiments should be done. For example, how large must the intracellular selection coefficient be before it significantly changes the effective cellular mutation rate? How can this kind of selection be measured? Even neutral theory has some unresolved differences between the equilibrium equations of different authors. And perhaps most important, neutral theory has shown that there are several parameters about which we are abysmally ignorant, notably mutation rates. There is a lot of fun to be had at the bench as well as at the computer.

CHAPTER 7

MOLECULAR EVOLUTION OF CHLOROPLAST DNA

Michael T. Clegg, Gerald H. Learn, and Edward M. Golenberg

Owing to the central role of the chloroplast in energy metabolism and the relative ease with which its genes can be cloned and characterized, the molecular biology of this photosynthetic organelle has been an area of intensive research activity during the past decade. The result has been the accumulation of a rich background of molecular information that has stimulated and facilitated evolutionary studies. As a consequence, much more is currently known about chloroplast molecular evolution than about the evolution of any other component of plant genomes.

The preponderance of evidence points to a symbiotic origin of the chloroplast from a cyanobacteria-like progenitor. In the course of evolution, many of the gene functions that are essential for photosynthesis are thought to have been transferred from the chloroplast genome to the nuclear genome. This process apparently slowed markedly before the origin of land plants, because the chloroplast genome of all land plants is relatively conserved in both gene content and organization. Chloroplast genes evolve slowly in comparison to nuclear genes and are not subject to the complex processes of conversion and intragenic recombination that often affect nuclear sequence evolution. Structural stability and simple patterns of mutational change make the chloroplast genome especially well suited for studies of plant phylogeny.

Our goal in this chapter is to examine chloroplast gene evolution from several perspectives. First, we will review broad features of chloroplast DNA (cpDNA) structural evolution, including changes in gene content. Second, we will consider the origin and evolution of chloroplast introns. Third, we will discuss rates of chloroplast gene evolution and the use of gene sequences as molecular clocks. Along the way, we will identify some intriguing evolutionary problems posed by chloroplast gene evolution.

CHLOROPLAST GENE AND GENOME EVOLUTION

Chloroplast genome structure

The chloroplast genome is a circular molecule that ranges in size from about 120 to about 220 kb in higher plants (Palmer, 1987). A large region that is duplicated and inverted divides the genome into a large single copy (LSC) region, a small single copy (SSC) segment, and an inverted repeat region (IR) (Figure 1). Much of the size variation in cpDNA is the result of variation in

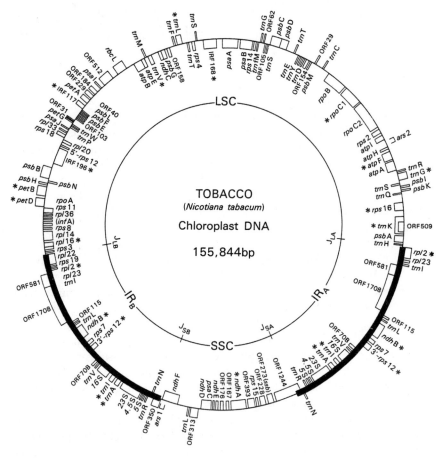

FIGURE 1. Map of the chloroplast DNA from *Nicotiana tabacum*, provided by M. Sugiura.

the length of the IR region. The IR region is typically about 20–25 kb long, but the repeat structure is absent in conifers (Strauss et al., 1988) and some legumes (Palmer, 1987). Remarkably, the IR region has expanded to 78 kb in *Pelargonium* and has engulfed a number of genes that are found in the large single copy region in most plant species (Palmer, 1987).

Data on algal cpDNA organization are still sketchy, although the chromo-phytic algae (those with chlorophyll a and c) and chlorophytic algae (those with chlorophyll a and b), except *Euglena*, have the IR structure, which is absent in the rhodophytic algae (those with chlorophyll a and phycobilins) (Cattolico, 1987). Although the IR structure arose early in plant evolution, it appears that the chloroplast can function without it. Absence of the IR region is associated with relatively high frequencies of rearrangements, an observation that has led to the hypothesis that it stabilizes the chloroplast genome (Palmer, 1987).

Knowledge of chloroplast gene organization and evolution in land plants has been greatly advanced by the determination of complete chloroplast DNA sequences for a liverwort, *Marchantia polymorpha* (Ohyama et al., 1986); tobacco, *Nicotiana tabacum* (Shinozaki et al., 1986), a dicot; and rice, *Oryza sativa* (Hiratsuka et al., 1989), a monocot. Because the monocot–dicot separation occurred at least 150 million years ago, and the divergence between liverworts and vascular plants occurred 350–400 million years ago, comparisons among these three data sets provide a broad view of the evolution of the chloroplast genome. Comparative analyses of the rice and tobacco sequences have been particularly helpful in elucidating the mechanisms responsible for the origin of major inversions in chloroplast gene order (Hiratsuka et al., 1989).

Several large inversions have arisen in the course of cpDNA evolution in land plants. One inversion of approximately 50 kb is associated with legumes, and a second inversion of about 20 kb is found in grasses (Palmer, 1987). Because these inversions are confined to specific taxa, each is assumed to be monophyletic, each having arisen only once in plant evolution. As phylogenetic markers, inversions have been exploited by Jansen and Palmer (1987) to define an ancient evolutionary split in the sunflower family.

Comparison of the rice and tobacco cpDNA sequences has suggested a mechanism for the origin of the major inversion characteristic of the grass family. Because the chloroplast genome has multiple copies within organellar structures known as nucleoids and cpDNA is often present as multimeric com-plexes (Deng et al., 1989), there is opportunity for occasional illegitimate recombination between cpDNA molecules. Hiratsuka et al. (1989) proposed that a series of recombination events occurred between two different transfer RNA genes on different cpDNA molecules to first generate two pseudo-tRNA genes and a duplicated portion of the genome as a complex multimer involving both genomes. Next, a deletion removed one pseudogene and the duplicated material, leaving the gene order inverted between the original sites of inter-molecular recombination. Two subsequent inversions must be postulated to arrive at the gene arrangement presently observed in grasses. Although inversions

TABLE 1. Comparisons of liverwort, tobacco, and rice chloroplast genes[a]

Category	Liverwort	Tobacco	Rice
tRNA	32	30	30
rRNA	4	4	4
ORFs	83	80	77
Introns			
tRNA	8	8	8
ORFs	12	12	9
Total size	121,024	155,844	134,525

[a]The total number of each category of gene is shown.

arising from intermolecular recombination do occur, these events must be either extremely rare, selectively disfavored, or structurally unstable, because they have rarely been fixed in plant evolution.

Chloroplast gene content

Table 1 compares the gene content of the *Marchantia*, tobacco, and rice chloroplast genomes. Approximately 80 open reading frames (ORFs), 30 tRNA genes and 4 rRNA genes (the *rrn* operon within the IR region, which includes the 16 S, 23 S, 4.5 S, and 5 S sequences) make up the chloroplast genome (Shinozaki et al., 1986; Ohyama et al., 1986; Hiratsuka et al., 1989). Approximately 27 ORFs, some of which are quite small, have not been assigned a coding function. Four ORFs show no sequence similarity between tobacco and *Marchantia*, and five ORFs are unique to one or the other cpDNAs (Wolfe and Sharp, 1988). Specifically, the gene *rps16* is absent in *Marchantia* and the genes *mbpX*, *infA*, *rpl21*, and *frxC* are absent in tobacco and rice (Table 2). In addition, at least five ORFs found in the tobacco or liverwort genomes are

TABLE 2. Gene functions lost in liverwort, tobacco, and rice genomes

Gene	Liverwort	Tobacco	Rice
infA	+	Pseudogene	+
frxC	+	−	−
mbpX	+	−	−
rpl21	+	−	−
rps16	−	+	+

absent or truncated in rice (Hiratsuka et al., 1989). These comparisons reveal a loss of some gene sequences in all three genomes. Whether these genes now reside in the nucleus is unknown, but according to the symbiotic origin hypothesis, many gene functions, such as those involved in photosynthesis, have been transferred from the chloroplast genome to the nuclear genome.

An instructive example of the coordinate function of nuclear- and plastid-encoded genes is the enzyme ribulose-1,5-bisphosphate carboxylase/oxygenase (RUBISCO). The active holoenzyme of all land plants and some algae is composed of eight identical large subunits that are encoded in the cpDNA (*rbcL* gene) and eight small subunits that are encoded as a small gene family in the nucleus (*rbcS* genes). The small subunit has a transit peptide that targets the protein to the chloroplast following synthesis on cytoplasmic ribosomes (Schmidt and Mishkind, 1986). After import to the chloroplast, the transit peptide is cleaved, leaving a mature protein that joins the holoenzyme complex. In the cyanobacterium *Anabaena*, the small subunit coding sequence is located immediately upstream from the large subunit sequence and both are transcribed as a dicistronic message (Nierzwicki-Bauer et al., 1984). Interestingly, the small subunit is encoded by genes adjacent to the large subunit gene on the cpDNA in some algae (Reith and Cattolico, 1986) and in the cyanelle genome of *Cyanophora paradoxa* (Heinhorst and Shively, 1983). These observations suggest that the small subunit coding sequences were transferred to the nuclear genome in the course of plastid evolution. Such a transfer would require the nuclear copy of the small subunit sequence to have come under the control of an appropriate eukaryotic promoter and to acquire a transit peptide mediating its transport back to the organelle.

There is now good evidence for recent incorporation of chloroplast sequences in the nuclear genome (Cheung and Scott, 1989). Presumably, sequence transfers occurred at a slow but continuous rate over the course of hundreds of millions of years of evolution. Out of the large number of small subunit transfers, one or more random integrations adjacent to an appropriate promoter, and in frame with a leader sequence that could function as a transit peptide, must have occurred. Palmer et al. (1990) present clear evidence for a more recent transfer of a *tufA* gene, which is present in the cpDNA of green algae but is absent in that of land plants. Instead, a *tufA* sequence very similar to the green algal cpDNA sequence occurs in the nuclear genome of the higher plant *Arabidopsis thaliana*. Variation in gene content among the rice, tobacco, and liverwort cpDNAs suggests that the process of gene transfer is a continuing one.

The evolution of chloroplast introns

Although chloroplast genomes are similar to prokaryotic genomes, they differ in that a number of cpDNA genes are interrupted by introns. There are many intriguing and unresolved questions concerning the evolution of cpDNA introns. Chloroplast introns are related to groups of other introns both in fungal mitochondrial genomes (Keller and Michel, 1985) and in the nuclear genomes of certain protozoans (Bonnard et al., 1984). All these introns are conserved in secondary structure (Waring and Davies, 1984; Michel and Dujon, 1983; Michel and Jacquier, 1987), which provides a basis for their classification. Chloroplast introns have been classified into two or three groups, based on secondary structure and conserved boundary sequences (Gruissem, 1989). Group I introns are found in a number of genes in *Chlamydomonas* but occur in only a single gene [*trn*L(UAA)] in the cpDNA of land plants. The remaining introns in the plastid genomes of land plants are classified as group II on the basis of secondary structure (Michel et al., 1989), although Shinozaki and co-workers (1986a) split group II into two groups (II and III) on the basis of boundary sequences.

Introns are common in algal cpDNA genes, and they have been found in some cyanobacterial tRNA genes (J. D. Palmer, personal communication). We therefore assume that these elements had an ancient origin. The evolutionary origins, relationships, and roles of various classes of introns have been recently discussed (Cech, 1986; Padgett et al., 1986; Doolittle, 1987; Krainer and Maniatis, 1988). One perspective emphasizes the potential catalytic properties of introns, their interactions with exons, and their possible role in evolution in a precellular, "RNA world" (Gilbert, 1986, 1987; Joyce, 1989). Another view focuses on recent evidence of the activity of certain group I and, possibly, group II introns as mobile genetic elements (Lambowitz, 1989). This view arises from the observation that the products of ORFs in particular introns are related to retroviral reverse transcriptases (Michel and Lang, 1985; Xiong and Eickbush, 1988b; Doolittle et al., 1989). Lambowitz (1989) suggested that the evolution of introns into, or from, mobile genetic elements was an ancient event and that many present-day introns represent degenerate mobile elements. Although fungal mitochondrial introns have several properties that enhance the capacity of self-propagation and insertion (see references in Lambowitz, 1989), the evidence suggests that mobility of chloroplast introns is rare.

In some cases, introns have been lost in particular plant lineages (Table 1). The gene encoding ribosomal protein L2 provides the best documented case of intron loss. The L2 gene lacks a group II intron in spinach and nine other species in four families in the Caryophyllales (Zurawski et al., 1984b; Zurawski and Clegg, 1987; Palmer et al., 1988). Recently, Downie et al. (submitted) have shown that the L2 intron may have been lost as many as five times in flowering plant evolution. The gene *trn*I(CAU) in *Chlamydomonas* and *Chlorella* chloroplasts lacks introns found in land plants (Schneider and Rochaix, 1986; Yamada and Shimaji, 1986), whereas a number of introns occurring in the chloroplast genome of *Euglena gracilis* are absent in land plants (Christopher et al., 1988). In some of these examples, the observed pattern may be explained by the loss of the intron in particular lineages. Obviously, mechanisms other than transposition may be responsible for intron loss. The case of *Euglena* is more difficult to explain in terms of intron loss, and it is tempting to invoke transposition, perhaps involving mobile elements related to introns.

The apparent conservation of intron location across the eukaryotic–prokaryotic boundary has been advanced as evidence that introns predated the origin of eukaryotes (Quigley et al., 1988; Shih et al., 1988). The specific location of introns in a number of chloroplast tRNA genes similarly suggests that introns are extremely old. It is remarkable that the introns in five chloroplast tRNA genes [*trn*V(UAC), *trn*K, *trn*L(UAA), *trn*I(GAU), and *trn*A] are all in the anticodon loop. Of these, *trn*L is interrupted by a group I intron, whereas the others are split by group II introns. [A sixth chloroplast tRNA gene, *trn*G(UCC), has a group II intron with its splice site in the D helix (Deno and Sugiura, 1984; Quigley and Weil, 1985); as far as is known this location is unique.] The location of introns in these five genes is especially interesting in light of the observation that introns in both eukaryotic nuclear and archaebacterial tRNA genes are also in the anticodon loop. Of the 981 tRNA genes sequenced to date, 65 contain introns (Sprinzl et al., 1989) (Figure 2).

Based on the occurrence of introns in archaebacterial tRNA genes, Kaine (1987) suggested that the presence of introns predates the archaebacterial–eukaryotic divergence. One could infer from the fact that tRNA genes are split by at least three different classes of introns that the presence of introns in these tRNAs predates the divergence of group I, group II, and nuclear tRNA introns as well as that of the presumed tRNA ancestor (Nicoghosian et al., 1987). This may not be the case, however. It is possible that the anticodon loop or its subtending helix serves as a recognition site for insertion of genetic elements.

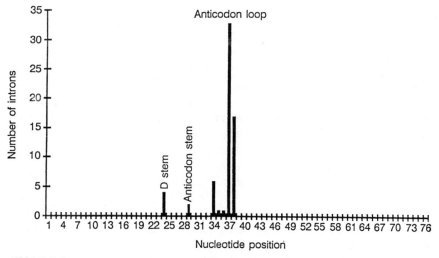

FIGURE 2. Locations of 65 introns in published tRNA gene sequences. On the abscissa are the 76 conserved nucleotide positions in tRNA molecules.

A recent survey found that the anticodon stem and loop of several prokaryotic tRNA genes are attachment sites for a number of mobile elements (Reiter et al., 1989). Dibb and Newman (1989) also presented evidence that nuclear introns were acquired at what are now splice sites in actin and tubulin genes. These observations suggest that the introns in chloroplast tRNA genes (as well as others) arrived at their location via transpositional insertion.

Strong functional constraints on the evolution of cpDNA introns are evident from detailed investigations of intron sequences. Clegg et al. (1986) studied the group II intron in the *trn*V(UAC) gene from four taxa (barley, maize, tobacco, and pea). Using a clustering algorithm based on frequency of nucleotide changes, they divided the intron into 12 regions that varied markedly in their acceptance of mutational change. The analysis of the *trn*V(UAC) intron has recently been extended by Learn et al. (unpublished data) to a total of seven taxa (the four mentioned above as well as *Marchantia*, *Pennisetum*, and *Cenchrus*). Structural models showing six stem-and-loop domains similar to those inferred for other group II introns (Michel and Dujon, 1983; Michel and Jacquier, 1987) were obtained by the methods of Zuker and Stiegler (1981). Domain II showed the greatest rates of sequence divergence, while other structural features showed rates an order of magnitude lower. These results are compatible with the molecular biology of self-splicing group II introns in yeast mitochondria, in which some structural features have been shown to be required for proper intron processing and domain II appears to be dispensable.

RATES OF CHLOROPLAST GENE EVOLUTION

Heterogeneities in rates of cpDNA evolution

Evolutionary rates vary among cpDNA genes. With regard to the tobacco–liverwort comparison, the most conservative genes appear to be the photosystem II genes *psbA*, *psbF*, and *psbD* and the ATPase subunit *atpH* (Wolfe and Sharp, 1988). Average rates of synonymous substitution have been estimated to be about 1×10^{-9} per synonymous site per year (Zurawski and Clegg, 1987; Wolfe et al., 1987). The synonymous substitution rate varies among genes from 0.2 to 1.0×10^{-9} and is positively correlated with variation in estimated rate of amino acid substitution for the same genes (Wolfe, et al., 1987; Wolfe and Sharp, 1988). It is at least twofold, and probably fourfold, slower than estimated rates of synonymous substitution for plant nuclear genes (Wolfe et al., 1987, 1989a; Meagher et al., 1989). This rate differential means that chloroplast genes are quite conservative and, as a consequence, are especially well suited to the construction of molecular phylogenies among plant taxa, especially at the deepest levels of evolution.

For reasons that are not entirely clear, the IR region evolves at a slower rate than the LSC or SSC regions. This differential was first observed in studies of restriction site variation (Clegg et al., 1984; Jansen and Palmer, 1987; Doebley et al., 1987), and the slower rate is even more pronounced at the DNA sequence level. Wolfe et al. (1987) found that noncoding DNA evolves at a threefold slower rate and that the synonymous rate for protein-coding genes is retarded by ninefold relative to genes in the single-copy regions. A comparison of functionally equivalent sequences also demonstrates the rate retardation. There are two group II introns in the inverted repeat of *Marchantia* (Kohchi et al., 1988); these are in two tRNA-encoding genes, *trnI* and *trnA*. Sequences for these introns are also available for tobacco and maize (Takaiwa and Sugiura, 1982; Koch et al., 1981). Group II intron sequences are available for *Marchantia*, tobacco, and a monocot for three genes in the single copy portion of the genome, two tRNA encoding genes, *trn*V(UAC) (Umesono et al., 1988; Deno et al., 1982; Krebbers et al., 1984) and *trn*G (Umesono et al., 1988; Deno and Sugiura, 1984; Quigley and Weil, 1985), and a protein-coding gene *atp*F (Umesono et al., 1986; Shinozaki et al., 1986b; Bird et al., 1985).

Sequence comparisons among these introns indicate that group II introns in the inverted repeat appear to be evolving more slowly than those in the large single-copy region (Table 3). When *Marchantia* is compared to the angiosperms, sequence identity for IR introns ranges from 74 to 79% while for LSC introns it varies between 60 and 67%. A similar relationship is seen in comparisons between the monocots and the dicot, tobacco: identities for IR introns are about 93% and those for LSC introns fall between 74 and 80%.

Noncoding regions of the cpDNA tend to accumulate additions and deletions that eventually obliterate sequence similarity (Zurawski and Clegg, 1987). Thus, for example, there is no sequence similarity between liverwort and tobacco noncoding regions (Wolfe and Sharp, 1988). Detailed examination of the noncoding region between the genes *rbcL* and *atpBE* reveals conserved promoter and ribosome binding sites; as a consequence of these conserved features, the total rate of nucleotide substitution is about equivalent to that estimated for the *rbcL* coding region (Zurawski and Clegg, 1987; Zurawski et al., 1984a). Many additions and deletions in noncoding regions involve short

TABLE 3. Percent identity between group II introns in the IR and LSC regions

Comparison		IR region		LSC region		
		trnI	*trnA*	*trnV*	*trnG*	*atpF*
Liverwort	Tobacco	76	79	63	66	67
	Maize	74	78	62	—	—
	Wheat	—	—	—	59	67
Tobacco	Maize	93	93	80	—	—
	Wheat	—	—	—	74	79

direct repeats that are likely to be the result of slipped-strand mispairing during replication. Analyses of the evolution of the noncoding region suggest that length mutations occur at least as often as nucleotide substitution mutations among closely related taxa (unpublished results).

Rates of molecular evolution also appear to be heterogeneous among plant lineages. Although the data are still few, several studies of restriction site variation suggest that cpDNA evolution is slowed considerably in long-lived tree species (Smith and Doyle, 1986; Wilson et al., 1990). In one case, rates estimated from restriction site variation and complete DNA sequences for the gene *rbcL* were in good agreement and showed an eightfold reduction in overall rate of substitution in the palm family relative to annual plant species (Wilson et al., 1990). Variations in rate of this magnitude are quite large and must be accounted for by a strong dependence of mutation rates on generation time.

Detailed analyses of evolutionary trees have revealed more subtle variations in rate of cpDNA evolution. For example, Ritland and Clegg (1987) found that the rate of evolution of a chloroplast intron and the third position rate for the *atpB* gene were accelerated in the lineage leading to pea. Wolfe et al. (1987, 1989b) also found accelerated rates of nonsynonymous substitution in the pea lineage for a number of chloroplast protein-coding genes. In relative rate tests involving the *rbcL* gene in nine grass taxa, Doebley et al. (1990) found that the lineage leading to maize had experienced an accelerated rate of evolution relative to the other grass taxa. It is clear from these examples that the rates of evolution of chloroplast-encoded genes are not constant across genes or lineages. This rate heterogeneity has implications for the use of cpDNA sequences as evolutionary clocks in phylogenetic analyses.

Applications of the molecular clock hypothesis

According to the molecular clock hypothesis, genetic divergence is a linear function of evolutionary time. Under this hypothesis phylogenetic events can be reconstructed by measuring the number of nucleotide substitutions and calibrating this measure against external estimates of divergence times for two or more taxa. Divergence times for other taxa can then be estimated directly from molecular distances. We have just seen, however, that chloroplast genes do not exhibit uniform rates of nucleotide substitution over different plant lineages, and, obviously, variation in rates adds considerable uncertainty to molecularly based time estimates. Nevertheless, an attempt has recently been made to use cpDNA sequence data to estimate the divergence times of major plant lineages.

Wolfe et al. (1989b) used 12 cpDNA protein-coding genes to estimate molecular distances between several monocot and dicot lineages. They also used a smaller set of genes to estimate distances among monocot, dicot, liverwort (Bryophyte), and algal lineages. In the case of the monocot–dicot contrasts, they based distance estimates on synonymous substitutions that are more likely to satisfy assumptions of selective neutrality. In the case of the flowering plant, bryophyte, and algal contrasts, nonsynonymous substitutions were employed, because synonymous substitutions were close to mutational saturation and were, hence, judged to be unreliable. Two rate estimates were obtained from palaeontological estimates of separation times: one for the origin of the grass family and one for the origin of land plants. Based on these rate estimates, the time of the monocot–dicot split was placed at 200 million years ago. This is somewhat earlier than contemporary estimates derived from palaeontological and evolutionary evidence (Stewart, 1983), but it is reasonable, given the many uncertainties in the plant fossil record.

Interestingly, a second group has also attempted to estimate the time of the monocot–dicot divergence based on nuclear gene sequences (Martin et al., 1989). In this case, an estimate of 320 million years was obtained. There are many points of difference between these two studies. Based on size of data set, choice of calibration times, and other considerations, the Wolfe et al. (1989b) estimate is probably more reliable (Clegg, 1990). Nevertheless, one cannot escape a sense of unease when a method leads to such divergent results. The simple fact is that assumptions of constant rates of mutational substitution are not satisfied by cpDNA-encoded genes, and are probably not satisfied by plant nuclear-encoded genes either.

CHLOROPLAST DNA IN STUDIES OF PLANT EVOLUTION AND SYSTEMATICS

Heteroplasmy and chloroplast DNA polymorphism

Recent studies (including those using interspecific crosses) indicate that although maternal inheritance of chloroplasts may be the rule (e.g., Soliman et al., 1987; Erickson et al., 1983), paternal inheritance (in conifers, Neale and Sederoff, 1989; Neale et al., 1986; Szmidt et al., 1987), and exceptional cases of biparental inheritance do occur. More than 35 years ago, Schötz (1954) demonstrated biparental inheritance in sexual crosses of *Oenothera*, and, more recently, Metzlaff et al. (1981) observed both biparental inheritance and subsequent sorting out of chloroplast genotypes in *Pelargonium zonale*. In addition, Moon et al. (1987) in *Oryza*, Rose et al. (1986) and Johnson and Palmer (1989) in *Medicago sativa*, and Govindaraju et al. (1988) in *Pinus banksiana* × *P. contorta* crosses found heteroplasmy within individual plants, which, in the latter two cases, was demonstrated to sort out in cuttings or on branches.

With biparental transmission, intermolecular recombination can generate new chloroplast genotypes. However, for recombination to occur, different cpDNA molecules in heteroplasmic cells must become intimately associated within a single plastid. At present, some population data hint at the possibility of intermolecular recombination generating new genotypes, but the data are far from compelling. Novel cpDNA genotypes observed in hybrid swarm populations may be the result of biparental inheritance and recombination (Neale et al., 1986; Wagner et al., 1987; Govindaraju et al., 1989; Szmidt et al., 1987). However, multiple mutations could also explain the patterns of variation. In the single study carefully designed to detect recombination in chloroplast DNA of *Oenothera*, there was no evidence for recombination in the progeny of crosses in biparentally inherited chloroplasts (Chiu and Sears, 1985).

Anomalous patterns of chloroplast inheritance have been noted in interspecific crosses [biparental or paternal inheritance in *Secale* × *Hordeum* hybrids (Soliman et al., 1987), and maternal inheritance in the conifers *Larix decidua* × *L. leptolepis* (Szmidt et al., 1987)]. These results suggest that specific nuclear–cytoplasmic interactions determine the rules of chloroplast transmission. Most investigators assume uniparental (maternal) inheritance of chloroplasts and a one-to-one correspondence of restriction site differences with individual mutations. Although these assumptions are usually correct, it is prudent to establish transmission patterns and to map site changes as adjuncts to systematic investigations on new plant groups.

Population studies of restriction site variation in cpDNA

There are few studies of cpDNA variation in natural plant populations. Banks and Birky (1985) found very limited restriction fragment length variation in *Lupinus texensis*. Although their sample may have been unintentionally nonrandom, their results, along with those from cultivated species and their relatives (e.g., Kung et al., 1982; Palmer and Zamir, 1982; Timothy et al., 1979), indicate that intraspecific or intrapopulational chloroplast variation is rare. Several recent studies, however, have documented both intrapopulational and intraspecific variation. For example, Rieseberg et al. (1988) reported chloroplast differentiation between two ecotypes of *Helianthus bolanderi* as well as one variant genotype within an ecotype. Sytsma and Gottlieb (1986) found a case of intraspecific variation even in a very limited sample of localized endemics. In studies where a larger number of populations was sampled, Soltis et al. (1989a,b) documented considerable intraspecific variation in *Heuchera micrantha* and *Tolmeia menziesii*. Intrapopulational variation was first reported in *Hordeum spontaneum* by Clegg et al. (1984), and studies with more extensive sampling (Neale et al., 1989; Golenberg and Clegg, unpublished data) have further corroborated these findings.

Uses of chloroplast restriction site variation in systematics

The use of cpDNA variation in systematic studies has grown rapidly in the past few years (Table 4). It has generally been assumed that lower taxonomic groups are monomorphic and, therefore, that a given mutation will be diagnostic. Sytsma and Schaal (1985) distinguished six monomorphic populations in the *Lisianthius skinneri* complex based on cpDNA restriction patterns and used this

TABLE 4. Studies of restriction fragment length polymorphism in the chloroplast genome in land plants

Taxon	Crop plant[a] or relative	Intraspecific[b] cpDNA variation	Reference
Aegilops spp.		X	Murai and Tsunewaki (1986)
Asteraceae taxa			Jansen and Palmer (1988)
Brassica spp.	X	X	Erikson et al. (1983)
Brassica spp.	X		Palmer et al. (1983)
Clarkia spp.		X	Sytsma and Gottlieb (1986)
Coffea spp.	X		Berthou et al. (1983)
Cucumis spp.	X		Perl-Treves and Galun (1985)
Daucus spp.	X		DeBonte et al. (1984)
Dryopteridaceae taxa			Yatskievych et al. (1988)
Eleusine spp.	X		Hilu (1988)
Festucoideae taxa			Lehväslaiho et al. (1987)
Glycine spp.	X	X	Close et al. (1989)
Helianthus spp.		X	Rieseberg et al. (1988)
Heuchera grossulariifolia		X	Wolf et al. (1988)
Heuchera micrantha		X	Soltis et al. (1989a)
Hordeum spp.	X	X	Clegg et al. (1984a)
Hordeum spp.	X	X	Holwerda et al. (1986)
Hordeum spp.	X	X	Neale et al. (1988)
Linum spp.	X		Coates and Cullis (1987)
Lycopersicon spp.	X		Palmer and Zamir (1982)

[a]Studies involving cultivated species or related species.
[b]Studies reporting intrapopulational or intraspecific cpDNA variation. Studies involving hybrid swarms are not included.

information, in conjunction with extensive morphological divergence data, to justify species status for each of the populations. Conversely, Lehväslaiho et al. (1987) argued for reconsideration of the taxonomic distinction between the genera *Lolium* and *Festuca* based on similarities of cpDNA restriction patterns. Similarly, Palmer and Zamir (1982) in *Lycopersicon* and Jansen and Palmer (1988) in an exhaustive study in the Asteraceae argued for changes in the previously suggested taxonomic groupings. Milo et al. (1988) found inconsistencies in the patterns of relationships between diploids and polyploids in the genus *Papaver* section *Oxytoma*, based on cpDNA and isozyme patterns, although small sample sizes may have influenced their results.

Many of the above studies (cited in Table 4) deal explicitly with the rela-

TABLE 4. (Continued)

Taxon	Crop plant[a] or relative	Intraspecific[b] cpDNA variation	Reference
Nicotiana spp.	X		Kung et al. (1982)
Nicotiana spp.	X		Salts et al. (1984)
Oenothera spp.			Gordon et al. (1982)
Papaver spp.			Milo et al. (1988)
Pelargonium spp.			Metzlaff et al. (1981)
Pennisetum spp.	X		Clegg et al. (1984b)
Pennisetum glaucum	X		Gepts and Clegg (1989)
Picea spp.			Szmidt et al. (1988)
Pinus spp.			Wagner et al. (1987)
Pinus spp.			Govindaraju et al. (1989)
Pisum spp.	X	X	Palmer et al. (1983)
Solanum tuberosum	X	X	Hosaka and Hanneman (1988a)
Solanum spp.	X	X	Hosaka and Hanneman (1988b)
Tolmiea menziesii		X	Soltis et al. (1989b)
Triticum–Aegilops spp.	X		Tsunewaki and Ogihara (1983)
Triticum–Aegilops spp.	X		Bowman et al. (1983)
Triticum–Aegilops spp.	X	X	Ogihara and Tsunewaki (1988)
Zea spp.	X		Timothy et al. (1979)
Zea spp.	X	X	Doebley et al. (1987)

tionships of cultivated species and their wild relatives. For example, Palmer et al. (1985) found that most accessions of the cultivated pea, *Pisum sativum*, grouped with individuals from northern populations of *P. humile*, thus supporting earlier classical models for the domestication of the pea. One accession of the cultivated pea, however, aligned with chloroplast genotypes found in southern populations of *P. humile* and *P. elatus*, suggesting the presence of some introgression into the cultivated pea gene pool. Hosaka and Hanneman (1988b) also argued for introgression of diploid species into the cultivated tetraploid potato, *Solanum tuberosum* ssp. *andigena*. In contrast to the findings in peas, the variability within the tetraploid potato was so great that Hosaka and Hanneman (1988b) argued for multiple origins of the Andean cultivar. Similarly, in a study of the origins of amphiploid *Brassica* species, Erickson et al. (1983) postulated multiple origins of *B. napus* based on the finding of the chloroplast genotypes of both elementary species, suggesting that the amphiploid arose independently from reciprocal hybridizations.

Chloroplast variation has been used to study evolutionary interactions such as species introgression and hybridization, and the origins of polyploidy. As in the examples of *Solanum tuberosum* and *Brassica*, chloroplast variation found in polyploid species that represent a subset of putative diploid progenitors suggests polyphyletic origins of polyploids in *Aegilops triuncialis* (Murai and Tsunewaki, 1986) and in *Heuchera micrantha* (Soltis et al., 1989b). Similarly, chloroplast polymorphisms have been used extensively in studies of conifers to identify hybrid populations and determine parentage (Szmidt et al., 1988; Wagner et al., 1987; Govindaraju et al., 1989).

CONCLUSIONS

Research on chloroplast DNA molecular evolution falls into two broad categories. One category includes the study of mechanisms that govern cpDNA evolution, and the other involves the use of cpDNA variation as a tool to determine evolutionary relationships among plant species. Knowledge of the structure, gene content, and evolution of the chloroplast genome has been greatly advanced by the determination of the complete DNA sequences for three chloroplast genomes (rice, tobacco, and a liverwort). Sequence comparisons have revealed that the chloroplast genome is a dynamic entity, with inversions and gene deletions playing a major role in structural changes in the genome. Complete gene sequences also demonstrate that the chloroplast genome evolves slowly relative to nuclear-encoded genes. Detailed analyses have disclosed several heterogeneities in rate of evolution among different cpDNA genes, and across major plant lineages. If there is a cpDNA molecular clock, it is variable.

Although the chloroplast genome has a prokaryotic-like structure, many cpDNA genes are interrupted by introns. Detailed analyses of intron sequence evolution clearly indicate constraints in the acceptance of mutational change. Two hypotheses concerning the origins of cpDNA introns are currently under debate; one hypothesis holds that cpDNA introns predate the prokaryote–eukaryote divergence, whereas a competing hypothesis maintains that cpDNA introns result from ancient transpositional insertions. At present we do not have adequate data to discriminate between these hypotheses.

It is generally assumed that chloroplasts are transmitted uniparentally and that different cpDNA molecules do not recombine. Although these assumptions appear to be justified in most cases, there are a few exceptional situations, especially among the progeny of interspecific crosses. Finally, because chloroplast genes evolve at a conservative rate, there has been a tendency to assume that intraspecific polymorphism is negligible. This is not always the case, and the number of studies that document moderate levels of intraspecific cpDNA variation is growing.

CHAPTER 8

POPULATION GENETICS OF TRANSPOSABLE ELEMENTS IN *DROSOPHILA*

Brian Charlesworth and Charles H. Langley

As a result of advances in molecular genetics, we know that the genomes of most organisms contain families of repeated DNA sequences with the properties of self-replication and movement to new locations within the genome (Berg and Howe, 1989; Shapiro, 1983). These sequences are variously known as jumping genes, mobile genetic elements, transposable elements, or transposons; we shall refer to them here as transposable elements (TEs). The discovery of transposable elements has stimulated a considerable amount of debate concerning the means by which they are maintained within their host populations and their significance in evolution (Campbell, 1983; Doolittle and Sapienza, 1980; Hickey, 1982; Nevers and Saedler, 1977; Orgel and Crick, 1980; Syvanen, 1984). Much of this debate was conducted in the absence of data on the distribution of transposable elements among host individuals in natural populations and of rigorous models of their population biology, but the situation has changed considerably in recent years as a result of theoretical and empirical studies of the population biology of TEs (Brookfield, 1986; Charlesworth, 1985; Charlesworth and Langley, 1989; Engels, 1986, 1989; Hartl et al., 1986). We are far from being able to give conclusive answers to all the questions that have been raised in this debate, but the theoretical and empirical basis for pursuing the answers has become clearer.

This chapter reviews both experimental evidence and theories concerning the mechanism of maintenance of transposable elements in their host populations. Most of the data we consider comes from studies of *Drosophila*, partly because of the feasibility of applying the powerful technique of in situ hybridization of transposable element probes to polytene chromosomes in this group of organisms. We will not consider dispersed repetitive sequences in other taxa, except where the data can be usefully compared with the results for *Drosophila*. We also omit any consideration of the taxonomic distribution of families of TEs within the genus *Drosophila*, except where it is directly relevant to population dynamics.

Our overall conclusion is that *Drosophila* TEs are maintained in populations as a result of transpositional increase in copy number, and that their spread is checked by one or more opposing forces. In other words, the concept that TEs are essentially intragenomic parasites (Doolittle and Sapienza, 1980; Orgel and Crick, 1980) is supported. The processes responsible for stabilizing elements at relatively low frequencies are, however, still a subject of active research. We also find that the genetic and evolutionary mechanisms influencing the distribution of TEs within populations may affect their abundance in different regions of the genome.

BASIC BIOLOGY OF *DROSOPHILA* TRANSPOSABLE ELEMENTS

Structure and properties

The diversity of structure of *Drosophila* TEs (Figure 1) approaches that observed among the TEs of all organisms. Fifty to 100 families of *D. melanogaster* TEs, distinguishable both by their general structure and detailed DNA sequences, are known (Finnegan and Fawcett, 1986; Rubin, 1983). The structural variation among TEs is thought to reflect important differences in their biology and mode of transposition. A typical *Drosophila melanogaster* element is between 1 and 9 kb in length and often has direct or inverted repeats at each end. Numbers of copies of elements per haploid genome range from a few to 60. Little is known about the structural characteristics of TEs in other *Drosophila* species, other than the fact that species related to *D. melanogaster* share many elements that show substantial DNA sequence homology with those in *D. melanogaster*, while the elements in increasingly distant species show less similarity (Brookfield et al., 1984; Dowsett and Young, 1982; Martin et al., 1983).

Most *D. melanogaster* TEs are probably either retroviruses or retroposons (Berg and Howe; 1989; Boeke, 1988; Finnegan and Fawcett, 1986). They are presumed to "retrotranspose" via an RNA intermediate that is reverse transcribed into DNA and integrated at a new site in the genome of the host. The primary distinction between retroviruses and retroposons is that the former are capable of infecting a different cell or host, whereas the latter replicate only through insertion into new chromosomal sites within a cell. No *Drosophila* TE is known to be a retrovirus, despite the fact that many families possess characteristics of retroviruses, including direct long terminal direct repeats (LTRs) and genes coding for reverse transcriptase. Some *Drosophila* elements (*I*, *F*) resemble mammalian retroposons in lacking LTRs but contain the genes needed for

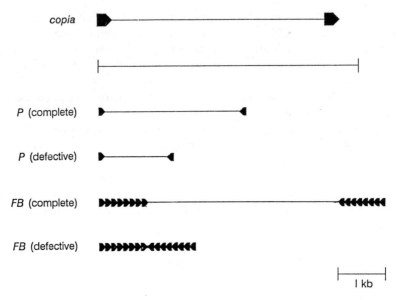

FIGURE 1. Examples of the four main classes of transposable elements (TEs) found in *D. melanogaster*: a retrovirus-like element with LTRs (*copia*), a retroposon without LTRs (*I*), a transposon with short inverted repeats (*P*), and a transposon with multiple inverted repeats (*FB*).

replication via reverse transcription. Other *Drosophila* TEs, such as the *P* and *hobo* elements, with inverted terminal repeats, and the foldback (*FB*) element, with its multiple internal repeats, are thought to transpose by direct DNA replication, possibly through the cointegration mechanisms known in bacteria (Berg and Howe, 1989). Except for the elements involved in the phenomenon of hybrid dysgenesis (*P*, *I*, and possibly *hobo*), in which elements mobilize at a high rate in the F_1 progeny of matings between flies lacking elements and flies that possess them (Engels, 1986, 1989), few of these distinctions among *Drosophila* TEs are known to be important in their population biology.

Most *Drosophila* TEs occur throughout the genome, as middle repetitive DNA, although many families tend to accumulate differentially in the proximal euchromatin and heterochromatin (see below). If there is site specificity for insertions, the number of sites available for occupation by a given family appears to be very large (Charlesworth and Langley, 1989; Finnegan and Fawcett, 1986; Rubin, 1983). An exception is provided by the type I and type II ribosomal insertion sequences of *Drosophila* and other insects, which show strong specificity for the ribosomal cistrons but occasionally are inserted elsewhere. They are retroposons without LTRs, and their reverse transcriptase genes show strong sequence similarity with the other *Drosophila* elements of this class (Xiong and Eickbush, 1988).

Rates of movement of elements

Many of the spontaneous morphological mutations used in *Drosophila* genetics are caused by the insertions of TEs into or near the coding regions of genes (Green, 1980; Rubin, 1983). Nevertheless, the low rate at which mutations occur, despite the fact that approximately 10% of the fly genome consists of middle repetitive, putatively transposable sequences, suggests that such insertion events are infrequent, except for the mobilization of *P*, *I*, and (possibly) *hobo* elements in crosses where hybrid dysgenesis is occurring. Similarly, the great stability of the majority of mutations caused by TE insertions indicates that the spontaneous complete loss of elements from their chromosomal locations is very rare (Rubin, 1983). Thus, although it is sometimes suggested that the genome is much more unstable than was thought to be the case by classical geneticists, in reality our knowledge of the existence of TEs does not qualitatively alter our view of the great stability of genetic information.

Because of the low frequency of the events concerned, relatively few experiments expressly designed to measure the rates of insertion of new elements into new chromosomal sites (transposition) or loss of elements from old sites (excision) have been performed. From the population point of view, one would ideally like to have estimates of the probability u that an element at a given site produces a new copy and the probability v that it is excised (see Table 1 and Figure 2). The experiments cited below almost certainly overestimate the true rate of excision, in the sense of the precise removal of elements (including their terminal repeats) from their chromosomal locations, because detailed molecular characterizations of the events have not been carried out. The experiments have either involved the use of genetic markers to follow insertions or excisions, or have followed the chromosomal locations of elements in a stock (as identified by in situ hybridization to polytene chromosomes) over many generations. Homologous recombination between the LTRs of a retrovirus-like element generates an extrachromosomal circle and leaves a copy of the LTR behind, mimicking excision (Finnegan and Fawcett, 1986). Crossing over between homologous elements located at different chromosomal sites (ectopic exchange) results in chromosomal rearrangements, such as deletions or inversions, and could cause the loss of much of the DNA of one or both of the elements concerned (Figure 3). There is experimental evidence for such ectopic exchange between TEs, and the concomitant production of rearrangements, in *D. melanogaster*. It appears to occur at a substantially higher rate than true excision (Davis et al., 1987; Goldberg et al., 1983).

Overall, experiments designed to measure rates of movement indicate that they are generally low, with u typically of the order of 10^{-4} and v an order of magnitude less (Charlesworth and Langley, 1989; Eggleston et al., 1988). These estimates are subject to considerable error, because of the small numbers of events on which they are based and the simplifying assumptions used to obtain them, but they are probably of the right order of magnitude (Charlesworth and

TABLE 1. Parameters for describing transposable element population biology

STATIC

m	Number of occupable sites in a haploid genome
n	Number of copies of a given family of elements in a given individual
$\bar{n}$	Mean number of copies of a given family of elements per individual in a population
V_n	Variance in copy number between individuals within a population
N	Number of breeding individuals in a population
x_i	Frequency of elements at the ith occupable site in a population
$\bar{x}$	Mean of x_i over all sites ($\bar{x} = \bar{n}/2m$).
σ_x^2	Variance of x_i between sites
D_{ij}	Coefficient of linkage disequilibrium in element frequency between the ith and jth occupable sites

DYNAMIC

u_n	The germline probability of transposition per generation of an element belonging to a given family, in a host individual carrying n elements from that family; the functional dependence of u on n, denoted by the subscript n, allows for possible regulation of the rate of transposition in response to copy number
v	The germline probability of excision per generation of an element of a given family
w_n	The fitness of a host individual carrying n members of a given family, relative to a value of one for an element-free individual
$\bar{w}$	The mean of w_n over all individuals in the population

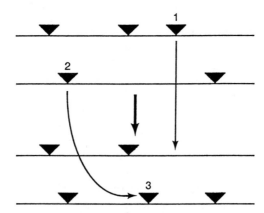

FIGURE 2. Examples of transposition and excision. The inverted triangles represent TEs, which are distributed over two different chromosomes of their host. Element 1 undergoes excision, whereas element 2 experiences replicative transposition such that a new copy (3) is inserted at a novel site while the parent copy remains at its original site.

(A) Interchromosomal ectopic exchange   (B) Intrachromosomal ectopic exchange

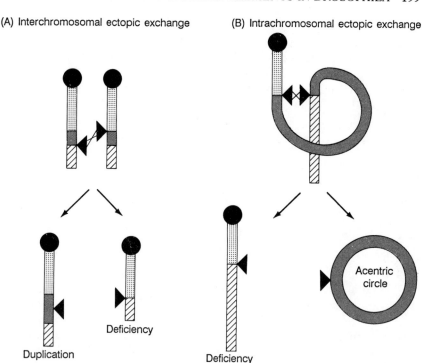

FIGURE 3. Interchromosomal and intrachromosomal ectopic exchange. (A) Pairing and exchange of two copies of a TE (triangles) inserted at different locations in the two homologues. The shading represents the three relevant homologous segments. One product of exchange is a chromosome bearing a duplication with one TE between the duplicated segments. The other product of this ectopic exchange is a deficiency with the other TE at the junction of the deficiency. (B) Pairing of two copies of a TE inserted at different locations in the same chromosome. Exchange between these elements produces a deficiency with one TE at the junction, and an acentric fragment bearing the other TE.

Langley, 1989). Exceptionally high rates of mobilization are found among elements responsible for hybrid dysgenesis, in which u values of the order of 10^{-3} have been reported in dysgenic crosses. There is no firm evidence that elements other than P are mobilized in P–M dysgenic crosses (Charlesworth and Langley, 1989; Eggleston et al., 1988).

There is extensive evidence in prokaryotic systems for regulation of the rate of transposition, such that u is a decreasing function of the number of elements of the family in question in the same cell (Berg and Howe, 1989; Charlesworth and Langley, 1986). The low rate at which *Drosophila* TEs typically transpose and excise has precluded the detection and characterization of any regulation of either process, with the exception of the P, I, and *hobo* elements involved in hybrid dysgenesis (Blackman et al., 1987; Engels, 1986, 1989; Yannopoulos et al., 1987). However, recent molecular studies of *P* elements suggest that repressor activity is a property of defective elements, which are widespread in natural populations with the *P* cytotype (Engels, 1989). The molecular basis of repression in the case of the *I* and *hobo* elements is unknown, although various kinds of defective *I* and *hobo* elements are often present in the genome (Blackman et al., 1987; Finnegan and Fawcett, 1986).

POPULATION MODELS OF TRANSPOSABLE ELEMENTS

General overview

Models of the population biology of transposable elements may be assigned to the following three categories.

1. Models that consider essentially only a single nuclear gene with non-Mendelian transmission biased in its own favor (Hickey, 1982; Ginzburg et al., 1984), or with an interaction between a nuclear gene and the cytoplasm (Uyenoyama, 1985).
2. Models of bacterial transposable elements, with elements dispersed over multiple genomic sites. These assume only limited possibilities of sexual transmission between individuals and a corresponding low frequency of recombination between elements at different genomic locations (Sawyer and Hartl, 1986; Hartl and Sawyer, 1988).
3. Models of randomly mating, diploid populations with elements dispersed over multiple genomic sites and with relatively high frequencies of recombination between different sites (Charlesworth and Charlesworth, 1983; Langley et al. 1983, 1988a).

Models of class 1 have been helpful in defining certain basic properties of the population behavior of transposable elements. In particular, Hickey (1982) pointed out that the spread of elements by a selfish DNA mechanism is crucially dependent on the existence of sexual or parasexual processes, such that there is a nonzero frequency of exchange of genetic material between individuals. With strictly clonal reproduction, any transpositional increase in copy number of a family of elements would be confined to lineages descended from individuals that possess at least one element belonging to the family in question. Transpositional increase in copy number thus could not in itself cause an element family that is initially present in only a few members of the population to spread to the majority of members of the population. Hickey suggested that this explains why transposable elements are absent from genomes with uniparental transmission, such as mitochondria and chloroplasts. Nonetheless, this type of model provides only a qualitative picture, in view of the fact that transposable elements are normally present as multiple, dispersed copies, and so cannot be expected to provide detailed predictions to test against population data.

Models of type 2 avoid this objection, but are specifically designed with bacterial populations in mind. Here, genetic exchange between individuals is mediated by plasmids or transducing phages. The frequency of such exchange is extremely low in most cases, so that the frequency of genetic recombination between elements located at different genomic sites is probably very low in most species (Hartl and Dykhuizen, 1984; Hartl et al., 1986). This has necessitated the development of mathematical models that take into account the possibility of extensive nonrandom associations between the frequencies of elements at different sites (Sawyer and Hartl, 1986; Hartl and Sawyer, 1988).

The analysis of the dynamics of bacterial elements thus encounters several formidable problems that are peculiar to the biology of their hosts. The question of the dynamics of elements in genomes with low frequencies of genetic exchange is, however, highly relevant to what may happen in plants with high frequencies of self-fertilization, where extensive nonrandom associations between alleles at different conventional loci are observed (Brown, 1979). At present, no work on partially self-fertilizing populations has been carried out.

These difficulties are largely absent from the analysis of randomly mating populations with sufficient recombination that linkage disequilibrium can be ignored. This is appropriate for outcrossing species of plants and most animal species. Models of class 3 apply here, and we shall now proceed to describe such models in more detail. They provide predictions concerning the population properties of TEs that can be tested against data.

Population statics

The first kind of prediction that can be made concerns the distribution of elements belonging to a given family among individuals within a population (Charlesworth and Charlesworth, 1983; Langley et al., 1983). The parameters used to describe this distribution are listed in the upper portion of Table 1. The state of a diploid population of size N at a given chromosomal site i can be described by the frequency x_i with which the $2N$ copies of that site are occupied by an element. If the element frequencies at different sites are independent (no linkage disequilibrium), a list of values of x_i for the entire array of sites ($i = 1$, $2, \ldots m$) provides a complete description of the state of the population.

The mean copy number n is equal to $2\Sigma_i x_i$. The variance in copy number is given by the following expression (Charlesworth and Charlesworth, 1983):

$$V_n = \bar{n} (1 - \bar{x}) - 2m\sigma_x^2 + 4 \sum_{i<j} D_{ij} \qquad (1)$$

If σ_x^2 and the sum of the D_{ij}s are sufficiently small, only the first term need be considered; this corresponds to a binomial distribution of n over individuals. If $\bar{x} << 1$, as frequently seems to be the case (see below), then there is a Poisson distribution of n, with $V_n \approx n$.

Element dynamics in an infinite population

A model of the transmission of a family of elements from generation to generation needs to incorporate the following features, summarized in the lower part of Table 1.

Transposition and excision. If transposition is not accompanied by excision of the parental elements, there is clearly an expected gain of copy number of nu_n in the germ line of an individual with n elements. Excision of elements at rate v in an individual with n elements reduces copy number by nv. The net change in mean copy number is approximately $\bar{n}(u_{\bar{n}} - v)$ (Charlesworth and Charlesworth, 1983). If there is no regulation of transposition rate, so that u is independent of n, transposition leads to fixation of elements at each site if $u > v$. If transposition is regulated, and if the the rate of transposition at low copy number (u_0) exceeds v, mean copy number will increase until $\bar{n} \approx \hat{n}$ (such that $u_{\hat{n}} = v$), or until all available sites have filled up with elements (Charlesworth and Charlesworth, 1983).

   The evidence on the values of u and v reviewed above suggests that *Drosophila* elements do indeed have the capacity to increase in copy number as a result of an excess of transposition events over excision.

Natural selection. The simplest models of the effects of natural selection on TE abundance assume that the fitness of a host individual with copy number n for a given family, w_n, is a decreasing function of n. Assuming no linkage disequilibrium between sites, the following equation is obtained for the change in copy number per generation (Charlesworth, 1985):

$$\Delta \bar{n} \approx \bar{n}(\bar{n} - \bar{x})\frac{\partial \ln \bar{w}}{\partial \bar{n}} + \bar{n}(u_{\hat{n}} - v) \qquad (2)$$

The approximate equilibrium value of n, $\hat{n}$, is given by setting this expression to zero. It is fairly easy to find functional forms that result in an equilibrium with low element frequencies at each site; for example, with $w_n = \exp[-(1/2)\,tn^2]$, we have $\hat{n} \approx (u_{\hat{n}} - v)/t$. An equilibrium is possible even in the absence of excision ($v = 0$). In the absence of regulated transposition, the stability of this equilibrium requires that the logarithm of fitness declines more steeply than linearly with increasing n (Charlesworth, 1985).

   Given that low frequencies of transposition seem to be the norm for *Drosophila* elements, even a weak pressure of selection (as measured by t, which is the slope of the relation between the logarithm of fitness and copy number at a copy number of 1) is capable of maintaining a balance with transpositional increase in copy number. For example, a mean copy number of 50 elements per diploid genome would be maintained if t is one-fiftieth of the excess of the rate of transposition over excision. Rates of movement are on the order of 10^{-4}

implying that selection coefficients of the order of 10^{-5} against individual insertions would be needed to maintain a balance. The mean fitness of the population at equilibrium, relative to the fitness of an element-free individual, is equal to $\exp[-\hat{n}(\hat{u}_n - v)]$, regardless of the form of the selection function (Charlesworth, 1985). There are approximately 50 families of elements, and the mean copy number for most of them is probably of the order of 10 per individual or less (Charlesworth and Lapid, 1989; Finnegan and Fawcett, 1986; J. Hey and W. F. Eanes, in preparation). With $\bar{n} = 500$ and $u_n - v = 10^{-4}$, $\bar{w} = 0.95$. This yields the conclusion that with the observed low transposition rates, the mean fitness of a *Drosophila* population is barely affected by the presence of elements.

Possible modes of selection

This section will consider modes of selection that could lead to the stabilization of element copy numbers in the face of transpositional increase in copy numbers.

It is well established that the insertion of elements into or near genes frequently alters their expression (Shapiro, 1983). Studies of *Drosophila* mutations affecting viability have shown that most of them have relatively small, detrimental effects that are expressed when heterozygous (Simmons and Crow, 1977). For example, *P* element insertions in *D. melanogaster* are frequently accompanied by such detrimental mutations (Engels, 1989). The studies of Mukai and colleagues on viability mutations indicate a mean selection coefficient of the order of 2% against homozygous detrimental mutations and 0.7% against heterozygotes (Simmons and Crow, 1977). If all TE insertions resulted in mutations with this magnitude of effect on fitness, it is clear that elements would never be able to spread in the face of selection, unless transposition rates are higher than seems realistic for most elements, or regulation of transposition is such that transposition rates are very high at low copy numbers.

A solution to this dilemma might be that a large fraction of insertions have little or no effect on fitness, because they involve sites that are sufficiently remote from active genes. The average impact on fitness of an element insertion is then the product of the probability that it involves a selectively significant site, times the selection coefficient against a mutation induced in such a site. Numerical analysis of a model with a mixture of neutral and selected sites showed that the majority of elements in an equilibrium population is found at the neutral sites, with element frequencies of 50% or more, reaching fixation if there is no excision (Charlesworth and Langley, 1989). Mean copy number may become very large in this case, if there is a large number of occupable neutral sites. However, this is inconsistent with most observations on *Drosophila* elements (see below).

A modification to the two-class model that could in principle explain the population data assumes that there are two classes of selected sites, one class being composed of the sites subject to relatively strong selection of the sort considered above, and the other subject to much weaker selection against insertions. These might correspond to insertions into transcribed sequences and nontranscribed flanking regions, for example. In this model, if weakly selected sites are sufficiently common, the dynamics of elements are essentially controlled by them, and copy numbers stabilize at values close to those predicted by Equation (2), with $\bar{n}$ and $\partial \ln \bar{w} / \partial \bar{n}$ corresponding to values for the weakly selected sites and the abundance of elements in the strongly selected sites being effectively negligible. Comparisons of DNA sequences among species suggest that the rate of nucleotide substitution in the 5' and 3' flanking regions of genes is higher than that for nonsynonymous changes in coding regions but considerably lower than the rate for pseudogenes (Li et al., 1985a). This suggests that there is indeed selection against insertions of elements into nontranscribed regions, but it is unclear whether neutral sites are so rare that the problems discussed above can be ignored.

The models suggest that equilibrium element frequencies will be negligibly low in sectors of the genome, such as coding sequences, where insertions have fitness effects comparable to those typical of spontaneous mutations (see below). Measurable frequencies in large populations will be observed only at sites where insertions have little or no negative effects on fitness. This is consistent with the patterns of restriction site variation found in surveys of segments of the *Drosophila* genome (see below).

These considerations lead to the conclusion that although selection against insertional mutations may be a factor in stabilizing element frequencies in natural populations, it may not be the only force involved. One possibility is that control of transposition rates occurs in addition to selection, but the evidence for such regulation is equivocal in *Drosophila* (see above). Another possible mechanism that could limit TE copy number is ectopic exchange between homologous elements located at different chromosomal sites (Langley et al., 1988a). As described above, this process leads to the production of deleterious chromosomal rearrangements, thus lowering the fitness of individuals as a function of the number of elements carried. Population models that take into account heterogeneity in rates of ectopic recombination over different parts of the genome (Langley et al., 1988a) show that, at equilibrium, elements tend to be more abundant in regions where ectopic exchange is less frequent.

Element dynamics in finite populations

The facts described above suggest that the deterministic forces affecting element frequencies in natural populations (transposition, excision, and selection) are normally rather weak. This means that the effects of random genetic drift must be taken into account in the development of models that are intended to aid in the interpretation of population data.

The probability distribution of element frequencies. Extensions of the models described above to finite populations yield the form of the probability distribution of element frequencies per site, for a given family, attained when the forces of transposition, excision, and selection come into statistical equilibrium with random changes in element frequencies (the stationary distribution of element frequencies) (Charlesworth and Charlesworth, 1983; Langley et al., 1983).

The following parameters provide an approximate description of the form of $\phi(x)$, the stationary probability density for element frequency x at a site: N_e is the effective size of a local population; $\hat{n}$ is the equilibrium value of $\bar{n}$ for a large population; $\alpha = 4N_e\hat{n}/(2m - \hat{n})$; $\beta = 4N_e(s + v)$, where s is the value of $-\partial\ln \bar{w}/\partial\bar{n}$ at $\bar{n} = \hat{n}$; α measures the effect of drift, and the effect of transposition in causing insertions into a given site; β measures the joint effects of drift, excision, and selection.

The distribution $\phi(x)$ is approximated by a beta distribution (Charlesworth and Charlesworth, 1983):

$$\phi(x) \approx \frac{\Gamma(\alpha + \beta)}{\Gamma(\alpha)\Gamma(\beta)} x^{\alpha-1}(1 - x)^{\beta-1} \tag{3}$$

This is the formula for a closed population; if $\hat{n} \ll 2m$, the effect of migration can be included by adding $4N_eM$ to β, where M is the the frequency of immigrants into a local population. If m is sufficiently large compared with $\hat{n}$, α can be neglected in Equation (3). This yields a simple formula for the expected number of sites per haploid genome with element frequency x, $\phi(x) = (1/2)\hat{n}$ $x^{-1}(1 - x)^{\beta-1}$ (Langley et al., 1983). As will be seen below, Equation (3) can be applied to population data to obtain estimates of the parameters α and β.

Stochastic loss of elements. Charlesworth (1985) and Kaplan et al. (1985) modeled the process of the chance loss of elements from finite populations, suggested by Engels (1981) as an explanation for the absence of the elements responsible for hybrid dysgenesis from old laboratory stocks of *D. melanogaster*. In a very small population, the expected time to loss of all elements is of the order of at least $1/v$, unless the initial mean copy number per individual is very small; the time is substantially longer than $1/v$ in larger populations (Charlesworth, 1985). This is difficult to reconcile with the fact that laboratory strains lacking active *P* elements may be only 30 years old or less, corresponding to 750 generations at 25°C (Bregliano and Kidwell, 1983). Hence, a high rate of excision of *P* elements from their chromosomal sites and a very small mean number of active elements per individual in the populations from which the laboratory strains in question were collected would be required to rescue the stochastic loss hypothesis for *P* elements. Since the mean copy number is of the order of 30 for natural populations (Ronsseray and Anxolabéhère, 1986), the latter condition is unlikely to be satisfied. It seems most probable that Kidwell's recent invasion hypothesis (Kidwell, 1983) is valid, particularly in view of evidence for the presence of *P* elements in unrelated species and their absence from the closest relatives of *D. melanogaster* (Brookfield et al., 1984; Daniels et al., 1984).

A striking feature of *P* and *hobo* elements is that a substantial fraction of elements is defective and unable to produce the enzymes needed for their own transposition (Blackman et al., 1987; Engels 1986, 1989). They can, however, transpose in the presence of complete elements. Kaplan et al. (1985) constructed a model in which defective elements are generated by mutation from functional elements. The equilibrium state of a large population is such that a large fraction of elements is defective, provided that the defective elements suffer no replicative disadvantage in host cells that contain a sufficient number of complete elements to supply them with transposase functions. In a finite population, stochastic loss of complete elements may occur relatively fast (in terms of evolutionary time), because of their low equilibrium numbers. This could lead to the evolution of an element family all of whose members are incapable of transposition. The ultimate fate of such a family would be mutational degeneration or extinction, the latter of course requiring a nonzero rate of removal by excision, selection, or ectopic exchange.

Muller's ratchet. This is a process by which insertions of elements into genomic regions that lack crossing over can cause a gradual increase in copy number over time, as a result of the effects of genetic drift. The theory of Muller's ratchet has been worked out for the case of conventional deleterious mutations (Haigh, 1978; Pamilo et al., 1987), based on an idea of Muller (1964), but essentially the same results apply to TEs, provided that excision rates are sufficiently small (Charlesworth, 1985). Consider a segment of chromosome in an infinitely large population. If there is some rate of insertion of elements into this segment, then (as we have seen) an equilibrium in the copy number distribution among individuals will be established under counterselection or regulated transposition. If the rate of insertion is high in relation to the pressure of selection, then the frequency of chromosomes lacking elements in the given segment is low. In a finite population, this means that this class of chromosomes is vulnerable to loss by drift. If there is crossing over in the segment in question, it is possible for the zero-element class to be regenerated by crossing over, and so the form of the distribution of element numbers will remain similar to that for an infinite population. If there is no crossing over, however, the zero class will be permanently lost in the absence of loss of elements by excision. Once it is lost, the class with one element in the region in question will be vulnerable to loss, and so on, such that the distribution of copy number moves steadily to the right in a ratchet-like process. Because the speed of this process depends on the inverse of the product of the species population size and the initial frequency of the zero class (Haigh, 1978; Pamilo et al., 1987), it will probably proceed very slowly in natural populations of *Drosophila*, especially if the size of the chromosome segment is small, so that the rate of input of new elements per generation is low.

The effect of the ratchet in producing a build-up of elements in regions where crossing over is suppressed is distinguished from the similar effects of ectopic exchange and transposition discussed above by the fact that the ratchet produces an unlimited build-up of elements, given a sufficiently long time, whereas the latter process will produce only a limited increase, unless exchange is totally suppressed and there is no regulation of copy numbers. The ratchet also fails to operate in the presence of even low frequencies of recombination or excision (Pamilo et al., 1987) and will thus produce an excess of elements that excise at low frequency only in genomic regions that almost totally lack crossing over, except in very small populations.

POPULATION DATA FROM IN SITU HYBRIDIZATION STUDIES

In *Drosophila*, the technique of in situ hybridization of labeled TE probes to the polytene salivary gland chromosomes permits identification of the sites where homologous elements are located. A survey of a set of stocks, each homozygous for chromosomes independently isolated from a natural population by standard genetic procedures, provides a picture of the haploid genomes present in the population from which the chromosomes were sampled. The resolution of this technique is somewhat coarse compared with restriction mapping of portions of the genome (see below), since at best the locations of elements can be determined only down to the level of the salivary chromosome bands shown in the Lefevre photographic map (Lefevre, 1976). However, a useful picture of the general properties of the distribution of elements among and within chromosomes has been obtained from such studies, despite their low resolution and limited number (Charlesworth and Lapid, 1989; Eanes et al., 1988; Leigh Brown and Moss, 1987; Montgomery and Langley, 1983; Montgomery et al., 1987; Yamaguchi et al., 1987).

Copy number distributions between homologous chromosomes

As shown above, if element frequencies for a given family are low and similar at individual chromosomal sites, and if linkage disequilibrium is negligible, a Poisson distribution of copy number is expected among individuals, with the variance in copy number approximating the mean. This result applies equally well to the haploid copy numbers determined for homozygous chromosomes or inbred stocks. Elements present in the centromeric heterochromatin cannot be detected by this technique, because of the underreplication of the heterochromatic DNA during polytenization, and so the data come from the counts of numbers of elements in the euchromatic regions of the salivary chromosomes. In addition, the tendency for elements to accumulate in the centromere-proximal regions of the euchromatin (see below) means that tests of fit to the Poisson distribution must be conducted with omission of the two proximal divisions of each chromosome arm, since inhomogeneity in element abundances between chromosomal regions will cause deviations from the Poisson expectation (Charlesworth and Lapid, 1989). Published data on such copy numbers for sets of isogenic chromosomes show generally good agreement with the Poisson distribution for 12 element families (Charlesworth and Langley, 1989). The one exception is the case of *roo* on the X chromosome in the study of Charlesworth and Lapid (1989), which had a mean of 11.4 and variance of 4.4, and deviated from the Poisson distribution, with $p < 0.01$. Detailed analysis suggested that this was a chance result, largely reflecting an effect of nonsignificant linkage disequilibrium terms.

Early studies of the distribution of *Drosophila* elements among inbred laboratory strains gave the impression of uniformity of copy number for a given family, in contrast to the wide variation in the identity of occupied sites. This led to the suggestion that there is selection for an optimal copy number (Young, 1979; Strobel et al., 1979). However, reanalysis of these data shows that the variance in copy number is approximately the same as the mean, consistent with the Poisson expectation (Charlesworth, 1988).

Distribution of elements between nonhomologous chromosomes

The hypothesis that frequencies of TEs are stabilized by the deleterious effects of insertional mutations predicts that the equilibrium element frequencies will be lower for X chromosomal sites than for autosomal sites, since X chromosomal mutational effects are expressed in the hemizygous state in males, compared with the predominantly heterozygous state of rare autosomal mutations. The partial recessivity of mutational effects on fitness (Simmons and Crow, 1977) thus means that there will be a greater selective impact of X-linked TEs relative to autosomal elements. A quantitative model of the relative equilibrium mean copy numbers for the X chromosomes and autosomes of *D. melanogaster* was developed by Montgomery et al. (1987) and Langley et al. (1988a) for the hypothesis of insertional mutation effects and for the null hypothesis that X-linked and autosomal elements are eliminated at the same rate. The model assumes that the sizes of the euchromatic regions of the two autosomes as targets for insertion are equal and are twice the target size presented by the X. This assumption seems reasonable in view of the genetic evidence for such equivalence, and would be violated only if there were a nonrandom distribution of insertion sites between the chromosomes. The results of scoring the numbers of copies of the retrovirus-like elements *copia*, *roo*, and *412* on sets of 20 X, 2nd, and 3rd chromosomes from a natural population showed no significant difference between the observed abundance of elements on the X and the expectation on the null hypothesis for *copia* and *roo*, whereas *412* showed a significant ($p < 0.001$) deviation from the null expectation, but agreed with the expectation on the insertional mutation hypothesis. There is a similar lack of evidence for a deficiency of P elements on the X compared with the autosomes (Eanes et al., 1988). Overall, these results suggest that although there may be some effect of insertional mutations on copy number, it is not necessarily the chief mechanism involved. This is in accord with the theoretical conclusions discussed above. It is unclear why the above differences between families should exist; additional data on a wider variety of families are clearly desirable.

Element frequencies at individual chromosomal sites

A convenient summary of the results of comparisons of the salivary chromosome band locations of members of a given element family among independently isolated homologous chromosomes is provided by the occupancy profile, which gives the numbers of sites at which hybridization is detected at that site in 1, 2, 3, . . . , S separate chromosomes of a sample of size S (Montgomery and Langley, 1983). Data from population surveys of 11 TE families in isogenic chromosome lines of D. *melanogaster* show clearly that element frequencies are low for sites in the distal 18 divisions of the salivary chromosomes (Table 2). Again, most of the data are for the X chromosome, but the data sets for the major autosomal arms suggest that the profiles are similar to those for the X

TABLE 2. Estimates of the parameters of the probability distribution of element frequencies for *Drosophila melanogaster* elements from natural populations[a]

Element	α	β	$\hat{x}$	m	Element	α	β	$\hat{x}$	m
roo[b]	0.8	12.5	0.060	191	2210[b]	25	550	0.044	15
roo[c]	3.4	32	0.096	128					
roo[d]	2.5	28	0.082	132					
2156[b]	0	∞	0	∞	2217[b]	0.4	40	0.009	204
2158[b]	0	∞	0	∞	297[e]	0.05	16.5	0.003	1340
					297[b]	0	5.5	0	∞
2161[b]	2.5	35	0.067	58	412[e]	0	30	0	∞
					412[b]	17	380	0.043	55
2181[b]	∞	∞	0.028	46	*copia*[e]	∞	∞	0.020	82
					copia[b]	∞	∞	0.019	63
I[f]	∞	∞	0.028	95					

[a]Note that $4N_eM$ is estimated to be equal to 8.8 for autosomal loci and 6.5 for X-linked loci (Singh and Rhomberg, 1987). The contribution of selection and excision to β can be found by deducting these from the appropriate entries in the table.
[b]Data for the X from Charlesworth and Lapid (1989).
[c,d]Data for *roo* for 3L and 3R, respectively, from Charlesworth and Lapid (1989).
[e]Data for the X from Montgomery and Langley (1983).
[f]Data for the X from Leigh Brown and Moss (1987).

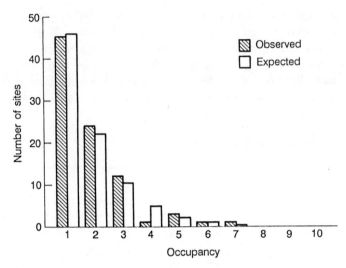

FIGURE 4. Observed and expected occupancy profiles for the retrovirus-like element *roo* in a sample of 14 X chromosomes from a natural population of *D. melanogaster* (data of Charlesworth and Lapid, 1989). The hatched columns of the histogram display the numbers of chromosomal sites at which *roo* is found to be represented in 1,2, . . . chromosomes in the sample. The clear columns display the numbers of sites expected with the best fitting estimates of the α and β parameters (see text for explanation).

(Charlesworth and Lapid, 1989, B. Charlesworth and A. Lapid, unpublished data). Figure 4 displays the observed occupancy profile and the best fitting expected profile (see below) for the element *roo* in a sample of 14 X chromosomes (Charlesworth and Lapid, 1989). Sites are mostly occupied only once by a member of a given family in samples of size 10–20. With high-copy number elements, such as *roo*, occupancies of up to seven have been observed; low-copy number families rarely show occupancies higher than two (Charlesworth and Langley, 1989).

A very different result was reported by Hey (1989) for four uncharacterized families of repetitive DNA in a small number of inbred stocks of *D. algonquin* and *D. affinis*. Copy numbers per haploid genome were much higher (in the hundreds) than those typical of *D. melanogaster*, and many sites showed occupancies of 50% or more. A similar result has been reported for an element in the Hawaiian species *D. heteroneura* and *D. silvestris* (Hunt et al., 1984). Detailed population surveys have not been carried out for species other than *D. melanogaster*, and these isolated reports suggest that it is important that such work be done.

Equation (3) can be applied to the analysis of in situ hybridization data on element frequencies using the sampling theory developed by Charlesworth and Charlesworth (1983), Kaplan and Brookfield (1983), and Langley et al. (1983).

Estimates of α and β for a number of elements scored on chromosomes sampled from natural populations of *D. melanogaster* are presented in Table 2, which shows estimates of the mean element frequency per occupable site $[\hat{x} = \alpha/(\alpha + \beta)]$ and the total number of occupable sites on the chromosome concerned $(m = \bar{n}/2\hat{x})$. It should be stressed that the estimates obtained with a model fitting α and β jointly (Charlesworth and Charlesworth, 1983) are usually not significantly better than those obtained with the assumption that $\alpha = 0$ (Langley et al., 1983) or that the expected element frequencies are equal at each site (Kaplan and Brookfield, 1983). The estimates of α and β are subject to considerable uncertainty, except for the high-copy number elements such as *roo*, *2161*, and *297*. Nevertheless, element frequencies per occupable site are at most a few percent, and the generally large values of β obtained are highly significant when compared with small values, reflecting the low frequencies of multiple occupation. *P* elements have not been included in this analysis, because these elements are probably far from equilibrium, but low *P* element frequencies seem to be the norm (Ajioka and Eanes, 1989; Ronsseray and Anxolabéhère, 1986).

Consideration of data from natural populations requires the inclusion of effects of immigration in β [see Equation (3)]. Estimates of the value of $4N_eM$ in *D. melanogaster* are available from studies of the geographic distribution of electrophoretic alleles (Singh and Rhomberg, 1987) and can be deducted from the estimated values of β to obtain the contribution from excision and selection. Except for *roo* on the X chromosome, these values are all of the order of 10 or more. If the expected mean number of elements per individual is unchanging over time, then the condition for equilibrium derived from Equation (2) implies that $\beta \approx 4N_eu$. Estimates of N_e for East Coast *D. melanogaster* populations from the frequencies of allelism between recessive lethals suggest values of the order of 2×10^4 (Mukai and Yamaguchi, 1974). With a β value of 20, this yields an estimate of 2.5×10^{-4} for *u*. The estimates of N_e are subject to considerable uncertainty, so that this value should not be taken too seriously. For the more abundant elements, the number of occupable sites is either infinite or of similar magnitude to the total number of bands on the photographic maps of the salivary chromosomes (Charlesworth and Lapid, 1989). This suggests that the elements in these studies do not have a high degree of insertional site specificity, at least at the level of salivary chromosome bands.

Variation among elements in the values of β suggests that there are real differences between elements in rates of transposition. Unfortunately, there are no reliable quantitative estimates of the rates of transposition for these elements for comparison with these estimates. The magnitude of β in *D. melanogaster* is so large that drift is clearly rather ineffective in relation to the deterministic forces responsible for reducing element frequencies; for element frequencies to be predominantly controlled by drift, β values of the order of one would be required (Wright, 1931). Values of this magnitude were, however, found for the elements of *D. algonquin* and *D. affinis* (Hey, 1989).

Accumulation of elements in regions of restricted crossing over

Ectopic exchange. A test of the role of selection against chromosome rearrangements produced by ectopic exchange between elements is provided by asking whether or not the abundance of elements tends to be higher in regions where exchange is reduced in frequency. Since there is no information about the distribution of ectopic exchange, the comparison must be made with data on regular meiotic exchange. *D. melanogaster* again provides useful material for such a test, since there is pronounced suppression of meiotic crossing over in the telomeric and centromeric regions of the euchromatin of the chromosome arms (Lindsley and Sandler, 1977). It is straightforward to use in situ hybridization to determine the distribution of the numbers of elements over the salivary chromosome maps in chromosomes sampled from natural populations and to compare this with the distribution expected under a null hypothesis of no variation in element abundance with respect to the rate of meiotic crossing over. The results of a study of *roo* by Langley et al. (1988) showed clear evidence of an excess of elements at the base of the X chromosome, of the order of 3-fold over the number of elements that would be expected if they were inserted in proportion to the physical size of the region. The picture was less clear for the other chromosomes, although the data suggested a slight tendency for elements to accumulate at the base (proximal end) of the euchromatin. There was no evidence of an accumulation of elements at the distal ends. A more recent study of 10 elements, including *roo*, on a sample of X chromosomes provided highly significant evidence for an accumulation of elements at the base of the euchromatin (Charlesworth and Lapid, 1989). In some cases, elements appear to attain high frequencies at individual salivary chromosome band positions in the basal euchromatin. There was no evidence of an excess of elements at the tip. Unpublished data of B. Charlesworth and A. Lapid indicate that there is also a tendency for many of these element families to accumulate at the bases of the autosomes.

These observations should not be taken as clear evidence in favor of the ectopic exchange hypothesis, since an alternative mechanism exists for creating a build-up of elements in regions where crossing over is suppressed (see discussion of Muller's ratchet above). Furthermore, the failure to detect an excess of *roo* near the chromosome tips, particularly that of the X, where crossing over is strongly reduced, presents problems of interpretation for either model. One possible solution is that crossing over near the tip is induced by an interchromosomal effect on crossing over caused by heterozygosity for polymorphic inversions (Langley et al., 1988a). Other evidence for an accumulation of elements in regions of restricted crossing over of the *D. melanogaster* genome is provided by the results of Montgomery et al. (1987) on laboratory cultures of balanced lethal inversions and unpublished data of W. F. Eanes and C. Wesley on *P* elements in naturally occurring inversions.

An extreme example of the differential concentration of elements in genomic regions where crossing over is suppressed is provided by the numerous *Drosophila* elements that appear to be disproportionately abundant in the β-heterochromatin that forms the boundary between the euchromatin (which gives a regular banding pattern in the polytene chromosomes) and the α-heterochromatin (which is not polytenized and consists mainly of highly repeated, satellite sequences). Clones containing material that hybridizes to the β-heterochromatin or to the euchromatin/heterochromatin junction often contain many sequences homologous to known TEs or to sequences that are present as dispersed copies in the euchromatin and hence represent presumptive TEs (Miklos et al. 1984, 1988; B. Wakimoto, personal communication). Detailed characterization of some of these clones has revealed the presence of multiple insertions of TEs, sometimes with several overlapping insertions of elements of different types, presumably as a result of a sequence of independent insertion events (di Nocera and Dawid, 1983). It is interesting that certain of these clones also show hybridization to chromosome 4, another region where crossing over is restricted, although it is not entirely clear whether this reflects the presence of TEs or tandemly repeated arrays that are also abundant in these regions (Miklos et al., 1984, 1988). The defective forms of the *I* element, which are responsible for the R cytotype characteristic of *D. melanogaster* lines that lack complete *I* elements, are also concentrated in the β-heterochromatin (Bucheton et al., 1984), and the same appears to be true for the defective *P* elements found in species of *Drosophila* other than *D. melanogaster* (Engels, 1989)

Muller's ratchet. Charlesworth and Langley (1989) have argued that it is unlikely that Muller's ratchet can account for the build-up of elements in the proximal euchromatin, since the values of the relevant parameters (rate of insertion and selection coefficient against insertions) are close to the boundary values for which the ratchet can operate in the total absence of recombination (Haigh, 1978; Pamilo et al., 1987), even in very small populations. *D. melanogaster* is known to have reasonably large effective local population sizes and rates of migration between populations (Mukai and Yamaguchi, 1974; Singh and Rhomberg, 1987), and there are significant frequencies of crossing over in the basal euchromatin (Lindsley and Sandler, 1977). It is, therefore, implausible that a ratchet mechanism has been responsible for the build-up of elements described above. It is possible, though, that the concentration of *Drosophila* elements in the heterochromatin discussed above is due to Muller's ratchet, since exchange is virtually absent here (Schalet and Lefevre, 1976). Similarly, the excess of middle repetitive DNA in the neo-Y chromosome of *D. miranda*, reported by Steinemann (1982), could be due to the operation of the ratchet. This chromosome is the result of a centric fusion between an autosome and the original Y chromosome, and so has been maintained strictly through the male line, without crossing over, since the establishment of the fusion. Finally, the small size of laboratory populations means that the ratchet will operate much more rapidly than in large natural populations. The apparent build-up of elements near the breakpoints of balanced lethal inversions in laboratory stocks (Montgomery et al., 1987) could also be due to the ratchet.

POPULATION DATA FROM RESTRICTION MAP SURVEYS

General findings

Much finer resolution concerning the identity of the sites occupied by TEs in a sample of chromosomes from a natural population is provided by restriction mapping surveys of defined genomic regions. In these studies, a cloned section of the genome is used as a probe against genomic DNA from a set of homozygous chromosomes extracted from a natural population, enabling comparisons of the restriction maps of the region homologous to the probe on each chromosome to be made. Several such surveys have now been carried out in a number of *Drosophila* species (Charlesworth and Langley, 1989). Table 3 summarizes the findings for *D. melanogaster*. Variation due to the presence of large insertions (more than 300 bases long) is readily distinguishable from nucleotide site polymorphism and variation due to small deletions and insertions. In two surveys of specific gene regions (*Hsp70*, Leigh Brown, 1983; *Adh*, Aquadro et al., 1986) in *D. melanogaster*, the large insertions found segregating in natural populations were cloned and characterized. Each of these insertions was similar to sequences of a previously cloned and characterized TE. Based on these results, the large insertions detected in the other studies listed in Table 3 are assumed to be TEs. The results for *Adh* are summarized in Figures 5 and 6.

Overall, there is remarkable concordance between the results of these surveys and the results from in situ hybridization studies. The general conclusion is that each particular TE insertion is very rare at a particular chromosomal site in a natural population, while the density of TEs is sufficiently high that one or more element insertions are commonly found in genomic regions more than 10 kb in length, if 20 or so independent chromosomes are sampled.

Results of surveys of *D. melanogaster* populations

Relationship of insertion sites to transcriptional units. Element insertions are rarely found in transcriptional units in *D. melanogaster* genes sampled from nature. Among the surveys listed in Table 3, there was only one case of a TE that was unambiguously mapped within a transcriptional unit (in the intron of *Ddc*, with no associated phenotype). Since such large insertions are usually in or very near the transcriptional unit in most spontaneous mutations used in *Drosophila* genetics (Rubin, 1983), it is likely (as discussed above) that such insertions are deleterious and rapidly eliminated from the population (Langley et al., 1988a). Naturally occurring insertions outside the transcriptional unit

TABLE 3. Summary of insertional variation found in natural populations of *Drosophila melanogaster*

Region	Size of region (kb)	Sample size	Density per kb	Proportion homozygous[a]	Reference or source
X Chromosome					
forked	25	64	0.002	0.000	C. H. Langley and N. Miyashita (in preparation)
Notch	37	60	0.002	0.028	Schaeffer et al. (1988)
per	52	78	0.002	0.000	D. Stern et al. (in preparation)
vermilion	24	64	0.002	0.000	C. H. Langley and N. Miyashita (in preparation)
white	45	64	0.011	0.067	Miyashita and Langley (1988)
white	45	38	0.010	0.046	Langley and Aquadro (1987)
zeste-tko	34	64	0.001	0.000	Aguadé et al. (1989a)
Total or mean[b]	217	—	0.004	0.017	
Reduced crossing over					
su(f)	24	64	0.004[c]	0.011	C. H. Langley and N. Miyashita (in preparation)
y-ac-sc	106	64	0.006	0.085	Aguadé et al. (1989b)
y-ac-sc	120	49	0.003	0.056	Beech and Leigh Brown (1989)
y-ac-sc	31	109	0.003	0.000	Eanes et al. (1989b)
Zw	13	64	0.029	0.063	N. Miyashita (in preparation)
Zw	13	127	0.015	0.041	Eanes et al. (1989a)
Total or mean[b]	143	—	0.006	0.041	

[a]The proportion of homozygous large insertions is the probability that a large insertion found in the sample would be homozygous in a zygote formed by two random (but distinct) chromosomes from the sample.
[b]The means were calculated by weighting by the size of the region surveyed (kilobases). Where the same region was surveyed more than once, the unweighted average of these surveys was calculated and then weighted by the size of the region.
[c]The *su(f)* transcriptional unit is surrounded by repetitive sequences that may be TEs fixed in the population (A. Mitchelson and K. O'Hare, personal communication). In that case, the density of TEs per kilobase would be greater than 0.5, and the proportion of homozygous TEs would approach 1.0.

TABLE 3. (*Continued*)

Region	Size of region (kb)	Sample size	Density per kb	Proportion homozygous[a]	Reference or source
Autosomes					
Amy	15	85	0.009	0.009	Langley et al. (1988b)
Adh	13	48	0.018	0.000[d]	Aquadro et al. (1986)
Ddc	65	46	0.003	0.000	C. F. Aquadro et al. (in preparation)
rosy	100	60	0.002	0.030	Aquadro et al. (1988), C. F. Aquadro et al. (in preparation)
Hsp70	25	29	0.006	0.000	Leigh Brown (1987)
Total or mean[b]	218	—	0.004	0.014	
Grand total or mean[b]	578	—	0.005	0.022	

[d]Subsequent sequencing of apparently identical insertions demonstrated that the TE involved (*2161*) is actually inserted at different sites in each chromosome (W. Quattlebaum, C. H. Langley, and C. F. Aquadro, unpublished data). Each TE in the *Adh* region is thus unique.

have only rarely been associated with reduced gene expression [e.g., *copia* in *Adh* (Aquadro et al., 1986)].

X chromosomes versus autosomes. If the slightly deleterious effects of TE insertions in the flanking regions of genes were responsible for keeping the frequencies of these insertions low, then (as discussed above) a substantial difference in the densities of TEs on the X and autosomes would arise, provided that the mutations concerned were partially recessive (Langley et al., 1988a; Montgomery et al., 1987). As with the in situ hybridization work described above, restriction map surveys provide little evidence to support such a hypothesis. In Table 3, the average density of large insertions on the X chromosome is not significantly lower than in autosomal regions. Neither is the expected proportion of homozygous insertions lower for the X, as might be expected under more intense purifying selection.

Regions of reduced crossing over. As discussed above, there are reasons to expect TEs to accumulate in the distal and proximal regions of chromosomes, where meiotic crossing over is reduced in frequency. Table 3 presents the results of surveys of restriction-map variation in the *yellow-achaete-scute* region at the tip of the X chromosome of *D. melanogaster*. The average density of large insertions in this region (0.004) is the same as in the central part of the X

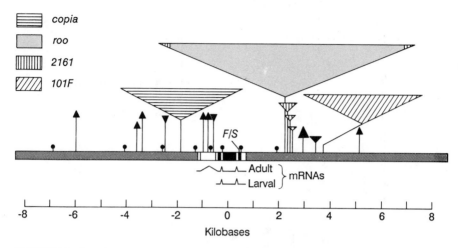

FIGURE 5. Restriction map summarizing the variation found in and around the *Adh* locus among 49 second chromosomes isolated from natural populations of *D. melanogaster* (data of Aquadro et al., 1986). The *Adh* gene is in the center. The filled boxes within the chromosome designate the exons. Below the gene are the two transcripts, larval and adult. The lollipops indicate the positions of polymorphic restriction sites. "F/S" shows the position of the nucleotide polymorphism responsible for the allozyme polymorphism. The small filled triangles show the approximate positions of unique sequence insertions and small deletions (not exactly to scale). The large triangles are TE sequences found inserted into the *Adh* region.

chromosome. The lack of an accumulation of elements at the telomere is consistent with the in situ hybridization surveys discussed above.

The restriction map survey of the proximal su(f) region of the X chromosome cannot be compared to the in situ surveys directly, since this region is located at 20F4, well into the β-heterochromatic region of the X where discrete sites of hybridization are difficult to identify. As would be expected from the results reported by Miklos et al. (1984, 1988), described above, the su(f) transcription unit is flanked by repetitive DNA (A. Mitchelson and K. O'Hare, personal communication). Restriction map results (C.H. Langley and N. Miyashita, in preparation) indicate that all sampled chromosomes characterized for the su(f) region are surrounded by identical repetitive sequences that may be TEs. The Zw region, located more distally in the basal euchromatin of the X, also shows a higher than average density of TEs. Table 3 also shows that the average expected proportion of homozygous insertions is somewhat higher for loci in regions of restricted recombination [even when the uniform repetitive region of su(f) is excluded]. This reflects a tendency for large insertions in these regions to be found at higher frequencies. Thus, the limited data on insertional variation at the base of the X chromosome are consistent with the in situ hybridization results.

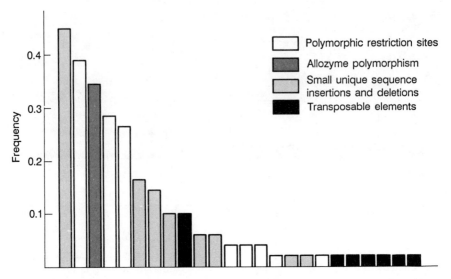

FIGURE 6. The frequency spectrum of the variation found in and around the *Adh* locus among 49 second chromosomes isolated from natural populations of *D. melanogaster* (data of Aquadro et al., 1986). Each column displays the population frequency of a polymorphism at a particular chromosomal site, in rank order decreasing from left to right. In each case the rarer variant is plotted, so that the frequencies are all less than 0.5. Each individual TE insertion was found to be rare, as is typical of such surveys. Subsequent sequencing of the element (*2161*) apparently present at a frequency of 0.1 has shown that it is really inserted into different but nearby sites (Quattlebaum et al., unpublished data).

Results for species other than *D. melanogaster*

D. ananassae. Three of the proximal regions of the metacentric X chromosome of *D. ananassae* have been surveyed (Stephan, 1989; Stephan and Langley, 1989). The *vermilion* region had two unique insertions and one highly polymorphic insertion, whereas no large insertions were observed in the *forked* region. Many insertions (large and small) were found in the *Om* region, and their frequencies were generally low. There are not enough data as yet to determine if the distribution and densities of TE-sized insertions in *D. ananassae* are different from those found in *D. melanogaster*. The *Om* and *forked* loci are in regions of relatively high rates of crossing over, whereas *vermilion* is in a region of reduced crossing over. The low rate of recombination in the *vermilion* region is of interest in view of the appreciable element frequencies observed here (see above).

D. pseudoobscura. Two surveys of restriction map variation in this species have yielded very little evidence of large-insertion polymorphism (Schaeffer et al., 1987; C. F. Aquadro et al., in preparation).

D. simulans. The *rosy*, *per*, and *Adh* regions of the *D. simulans* have been surveyed (Aquadro et al., 1988; C.F. Aquadro et al., in preparation; Stern et al., in preparation). Consistent with earlier reports (Dowsett and Young, 1982), the densities of large insertions (0.001) were dramatically lower in this species than in *D. melanogaster*. The few large insertions found were unique. The reason for this difference, which is in sharp contrast to the lower level of nucleotide-site variation in *D. melanogaster*, is unknown.

Future prospects

Unfortunately, there are no restriction map surveys of *D. affinis*, *algonquin*, *heteroneura*, or *silvestris*, in which in situ hybridization studies have suggested much higher abundances and element frequencies, at least for some types of middle repetitive DNA families (Hey, 1989; Hunt et al., 1984). If the findings of the two initial in situ surveys of these species are confirmed by further, more detailed studies, then the range of variation in the distribution of TEs among *Drosophila* species may be large and affords opportunities to study situations quite different from that in *D. melanogaster*.

CONCLUSIONS

The main conclusion of this chapter is that the data and models presented are consistent with the view that TEs are maintained in *Drosophila* populations as a result of transpositional increase in copy number, balanced by some opposing force or forces. This interpretation explains the fact that elements in *D. melanogaster* (and several other species) are usually present at low frequencies at individual chromosomal sites into which they can insert. Similar results have been obtained for yeast (Cameron et al., 1979) and bacteria (Hartl et al., 1986). These data are almost impossible to accommodate on the hypothesis that elements persist as a result of favorable mutations associated with their transpositional activities (Chao et. al., 1983; Nevers and Saedler, 1977; Syvanen, 1984). It is possible, however, that certain classes of mutation, such as chromosome rearrangements, are generated as a result of the excision of elements (Engels, 1989) or ectopic exchange between elements at different locations (Davis et al., 1987; Goldberg et al., 1983; Langley et al., 1988a) and become fixed in populations by drift or selection without leading to an increase in element frequency at the sites in question. Thus TEs could act indirectly as a source of mutational variation for evolutionary change, but this would not have much influence on their distribution within populations.

The DNA of *D. melanogaster* contains fewer repetitive dispersed sequences than the genomes of typical vertebrates and many plants (Spradling and Rubin, 1981). This difference is probably due to the low frequencies of TE insertions in *Drosophila* populations. Despite the extensive efforts exerted to discover restriction site polymorphisms in human genetics and the abundance of TEs in the human genome, it is remarkable that so few such polymorphisms are known to be attributable to the presence versus absence of a TE. Virtually all middle repetitive dispersed sequences are fixed in the human population (Schmid et al., 1985). The reasons for these differences between taxa are unknown; factors such as smaller effective population sizes in mammals compared with insects, and differences in the sizes of intergenic regions, may play a role.

The nature of the forces opposing the spread of TEs in *Drosophila* is also not known; regulation of rate of transposition in relation to number of element copies per individual, selection against insertional mutations, and the induction of highly deleterious chromosome rearrangements by ectopic exchange may all be involved. In view of the weak intensities of forces affecting element frequencies in natural populations, it seems clear that direct experimental tests to discriminate between the various possibilities discussed above will be difficult. The collection and analysis of quantitative population data should, however, continue to provide means of examining these questions, which are fundamental to an understanding of the biological significance of TEs.

Another major conclusion emerging from these studies is that there are theoretical reasons to expect TEs to accumulate in regions where ectopic exchange and/or regular meiotic exchange are suppressed. The data reviewed above suggest that there is indeed such an accumulation in the basal euchromatin and β-heterochromatin of *Drosophila* chromosomes, although not in the distal sections of the chromosomes where meiotic exchange (if not ectopic exchange) is also suppressed. The population genetic reasons for this accumulation are very different from that proposed for the accumulation of nontransposable, highly repeated sequences by gene amplification and genetic drift, which occurs more effectively in regions of reduced recombination (Charlesworth et al., 1986; Stephan, 1986; Walsh, 1987). Such sequences are also abundant in these regions (John and Miklos, 1988). Once TEs have inserted into regions of restricted genetic recombination and have accumulated there by the mechanisms we have described, it is possible that their abundance is increased even further by amplification processes. Recent data on repetitious, retroviral-related sequences on the mouse Y chromosome (Eicher et al., 1989) are suggestive of this. It is clear that the population dynamics of repeated DNA sequences have a considerable influence on the evolution of the structure of the chromosome.

CHAPTER 9

MOLECULAR EVOLUTION
OF ULTRASELFISH GENES OF
MEIOTIC DRIVE SYSTEMS

Chung-I Wu and Michael F. Hammer

The original notion of selfish genes, attributed to J. B. S. Haldane and widely popularized by Dawkins (1976), is that genes improve their own chances of survival by enhancing the fitness of their carriers and, hence, are the level at which natural selection takes place. Because the genes themselves are the ultimate beneficiaries of their own actions, all genes are selfish.

A more recent definition of selfish genes refers specifically to a class of DNA "with no phenotypic expression whose only function is survival within genomes" (Doolittle and Sapienza, 1980). In this view, selfish DNAs do not increase the survivorship of individuals carrying them but are merely genetic parasites. Werren et al. (1988) made this definition even more precise: Selfish genes are "those having characteristics that enhance their own transmission relative to the rest of an individual's genome, and that are either neutral or detrimental to the organism as a whole." In this chapter, the selfish genes discussed by Werren et al. (1988) are divided into two categories. The first includes transposable elements, B chromosomes, and organelle variants, which do not actively destroy other portions of the genome to promote their own transmission. For example, most transposable elements have deleterious effects on their hosts, as evidenced by their low copy numbers (Charlesworth and Langley, this volume), but there are notable exceptions, such as the *Alu* repetitive sequences in mammalian genomes (Hwu et al., 1986). In humans, there are more than half a million copies of *Alu*, comprising about 5% of the total DNA. Such a large number implies that the harmful effects of transposable selfish DNA can be very small. Similarly, B chromosomes do not actively destroy DNA within the same genome (Nur and Brett, 1987).

Selfish genes of the second category actively interfere with the functions of other genes in the same nucleus. We shall refer to these elements as "ultraselfish genes," a term first used by Crow (1988). Meiotic drive genes are an important class of ultraselfish elements, which destroy their homologues and thereby increase their own representation in the gametic pool. They first attracted the attention of geneticists because of their violation of the first law of Mendelian genetics—the law of equal segregation. Ultraselfish genes may also destroy DNAs other than their homologous partner in a diploid cell. For example, in the haplodiploid system of the wasp, *Nasonia vitripennis*, a paternal sex ratio factor, *psr*, is carried by the haploid male and acquires a transmission advantage by destroying the entire paternal genome after fertilization, thus ensuring all progeny are males carrying *psr*. Finally, even some transposable elements have an ultraselfish character. This happens when an element specializes in inserting itself into a particular gene and inactivating it. For example, the ribosomal insertional elements R1 and R2 in insects are very specific in their sites of insertion and are likely to usurp the transcription apparatus of the host gene for their own benefit (Xiong and Eickbush, 1988a).

In this chapter, we shall concentrate on genes that cause meiotic drive. Interested readers should consult Werren et al. (1987), Nur et al. (1988), and Xiong and Eickbush (1988a) for information on other types of "ultraselfish genes."

MEIOTIC DRIVE: WHY STUDY IT?

"Meiotic drive," a term coined by Sandler and Novitski (1957), is defined as an excess recovery of one allelic alternative in the gametes of a heterozygous parent. The first recorded case of meiotic drive is the *Sex-Ratio* (SR) condition in *Drosophila*, which was discovered in 1922 by A. H. Sturtevant. We shall discuss SR and *Segregation Distorter* (SD), another well-known meiotic drive system in *D. melanogaster* [see Temin et al. (1990) for a review]. A third case that will be briefly discussed is the *t*-complex in mice. Figure 1 is an illustration of the phenotypes of the three systems.

What can be learned from the study of meiotic drive? First, by studying the effect of modifiers on the degree of distortion, one gains insight regarding the broader question of what keeps Mendelian segregation so exact (Crow, 1979). Second, as meiotic drive genes will become fixed in the absence of counter-balancing selection, they present an opportunity to study natural selection. For example, variation in male virility appears to play an important role in main-taining the *Sex-Ratio* polymorphism in *Drosophila* (Wu, 1983a,b), and both interdemic selection (Lewontin, 1962) and female discrimination of male odors (Lenington and Egid, 1985) have been implicated in the maintenance of the *t*-complex polymorphism.

Third, in every system that has been analyzed in detail, meiotic drive involves interactions among several loci of a gene complex. The explicit char-

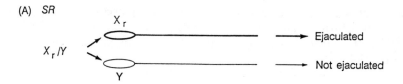

(A) *SR*

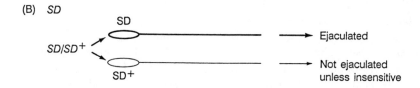

(B) *SD*

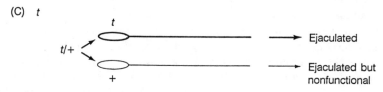

(C) *t*

FIGURE 1. The sperm phenotypes of *SR*, *SD*, and *t*. Sperm drawn in the thin outline are the nonfunctional ones.

acterization of these interactions makes meiotic drive an interesting example of coadapted gene complexes. Fourth, the cause of meiotic drive is often an aberrant form of a fundamental process. Like many other abnormal phenomena, it serves as a window for viewing the normal process. For example, meiotic drive in *Drosophila* is caused by improper condensation of wild-type chromatin, resulting in faulty spermiogenesis (Tokuyasu et al., 1972). Consequently, the study of meiotic drive may be useful in understanding chromosome inactivation, chromosome condensation, and sperm maturation. In mice, meiotic drive appears to be caused by a failure of certain sperm to function properly in the female's reproductive tract (Brown et al., 1989).

Knowledge of meiotic drive mechanisms may one day be applied to such practical problems as pest control. For example, it may be possible to "engineer" a Y chromosome that renders X-bearing sperm nonfunctional, resulting in all-male progeny. Such a driving Y chromosome would perpetuate itself at the expense of the normal Y, thereby reducing the number of females in the population and driving the population to extinction (Hamilton, 1967). Lyttle (1977) has already constructed a pseudo-Y drive system.

The role of many polymorphisms at the molecular level is hard to assess for want of a selectively significant phenotype. Although a neutral, or nearly neutral, polymorphism can be informative about certain processes, there is a need to study molecular variation that is manifested as variation in fitness. Meiotic drive is a phenomenon that can be studied at the population, cellular, and molecular levels.

SEX-RATIO IN DROSOPHILA

Drosophila males with the Sex-Ratio chromosome, X_r, transmit predominantly X_r-bearing sperm and produce nearly all female progeny. SR has been observed in two divergent subgenera of the genus Drosophila. In the subgenus Drosophila, it has been found in D. paramelanica (Stalker, 1961), D. mediopunctata (De Carvalho et al., 1989), D. quinaria, and D. testacea (James and Jaenike, 1990). In the subgenus Sophophora, SR occurs in at least three species in the affinis subgroup and four species in the obscura subgroup, including D. pseudoobscura and D. persimilis. The molecular nature of SR is completely unknown.

X_r is ultraselfish in that it causes sperm bearing the Y chromosome to degenerate after the completion of meiosis. As a consequence, most of the sperm produced by X_rY males carry X_r, whereas only half of the sperm of normal males carry X. Provided that sperm are not in limiting supply, X_r will be transmitted at twice the rate of the normal X and will rapidly replace it in the population. A problem presented by SR is why it remains polymorphic in populations rather than causing their extinction. There are two possibilities: either natural selection acts against it, or genes that suppress Sex-Ratio distortion have evolved.

Components of selection against SR

The theoretical conditions for the maintenance of the X_r–X–Y polymorphism were worked out by Edwards (1961). Many experiments have measured selection components, such as male virility, female fertility, and larval viability, that act against X_r in D. pseudoobscura (Curtsinger and Feldman, 1980; Beckenbach, 1983). An important issue is whether SR males are less virile than SR^+ males. Because half the sperm produced by these males degenerates, a fertility reduction may be expected, in which case, selection against X_r is linked to meiotic drive itself. However, because fecund males ejaculate more sperm than females can store, sperm degeneracy does not always reduce the fertility of males inseminating virgin females (Beckenbach, 1978; Wu, 1983a). Why, then, do Drosophila males ejaculate more sperm than needed for insemination? The evidence suggests that an excess ejaculate is more effective in displacing sperm from a previous mating stored in the female's receptacles. In Drosophila, multiple insemination in nature appears to be the rule (Cobbs, 1977). Thus, SR males may be less virile because of weaker sperm displacement (Wu, 1983a,b).

Suppressor genes of *SR*

In general, genes that suppress *SR* will have an advantage if they are either autosomal or Y-linked; the advantage of the latter condition is self-evident. Similarly, an autosomal suppressor will increase for the reason of rectifying a biased sex ratio: it distributes itself evenly between the two sexes, whereas its nonsuppressor allele tends to be associated with females, the more common sex. Such suppressors have been discovered in several species (Stalker, 1961), and it is conceivable that *SR* has disappeared in others because of suppression. Why, then, are suppressors absent in species such as *D. pseudoobscura* (Beckenbach et al., 1982)? A possible explanation is that selection acts against suppressors because gametic phase disequilibrium generated by meiotic drive often causes suppressors to occur in genotypes with reduced fitness. Theoretical calculations further suggest that strong selection against *SR* males, as opposed to selection against females carrying X_r, is highly effective against the emergence of autosomal suppressors (Wu, 1983b).

Genetics of *Sex Ratio*

SR, like all other meiotic drive systems, is associated with various chromosomal inversions. The condition is associated with XR (the right arm of X) in the *obscura* group, whereas in *D. testacea* it is associated with the telocentric X, which is homologous to XL in the *obscura* group. This suggests that *SR* evolved at least twice independently. It can be seen in Figure 2A that inversions are not needed for the *SR* trait. The *Sex-Ratio* chromosome of *D. persimilis* is homosequential to the standard X of *D. pseudoobscura*. Such peculiar arrangements illustrate the point that these inversions evolve to reduce recombination between "*Sex-Ratio* genes" on X_r. Inversions serve the same purpose on X or on X_r. Within each species, such inversions preclude genetic analysis by recombination, but because hybrid females between these two sibling species are fertile, it is possible to introgress X_r of *D. persimilis* into *D. pseudoobscura* and, reciprocally (Wu and Beckenbach, 1983). Wu and Beckenbach (1983) tested males carrying a recombinant X between X_r and X for the presence of the *SR* trait (see Figure 2B). Surprisingly, there were four different regions on X_r that, on replacement by the homologous region from X, resulted in the complete loss of the *SR* phenotype. This observation indicates the presence of at least one *SR* factor in each region and suggests very extensive differentiation between the two types of chromosomes.

Molecular studies of *SR*: A perspective

Many questions about the evolution of *SR* can be answered by examining nucleotide sequence variations in the X_r chromosome, as has been done for *SD* and the *t*-complex. It is possible to reconstruct the history of the system of

(A)

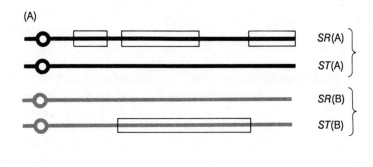

(B)

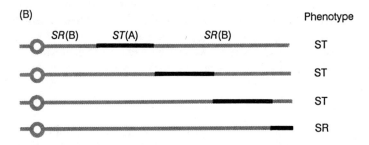

FIGURE 2. (A) The *Sex-Ratio* (X_r) and standard (X) chromosomes in two sibling species. Boxes represent inversions. (B) Phenotypes of X_rY males with a segment of the standard X chromosome (dark bar).

inversions because DNA sequences within the older inversion are likely to be more divergent. The divergence between the X_rs of different species will be most informative. Is it possible that the X_r of *D. pseudoobscura* came from *D. persimilis* by introgression, and subsequently acquired its inversions? The two species can hybridize, and there is some indication that they do so in nature (Powell, 1983). Later sections on the molecular evolution of *SD* chromosomes (Table 2) and the *t*-complex (Figure 10) will provide clues as to how *SR* research might benefit from molecular population genetic analyses.

SEGREGATION DISTORTER IN DROSOPHILA

Segregation Distorter (SD) in *Drosophila melanogaster* is one of the better understood systems of meiotic drive (Crow, 1979, 1988; Temin et al., 1990). In this chapter, we will emphasize the population and evolutionary genetic aspects of *SD*, taking advantage of knowledge gained from recent advances in molecular cloning.

SD refers to certain naturally occurring second chromosomes. Males heterozygous for *SD* and a wild-type chromosome usually transmit *SD* to 95–99% of their offspring (Hartl and Hiraizumi, 1977; Ganetzky, 1977; Crow, 1979;

Temin and Marthas, 1984; Lyttle et al., 1986). *SD* chromosomes have two major genetic loci that are closely linked: *Segregation distorter* (*Sd*) and *Responder* (*Rsp*). *Sd* exists in two allelic forms, corresponding to either the presence (*Sd*) or absence (Sd^+) of segregation distortion activity, while *Rsp* alleles are roughly distinguished as either sensitive (Rsp^s) or insensitive (Rsp^i). The structure of the gene complex is shown in Figure 3. A third locus, called *Enhancer of SD*, is also depicted. The *Sd* allele causes the dysfunction of Rsp^s-bearing sperm by interfering with chromatin condensation during spermiogenesis (Tokuyasu et al., 1972). Consequently, a $Sd\ Rsp^i / Sd^+\ Rsp^s$ male transmits more than 95% *SD*-bearing sperm. (Segregation in females, however, is normal.) The wild-type chromosome, $Sd^+\ Rsp^s$, is referred to as sensitive to distortion. With two loci, there can be two other kinds of chromosomes: insensitive ($Sd^+\ Rsp^i$) and suicidal (*Sd Rsp^s*). An insensitive chromosome is not distorted by *Sd* but has no distorting activities itself, whereas a suicidal chromosome distorts itself.

Certain genetic transmission properties are important for understanding both the mechanisms and the origin and evolution of this selfish gene system. First, the deletion of either *Sd* or Rsp^s completely eliminates the mutant phenotype, restoring normal segregation. Second, the actions of *Sd* and Rsp^s do not depend on their cytological locations, because they can be translocated to other chromosomes without changing their genetic properties (Lyttle et al., 1986; Lyttle, 1989). Third, many *SD* chromosomes have inversions that suppress recombination between *Sd* and *Rsp*.

Two aspects of segregation distortion were particularly puzzling. First, aberration in sperm formation first appears at the stage of sperm individualization, well after the completion of meiosis. Yet it is known that after meiosis no genes are transcribed (Olivieri and Olivieri, 1965) and no gene is needed (Lindsley

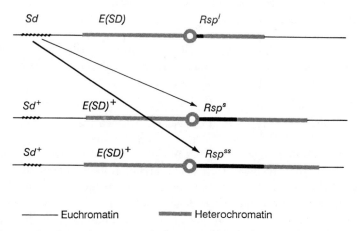

FIGURE 3. The structure of the *SD* gene complex. Weight of arrows indicates the strength of distortion. The sensitivity at the *Rsp* locus is proportional to the copy number of a satellite DNA repeat. *E(SD)* is an enhancer of distortion.

and Grell, 1969) for sperm maturation in *Drosophila*. Second, the existence of *Rsp*[s] alleles is an evolutionary enigma. Ganetzky (1977) demonstrated that deleting *Rsp*[s] renders the chromosome insensitive to segregation distortion, and homozygotes for the deletion are viable and fully fertile. Since *Rsp*[s] is a target of *Sd* action, one might expect deletions of *Rsp*[s] to be advantageous in the face of the worldwide occurrence of *SD* chromosomes (1–5%). The high frequency (60–80%, Temin and Marthas, 1984) of *Rsp*[s] in natural populations, therefore, needs to be explained.

Molecular cloning of the *Sd* locus

The cloning and characterization of the *Sd* locus are in progress (see Powers in Temin et al., 1990). The cytological location of *Sd* has been defined by deletion studies and mapped to the region of 37D2-6 on polytene chromosomes (Brittnacher and Ganetzky, 1983). Starting with a microdissection clone from that region, Powers obtained overlapping clones spanning most of the region. She identified a 5-kb tandem duplication that is always present in *SD* chromosomes but is absent in non-*SD*s. Interestingly, this duplication is only a few kilobases distant from the topoisomerase II gene (see Figure 6).

Molecular cloning of the *Rsp* locus

The cloning of *Rsp* relied on a major assumption that *Rsp*[s] consists of a large array of repetitive sequences that either are missing or exist in a much lower copy number on *Rsp*[i] chromosomes (Wu, in Temin et al., 1990). Wu et al. (1988) obtained an *Rsp* clone that contains an array of repeats; each of the repeats is about 120 bp long and rich in AT content, with short runs of As and Ts (see Figure 9). This *Rsp* repeat has most of the characteristics of satellite DNA (Miklos, 1985), which is a class of tandem repeats with very simple sequences, usually AT or GC rich, thus forming a satellite band on a CsCl gradient. Satellite DNAs are ubiquitous and abundant in higher eukaryotes.

The evidence that this array of repeats represents the *Rsp* locus is as follows:

1. *Rsp sensitivity of naturally occurring chromosomes correlates with copy number.* There is a nearly perfect correlation between the copy number of the *Rsp* repeat and sensitivity to distortion among 35 chromosomes, with a typical sensitive chromosome having about 700 copies and a supersensitive one having up to 2500 copies. One salient finding is the distinction between the *Rsp* allele on *SD* chromosomes and the corresponding allele on non-*SD*-insensitive chromosomes. The latter always have 100–200 copies of the repeats whereas *SD* chromosomes have fewer than 20. The observation is consistent with the hypothesis that *Rsp* has a cellular function, albeit a dispensable one. Given the low frequency of *SD* chromosomes in most natural populations (1–5%), a chromosome may gain an advantage by main-

taining a small array of repeats and remaining very weakly sensitive to distortion, thus balancing its need to retain some minimal level of the normal repeat function against the need to avoid significant segregation distortion by *SD*.

2. *The repeat array is deleted whenever* Rsp *function is deleted.* Ganetzky (1977) constructed five deletion chromosomes that lacked the *Rsp* sensitivity. Although the original chromosome had about 700 copies of the repeat, four of the deletions had fewer than 30 copies. The remaining one contained a full complement of the repeat array. On retesting, this chromosome turned out to be fully sensitive at the time of molecular cloning. A plausible hypothesis is that the residual *Rsp* repeats were amplified in a manner similar to that observed for the ribosomal genes of *Drosophila* (Tartof, 1974; Endow and Komma, 1986; see a later section on *Rsp* evolution).

3. *The repeat array is translocated to the Y chromosome whenever* Rsp *is.* Seventeen translocations constructed by Lyttle were tested (Lyttle, 1989). Ten of them moved *Rsp* to the Y chromosome (i.e., the Y chromosome became sensitive to *SD* distortion); the remaining seven had breakpoints near the *Rsp* locus but retained *Rsp* on the second chromosome. The repeat array cotranslocated with the *Rsp* locus in every case. In one case where *Rsp* was not completely translocated to Y, as judged by the reduced sensitivity, molecular data also indicate translocation of a partial array of repeats.

The evolution of autosomal drive

Although sex-linked meiotic drive is widespread in *Drosophila*, autosomal drive appears to be much less common. This may be attributed to a bias in detection, because sex-linked meiotic drive can be more easily observed, but theoretical studies and genetic/molecular analyses suggest at least two conditions that are conducive for the evolution of sex-linked drive. First, the Y chromosome, being entirely heterochromatic, may potentially be the largest target for meiotic drive (see the section on mechanisms). Second, there is virtually no recombination between X and Y chromosomes.

Recombination is important because gene complexes, rather than single gene loci, are the genetic basis of all meiotic drive systems analyzed. From a mechanistic point of view, meiotic drive results from the action of a distorting locus, such as *Sd*, on a target locus, such as *Rsp*. Such an interaction renders gametes bearing one allelic form of the target locus nonfunctional. Molecular studies of *SD* reveal a great difference in the nature of these two components. It is highly unlikely that true single-locus drive could exist; the locus would have to encode a product that not only recognizes itself but selectively interacts with one allelic form but not the other. As meiotic drive necessarily involves at least two loci, one can see immediately why linkage relationships between loci are important. The distorting gene has to be associated with a nonresponding allele at the target locus, whereas the responding allele has to be associated with the nondistorting gene. Tight linkage reduces the formation of "wrong" recombinants.

There is a rich body of theoretical studies on this subject (e.g., Prout et al., 1973; Thompson and Feldman, 1976; Eshel, 1985), some general features of which are as follows: (1) If the driving locus and the target locus are sufficiently tightly linked, depending on the fitness effects and drive strength, meiotic drive can evolve initially. (2) With tight linkage, the genes in the meiotic drive system can reach a stable equilibrium, again depending on fitnesses, drive strength, and linkage. (3) If there is a stable polymorphism at both the driving and target loci, a strong linkage disequilibrium will be observed, as noted above. (4) Following (3), inversions that reduce recombination between the two loci will increase in frequency, eventually giving rise to tightly bound gene complexes. (5) Genes that suppress meiotic drive will tend to increase in frequency, even if they are not linked to the target locus. This is an important contrast with enhancers of drive, which need to be tightly linked, as indicated in (1). An intuitive explanation for this contrast is given by Eshel (1985). Most of these conditions are qualitatively applicable to SR as well.

These conclusions explain why meiotic drive is not more commonly observed (or why Mendelian segregation is as stable as it is): the distorting locus has to be close to the target locus for meiotic drive to evolve (condition 1), but the suppressor locus does not even have to be linked (condition 5) to reverse the evolution of meiotic drive. The part of the genome that can potentially suppress the evolution of meiotic drive is far larger than the part that can promote its evolution. It is no coincidence that the two distorting loci, Sd and $E(SD)$ (Figure 3), are located within 2 cM of Rsp. There are also modifiers of SD on many naturally occurring X chromosomes (Hiraizumi and Thomas, 1984) and on the third chromosome (Trippa and Loverre, 1975). In this context, condition 1 is much more easily fulfilled by sex-linked drive because the absence of X–Y recombination precludes formation of "wrong" recombinants between any distorting gene on X and any target locus on Y.

It has sometimes been suggested that autosomal meiotic drive genes, if sufficiently potent, would sweep through a population, in which case no meiotic drive would be observed. One test of this hypothesis is to carry out interspecific crosses, on the assumption that driving chromosomes are fixed in one species and are absent in the other. But no hidden meiotic drive has yet been found by this method (e.g., Coyne, 1990). In the case of SD, interspecific comparisons of sequence data could also provide a test of this hypothesis.

Models of SD evolution

Most of the models discussed above are of a very general kind that impose no fitness constraint on the distorting locus but assume drive and selection at the target locus. However, experiments have now shown that selection occurs primarily at the distorting locus (see Table 1 and Figure 4). Charlesworth and

TABLE 1. Changes in the frequency of SD/SD among SD/SD, SD/S, and S/S (cages 1 and 2), or among SD/SD, SD/I, and I/I (cages 3 and 4) in laboratory populations[a]

Cage	Generations					
	0	2	3	4	5	8
1. SD vs. S	0.07	0.157	0.108	0.086	0.159	0.370
2. SD vs. S	0.12	0.143	—	0.15	0.124	0.179
3. SD vs. I	0.04	0.057	—	0.0	—	0.014
4. SD vs. I	0.136	0.014	0.021	0.071	—	0.007

[a]Each type of chromosomes, SD, S (for sensitive), or I (for insensitive), is represented by three different isolates from nature. These isolates were mixed among themselves for 8 weeks before being placed with chromosomes of another type for competition. Chromosome types were determined by squash-blotting (Wu et al., 1989).

Hartl's (1978) model was the first to consider the SD polymorphism with realistic parameters. They studied the changes in frequency of the three chromosomal types: SD (Sd Rsp^i), S (Sd^+ Rsp^s; S for sensitive) and I (Sd^+ Rsp^i; I for insensitive); the fourth type, Sd Rsp^s, is suicidal. Rsp^s/Rsp^i males transmit predominantly Rsp^i-bearing sperm when Sd is also present; otherwise segregation is normal. Selection is assumed to be strong against Sd and weak against Rsp^i. They also showed that, as expected, SD can be established only when linkage is sufficiently tight. This model goes a long way toward explaining the evolution of the SD system, as shown in Figure 5A. The frequencies oscillate dramatically and spiral toward an equilibrium, if it exists. Intuitively, it is like a game of scissor-paper-stone, where SD dominates S (due to distortion), I dominates SD (selection against Sd), and S dominates I (selection against Rsp^i).

An experimental demonstration of an SD/S/I equilibrium is provided by Hartl's (1977) analysis of a long-term population cage experiment and Temin and Marthas' (1984) investigation of two natural populations. In these studies, genotypes were determined by single-pair mating. The allele frequency dynamics have evaded analysis because of the tedious nature of genotype determination. Because of the large differences in Rsp repeat number among sensitive, insensitive, and SD chromosomes, Wu et al. (1989) were able to distinguish many genotypes in a large sample of individuals by squash-blotting adult flies directly onto nitrocellulose paper for hybridization (Table 1). Results of these studies by J. R. True and C.-I. Wu (unpublished results) can be summarized as follows: (1) Competition between SD and I shows clearly that selection against SD is very substantial. (2) SD, in competition with S, increases in frequency apparently because the drive advantage is sufficient to overcome the fitness defect. We also confirmed the high distortion in transmission ratio of SD/S males from the cages by squash-blotting many males and their progeny.

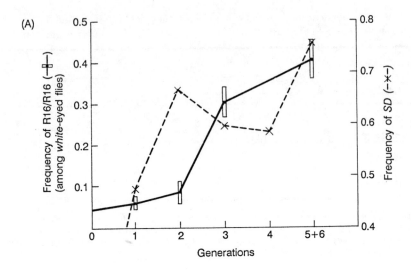

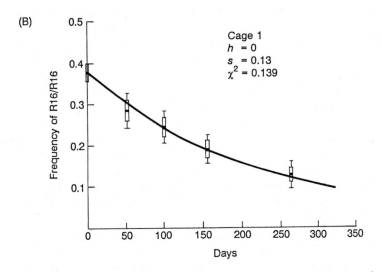

FIGURE 4. (A) Changes in the frequency of the insensitive chromosome (I, designated R16 in the figure) in the presence of SD. Homozygous non-SD flies are *white*-eyed, and among *white*-eyed flies, genotypes are determined by squash-blotting (Wu et al., 1989). (B) Changes in the frequency of the insensitive chromosome in the absence of SD. Note that the time scale is five times larger in (B) than in (A). *h* is the degree of dominance in fitness and *s* is the coefficient of selection against the insensitive chromosome, as determined by least-squares fitting.

A fitness comparison between S and I chromosomes, carried out by Wu et al. (1989), used two chromosomes that differed in amount of Rsp satellite (20 vs. 700) but were otherwise genetically identical. The I chromosome was derived from S by X-ray irradiation (Ganetzky, 1977) and was homozygous viable and fully fertile. As shown in Figure 4A, I replaces S in the presence of SD in a few generations, largely because SD/S males transmit mostly SD-bearing sperm (>95%) whereas the transmission in SD/I males is normal. The question is why I has not replaced S in nature, given such a large advantage in the presence of SD. Figure 4B offers an explanation: in the absence of SD, S gradually replaced I in large laboratory populations (size = 5000), with selection against I of about 15% per generation. Wu et al. (1989) showed that selection against I was largely due to a viability difference. Since segregation distortion takes place in the germline, these results suggest that the normal function of Rsp is independent of segregation distortion.

While Charlesworth and Hartl's (1978) model elegantly sketches the evolution of SD, we now know that the Rsp locus does not exist in only two states; rather, there is a continuum of sensitivity, correlated with copy number of the Rsp repeat. In addition, changes in the molecular structure of Rsp are very rapid because of unequal exchange (see the section on Rsp). Incorporation of these features into the model may have a stabilizing effect. Figure 5A and B illustrates this important point. In the Charlesworth and Hartl model, the oscillation in chromosome frequency can be so large that SD remains near zero for a long period of time and the trajectory approaches the equilibrium with dampened oscillation. If one changes the parameter slightly, the amplitude of oscillation can actually increase with time; the trajectory therefore goes away from, rather than toward, the equilibrium (Crow, personal communication). Recently, Takahata and Wu attempted to incorporate unequal sister chromatid exchange (USCE) at the Rsp locus into a model similar to that of Charlesworth and Hartl's (1978). Their preliminary results, as shown in Figure 5C, support the hypothesis that a high rate of USCE has a significant effect in dampening the oscillation in chromosome frequencies. When there is a stable equilibrium, SD avoids staying in a precariously low frequency and the equilibrium is approached much more quickly.

Molecular evolution of the Sd locus

DNA polymorphisms in the Sd region are intriguing for several reasons. First, their analysis reveals the level of divergence between SDs and SD^+s, shedding light on the evolution of SD, where meiotic drive and the counterbalancing selection are acting. Polymorphisms in this region will provide an interesting contrast with polymorphisms in regions that are weakly constrained by selection. Second, the region is expected to show very distinct linkage relationships among polymorphic sites. Referring to Figure 3, we can see that recombination between SD and SD^+ chromosomes proximal to the Sd locus will generate unfavorable

(A)

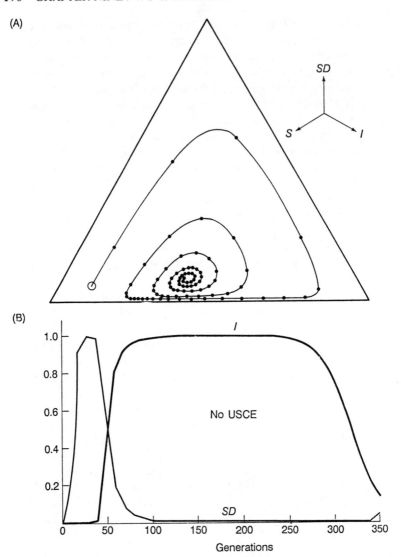

(B)

recombinants (such as suicidal chromosomes) whereas recombination distal to *Sd* should yield no deleterious products; hence, *Sd* is expected to sharply demarcate two contrasting regions. Third, combining the information on the evolution of *Rsp* with the polymorphism data for *Sd* should help reveal how the whole gene complex was formed.

Wu et al. (in preparation) studied sequence polymorphism of the *Sd* region based on the clones obtained and characterized by Powers (see Powers, in Temin et al., 1990), as shown in Figure 6. The region surveyed spans 70 kb around the *Sd* duplication, but here we will focus on three subregions marked in Figure 6. One covers 15 kb immediately surrounding *Sd* while each of the other two covers a region of similar size but approximately 30 kb on each side

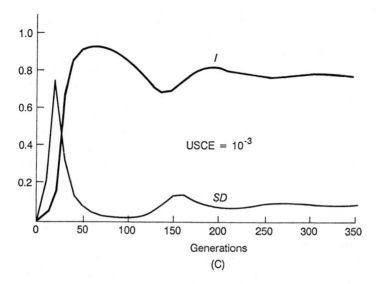

$USCE = 10^{-3}$

Generations

(C)

FIGURE 5. (A) Charlesworth and Hartl's (1978) model of the dynamics of the $SD/S/I$ changes. The frequency of each chromosome is 1 at the apex and 0 on the opposite side, as indicated by the axes. The blip interval is 10 generations. (B), (C) Changes in gene frequency under the assumption of no sister chromatid exchange (B) or sister chromatid exchange (C). Note that the trajectory in (B) is going into the next cycle at generation 350.

of Sd. The chromosomes studied here, including both SD and SD^+, are of worldwide origin (see legend of Table 2). Six six-cutter restriction enzymes were used in the analysis (Table 2).

On the basis of the variation in the proximal region, there are two distinct types of SD chromosomes: type I represented by SD_{Roma} and $SD_{Los\ Arenos}$, from Italy and Spain, respectively; and type II, represented by all other SDs from North America, Japan, Australia, and another Spanish chromosome, SD_{VO17B}. A simple explanation for this observation is that SD_{Roma} and SD_{LA} represent the "prototype," which retains the ancestral SD^+ character, whereas all other SDs have acquired SD-specific sites. This suggests that the SD polymorphism is old and provides further support for the view that the polymorphism is maintained by the counterbalancing forces of selection and drive. The fact that

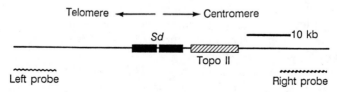

FIGURE 6. Structure of the Sd region.

SD_{Roma} is not associated with inversions and is one of the weakest driving chromosomes is consistent with it being ancestral. In addition, there must be a mechanism to maintain the coexistence of the two types of SD in the Mediterranean region, despite their differences in drive strength. There are many inversion karyotypes among type II SDs (Hartl and Hiraizumi, 1977) that must have had a recent origin in light of the low level of nucleotide divergence among them. Perhaps some inversion karyotypes are in the process of replacing others; thus, the inversion polymorphism is only transient. The observation of a gradual replacement of the SD_5 inversions by the SD_{72} inversions in Madison, Wisconsin, during the past 30 years is compatible with this interpretation (Temin and Marthas, 1984).

The presence of both type I and type II (SD_{VO17B}) SDs in the Mediterranean region, while only type II has been found elsewhere, suggests that SD chromosomes originated in that region. [The same rationale has been used to argue for the African origin of hominoids (Cann et al., 1987).] Temin (in Temin et. al., 1990) also suggests a Mediterranean origin of the SD chromosomes, based on the observation that SD_{Roma} and SD_{VO17B} are the only two SDs free of inversions. However, SD_{LA}, the prototype SD, carries an inversion commonly associated with SD chromosomes (Temin and Krieber, personal communication), which can best be explained by recombination between this inversion and Sd.

A second interesting feature shown in Table 2 is the contrasting patterns of divergence in the distal vs. proximal regions. The distal region in all but one SD appears to have the "consensus" SD^+ haplotype. In the proximal region, all type II SDs possess a haplotype unique to SD, which is quite different from the consensus SD^+ haplotype. Why? Recombination distal to Sd does not disrupt the SD gene complex (see Figure 3) and does not create unfavorable recombinants, but if an SD recombines with an SD^+ anywhere proximal to the Sd locus, it becomes a suicidal chromosome, which has never been recovered in natural or laboratory populations. The latter explains the very strong linkage disequilibrium between the Sd locus and the polymorphic sites covered by the right probe (28–43 kb proximal to Sd) in worldwide populations.

In the region distal to Sd, however, SD chromosomes must have been recombining with SD^+s and thus have not been able to maintain an SD-specific haplotype. This is supported by an SD chromosome from Australia, which appears to share a distinct HaeIII site with two local SD^+ chromosomes in this distal region, while maintaining identity of Sd as well as the proximal haplotype. The observations are not due to a difference in the rate of recombination between the two regions, which are only 60 kb from each other. None of the large scale effects on recombination in Drosophila, such as the distance to heterochromatin or inversion breakpoints, can adequately explain the contrasting patterns on such a fine scale.

There are other important features related to linkage in the Sd region. First, although SD is associated with recombination suppression, the suppression is

TABLE 2. RFLP patterns in the *Sd* region[a]

	Left (20–35 kb distal)						*Sd*	Right (28–43 kb proximal)					
	EcoRI	HindIII	ClaI	XhoI	BamHI	HaeII		EcoRI	HindIII	ClaI	XhoI	BamHI	HaeII
SD													
Roma	A	A	A	A	A	A	+	A	B	A	A	A	A
SPN(LA)	A	A	A	C	A	A	+	A	B	A	A	A	A
SPN(VO)	A	A	A	A	A	A	+	C	D	B	B	C	A
USA(5)	A	A	A	A	A	A,E	+	C	D,D′	B	B	C	A
JPN	A	A	A	A	K	A	+	C	D	B	B	C	A
AUS	A	F	B	C′	M	B′	+	C	D′	B	B	C	A
SD⁺													
CanS	A	A	A	A	B	A	−	A	A	A	A	A	B
AUS(2)	A	A	A	A	L	B′	−	A	B,E	A	A	A	A
CAL(9)	A	A,C,E	A	A	A,B,J	A	−	A	A,B	A	A,B′	A	A,D
WIS(3)	A	A	A,C	A,E	A,B	A	−	A	A,C	A	A,B′	A	B,A
NYS(7)	B,A,F	B,D	A,E	A,E	B,F,G	B,A	−	A	A,B	A	A	A	B,A
EUR(4)	A,D	A	A,C,E′	A	A,C	A,F	−	A,B	B	A	A,D	A	A
D. simulans	C	G	D	D	H	D	−	E	G	D	C	D	C

[a] The letters designate restriction patterns. Patterns designated by the same letter with a different prime may or may not be the same. Chromosomes are classified by their geographical origin: SPN (Spain), JPN (Japan), AUS (Australia), CAL (California), WIS (Wisconsin), NYS (New York State), EUR (Europe). The numbers in parentheses are the numbers of chromosomes surveyed.

far from complete. Therefore, meiotic drive and selection must have played a significant role in preserving these distinct patterns of linkage association. Second, the distance between *Sd* and other polymorphic sites in this study is more than 30 kb, a distance long enough to ensure linkage equilibrium in many regions of *Drosophila* (Aquadro et al., 1986; Langley and Aquadro, 1987). Third, as the frequency of *SD* is only about 1–5% in nature, it must have existed mostly as SD/SD^+, increasing the chance of shuffling polymorphic sites between them. Fourth, the polymorphism of SD/SD^+ has been actively maintained by balancing selection, a conclusion supported by nucleotide sequence data. Such a long-term coexistence further permits shuffling of polymorphic sites. In contrast, recombination in a neutral (or nearly neutral) region can occur only during the transient sojourn of neutral variants.

Molecular evolution of the *Responder* satellite DNA

Since none of several second chromosomes from *D. simulans* examined so far has the *Sd* duplication, we might have expected *Rsp* in this sibling species of *D. melanogaster* to be of a sensitive type, characterized by a large array of the *Rsp* repeat. To our surprise, the *Rsp* repeat exists in very low number (<10) in every *D. simulans* strain examined. This is in sharp contrast to the observations of the naturally occurring SD^+ chromosomes in *D. melanogaster*, which have between 100 and 2500 copies. The same low number of the *Rsp* repeat was observed in the other two sibling species, *D. mauritiana* and *D. sechellia*. Genomic DNA of none of the 12 other species from either the *Drosophila* or the *Sophophora* subgenus shows detectable hybridization to the *Rsp* repeat. The absence of *Rsp* in *D. pseudoobscura* suggests that *SR* and *SD* are not likely to be homologous at the target locus. We conclude that the expansion of the *Rsp* array took place after the divergence of *D. melanogaster* from its sibling species. Such an expansion provided a target for segregation distortion and appears to be a precondition for the evolution of *SD*.

A striking feature of the *Rsp* locus in *D. melanogaster* is the enormous variation among chromosomes. A typical example is shown in Figure 7. Genomic DNA was extracted from lines carrying one of many second chromosomes isolated from the same population at the same time, restriction digested, Southern-blotted, and hybridized to the *Rsp* repeat. Among 60 second chromosomes, we observed 59 restriction patterns; two chromosomes isolated from flies caught in the same bucket of bait appear identical.

Why is the *Rsp* locus so variable? A likely explanation is unequal crossover between either sister chromatids or homologous chromosomes (Figure 8A). One of the best studied examples is unequal crossover between the ribosomal RNA genes in *Drosophila*, which can lead to the rapid expansion of the rRNA gene array thus reverting the severe bobbed phenotype due to rRNA deficiency (Tartof, 1974; Endow and Komma, 1986). A possible case of expansion of an *Rsp* repeat array has been discussed in the section on *Rsp* cloning.

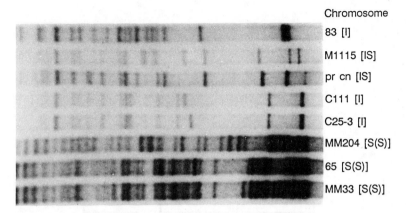

Chromosome

83 [I]

M1115 [IS]

pr cn [IS]

C111 [I]

C25-3 [I]

MM204 [S(S)]

65 [S(S)]

MM33 [S(S)]

FIGURE 7. The MspI restriction patterns in the *Rsp* region of eight naturally occurring chromosomes. MspI cuts the *Rsp* repeat occasionally. Note the extent of divergence among these chromosomes.

Analysis of nucleotide sequences of *Rsp* repeats suggests that unequal crossover occurs at the *Rsp* locus. Figure 9 shows some of the 40 *Rsp* repeats sequenced to date. As noted previously, the repeats are organized mainly as dimers consisting of pairs of 120-bp repeats (Wu et al., 1988). Dimers are delineated by an XbaI restriction site (TCTAGA) on each end while the site in the middle has changed to TCTACA. The sequence diversity among left repeats averages 4.2%, and diversity among right repeats averages 4.3%. Between a left and a right repeat, the average diversity, however, is about 20%. Since the nucleotide divergence between *D. melanogaster* and its sibling species is no more than 5% (Zwiebel et al., 1982), the age of this dimeric structure appears greater than the age of these sibling species. This is consistent with the presence of the *Rsp* repeat, albeit in very low number, in these sibling species.

An indication of unequal crossover is the divergence among the three dimers, CB5, CB6, and CB10. Their left repeats are identical whereas the right repeats differ by 4–8%. Direct evidence is shown in Figure 8A. When unequal crossover takes place between a left and a right repeat, the event is expected to yield two products—a trimer and a monomer. The monomer should resemble the left repeat type on its left side and the right repeat type on its right side whereas the second repeat of the trimer should be the opposite. Figure 9 highlights such changes. Three monomers, CB2, CB8, and LP24, possess all the characters of the left repeat between positions 1–60 but have the characters of the right repeat between position 101–120. The opposite pattern can be seen in LP14-2, LP16-2, and LP26-2, all being the middle repeat of a trimer. It is interesting to note that most of the crossover points fall between positions 60 and 100, a region exhibiting least divergence between the left and the right repeat. This is the region where a left repeat can be in nearly perfect register with a right repeat and undergo unequal exchange. Other derived products of

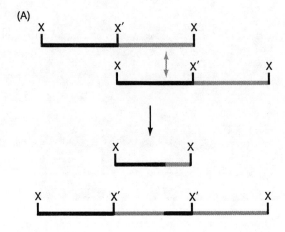

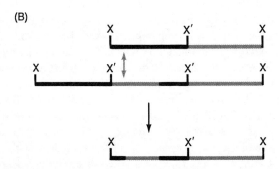

FIGURE 8. (A), (B) Models of unequal crossover between different *Rsp* repeats. The dark bar represents the left repeat type and the shaded bar is the right repeat type. X stands for an *Xba*I site (TCTAGA), while X′ stands for TCTACA. Two-headed arrows represent points of crossover.

such "odd-numbered" unequal crossover can be inferred as well. LP22-1 as shown in Figure 8B is another example that may have resulted from an exchange between a dimer and the second repeat of a trimer like LP16-2. (Note the extensive sharing of unique sites between LP22-1 and LP16-2 from positions 80 to 120.) A detailed analysis of the *Rsp* sequences from three chromosomes will be published elsewhere.

It has often been observed that tandem repeats within the same species resemble one another more than they resemble repeats from another species. The spread of repeats within the same genome by unequal crossover keeps repeats of the same species from diverging in a process known as concerted evolution (Brown et al., 1972). An interesting question is "Do repeats on a given chromosome also evolve in unison?" In other words, does concerted evolution occur below the species level (on a given chromosome). This is plausible if the rate of unequal sister chromatid exchange is high and if the repeats are in a region where recombination between homologous chromosomes is rare. We notice in Figure 9 that identical repeats (either the left or the right repeat of a dimer) have been found only on the same chromosome, and many rare variants are found more than once on the same chromosome but are not observed on others. The observations suggest that many unequal exchanges are between sister chromatids, consistent with earlier genetic studies (Schalet, 1969). Thus a new mutation on a repeat would initially spread on the same chromosome where it emerges. Nevertheless, occasional homologous recombination appears sufficient to prevent repeats on different chromosomes from diverging, as evidenced by the sharing of all common variants by different chromosomes.

Finally, the intraspecific variability of satellite DNA arrays, such as the *Rsp* locus, may be useful in population studies. The variation shown in Figure 7 resembles that of DNA fingerprinting (Jeffreys et al., 1985). Although DNA fingerprinting has now been widely used in genetic identification of individuals, the random assortment of bands representing unlinked loci limits its application in population studies, because the restriction patterns of two related individuals are often as different as two unrelated ones (Lynch, 1988). Since *Rsp* repeats are located entirely within the centromeric heterochromatin of 2R, with little recombination among them, all the bands in Figure 7 are tightly linked and segregate as a block. It can also be seen that some chromosomes share a large number of bands, presumably reflecting a close relationship. Such a high level of polymorphism may be useful for studying fine-scale population differentiation and migration. There is some evidence that localized satellite DNA arrays are present in species ranging from insects to man (Miklos, 1985). The identification of such an array on the Y chromosome may offer intriguing possibilities for tracking of the migration of males, provided the array is as variable as *Rsp*.

Molecular mechanisms of segregation distortion

Because the molecular structure of the key components of SD has been determined only recently, the molecular mechanism underlying SD is still unknown. The cloning of *Rsp* helps address several obvious explanations for SD. First, it has been hypothesized many times that as transcriptional activities are normally turned off during spermiogenesis, Rsp^s may remain turned on by Sd, and that such aberrant activities result in spermiogenic failure. However, we have detected no transcription from Rsp^s under either distorting or nondistorting conditions in any stage of development of the whole organism (Doshi and Wu, unpublished results). It is still possible that the aberrant transcription is restricted in the germ cells and escaped detection in our analysis. In situ hybridization of testis tissues may resolve the issue. Second, we can also rule out the possibility

```
              1                          31                        60
LEFT
Cb5-1     ---------- ---------- ---------- ---------- ---------- ---.----a-
Cb6-1     ---------- ---------- ---------- ---------- ---------- ---.----a-
Cb10-1    ---------- ---------- ---------- ---------- ---------- ---.----a-
Cb1       ---------- ---------- ---------- ---------- ---------- ---.----a-
Cb2       ---------- ---------- ---------- ---------- ---------- ---.------
Cb8       ---------- ---------- ---------- ---------- ---------- ---.------
Lp24      ---------- ---------- ---------- ..------.- ---------- ---.------
Lp22-1    ---------- a-c--a-c-- ----ca---- ----a----- --a------- a-g-------
Lp13-1    ---------- ---------- ---------- ---------- ---------- ---.----a-
Lp17-1    ---------- ---------- ---------- ---------- ---------- ---.----a-
Lp23-1    ---------- ---------- ---------- ---------- ---------- ---------a-
Lp14-1    ---------- ---------- ---------- ---------- ---------- -a-.------
Lp26-1    ---------- ---------- ---------- ---------- -----g---- ---.------
Lp16-1    ---------- ---------- ---------- ---------- ---------- ---.------

RIGHT
Cb5-2     ----c----- a-t-aa-c-- ----ca---- ----a----- --a------- a-g-------
Cb6-2     ----c----- a-t-aa-c-- ----ca---- --g-a----- --a------- a-g-------
Cb10-2    ----c----- a-t-aa-c-- ----c-ac-- --g-a----- --a------- a-g-------
Lp22-2    ----c--t-- a-t-aa-c-- ----ca---- ----a----- --a------- a-g-------
Lp13-2    ----c----- a-t-aa-c-- ----ca---- ----a----- --a------- a-g-------
Lp17-2    ----c----- a-t-aa-c-- ----ca---- ----a----- --a------- a-g-------
Lp23-2    ----a----- a-t-aa-c-- ----ca---- ----a----- --a----... ..........
Lp14-2*   ----c----- a-t-aa-c-- ----ca---- ----a----- --a------- a-g-------
Lp26-2*   ----c----- a-t-aa-c-- ----ca---- ----a----- -ta------- a-g-------
Lp16-2*   ----c----- a-t-aa-c-- -t--ca---- ----a----- --a-g---c a-g-------

Consen    TCTAGAGATT .CTGTTCAACT GGTAAGCAAA AACAGTAAAT TGCCTAAGTT TTATCATTTT


              61                         91                       120
LEFT
Cb5-1     ---------- ---t------ ---------- ---------- ---------- ----------
Cb6-1     ---------- ---t------ ---------- ---------- ---------- ----------
Cb10-1    ---------- ---t------ ---------- ---------- ---------- ----------
Cb1       ---------- ---------- ---------- ---a------ ---------- ----------
Cb2       ---------- ---------- ---t------ ---------- g-g------- --c-------
Cb8       ---------- ---------- ---t------ ---------- g-g------- --c-------
Lp24      ---------- ---------- ---t------ ---------- g-g------- --c-------
Lp22-1    ---------- --------ta ---a------ ------at-- g-g---tc-- ----------
Lp13-1    --cg------ ---ta----- ---------- ---------- ---------- ----------
Lp17-1    ----c----- ---tat---- ---------- -------t-- ---c------ ----------
Lp23-1    ---------- ---------- ---------- ---------- ---------- ----------
Lp14-1    ---------- ---------- ---------- ---------- ---------- --c-------
Lp26-1    ---------- ---------- ----t----- -------.-- ---------- ----------
Lp16-1    ---------- ---t------ ---------- ---------- ---c------ ----------

RIGHT
Cb5-2     ---------- -.-------- ---t------ --------g- g-g---g--- --c-------
Cb6-2     ---------- ---------a ---t------ ---------- g-gtt----- --c-------
Cb10-2    ---------- ---------a ---------- ---------- g-g------- --c-------
Lp22-2    ---------- ---------- ---t------ -------.-- g-g------- --c-------
Lp13-2    ---------- ---------- ---.------ --.------- g-g------. --c--.----
Lp17-2    ---------- ---------- ---t------ -------.-- g-g------- --c-------
Lp23-2    .......... .......--- ---t------ -------.-- g-g------- --c--c---
Lp14-2*   ---------- ---------- ---.------ ---------- ---------- ----------
Lp26-2*   ---------- ---------- .......... -------.-- ---------- ----------
Lp16-2*   ---t------ ---------a ---------- ------at-- g-g---tc-- ----------

Consen    AAGCGGTCAA AATGGGTGAT TTTCCGATTT CAAGTACCAG ACAAACAGAA GATACCTTCT
```

◀ **FIGURE 9.** Sequences of repeats from two different chromosomes (Cb and Lp). Repeats are of three kinds: monomers (Cb1, Cb2, Cb8, and Lp24), dimers, and trimers. The suffix, -1, indicates a left repeat of a dimer or a trimer, -2 indicates the right repeat of a dimer, and -2* indicates the second repeat of a trimer.

that the failure of Rsp^s-bearing sperm is due to nuclease activities that specifically cleave Rsp repeats. Genomic DNA isolated from the testes of distorting males does not appear to be cleaved at the Rsp locus. Nor is there any difference in the level of methylation of the Rsp repeats from testes of distorting and nondistorting males. This can be shown by restriction digestion with enzymes that recognize the same sequence but have different sensitivity to methylated bases.

Genetic data indicate that Sd interferes with the maturation of sperm bearing Rsp^s (Ganetzky, 1977) but do not tell us if such an interference is direct or indirect at the molecular level. Based entirely on genetic observations, many models have been proposed with regard to the possible molecular mechanisms (Hiraizumi et al., 1980; Brittnacher and Ganetzky, 1983; Lyttle et al., 1986). Some of these models postulate direct interaction between Sd and Rsp, while others invoke a product from a third locus mediating the interaction. As there are many enhancers and suppressors of SD drive, it is conceivable that any one of them might indeed be mediating the Sd–Rsp interaction. Assays of the ability of the Sd products to form complexes may provide a much needed clue to whether and how Sd interacts with Rsp. The consequence of the Sd–Rsp interaction, whether direct or indirect, is postulated to be a delay (as opposed to disruption) in the maturation of Rsp^s-bearing sperm (Nur and Wu, 1991).

An issue rarely addressed in the SD literature, and one that may be crucial to an understanding of the molecular mechanism of SD, is the true function of the Rsp locus. It seems inevitable to conclude that Rsp must have a normal function unrelated to SD, as explained previously. Population genetic experiments indeed show a fitness effect associated with the Rsp locus (Wu et al., 1989). (The identification of the Rsp satellite repeats poses another interesting question: Is this putative normal function associated with other satellite DNAs as well?) Since there are many nuclear proteins that bind specifically to heterochromatic DNAs (Levinger and Varshavsky, 1982; James and Elgin, 1986), a search for Rsp-binding proteins may shed some light on the true functions of the Rsp locus. Evidence from protein-DNA binding assays shows that proteins in the nuclear extract form complexes with the Rsp DNA (Doshi et al., 1991). These proteins are present in the extract from both the SD and SD^+ strains. If the products of Sd do not interact with the Rsp repeats directly, cloning and characterization of genes coding for these proteins may be the key to understanding the mechanism of SD.

The hypothesis that the Rsp repeat is a site for protein binding is in part based on its molecular structure. The Rsp repeat has a strong sequence-directed curvature, or DNA bending, as determined by its migration on a polyacrylamide gel. Many bent DNA regions have been shown to be complexed with proteins

(Zahn and Blattner, 1985; Hsieh and Griffith, 1988), and it has also been shown that the curvature of mouse satellite DNA is important in chromatin condensation (Radic et al., 1987). Interestingly, the failure of Rsp^s-bearing sperm during spermiogenesis is due to improper condensation of chromatin. We have found that the curvature on an Rsp repeat can be influenced by Hoechst stain, which has been shown to bind to Rsp cytologically (Pimpinelli and Dimitri, 1989). Such an effect may be useful for manipulating the property of Rsp and help in elucidating the mechanism of SD.

THE t COMPLEX IN MICE

The t-complex refers to the proximal third of the mouse chromosome 17, of which there are two forms in natural populations of house mice—the wild-type and the t-haplotype (for recent reviews see Silver, 1985; Klein, 1986; Lyon, 1989). t-Haplotypes are analogous to SD and SR and, hence, share a number of features with these meiotic drive systems. Heterozygous males carrying a t-haplotype and a wild-type chromosome transmit the t-haplotype to greater than 95% of their progeny (see Figure 1). Recently, Brown et al. (1989) showed that a high proportion of sperm from heterozygous $+/t$ male mice (presumably those bearing the wild-type chromosome) undergo a premature acrosome reaction that prevents such sperm from binding to the zona pellucida and fertilizing the egg.

Maintenance in natural populations

t-Haplotypes have been identified in several species of house mice. They are found in most populations at frequencies between 10 and 20% (Petras, 1967; Klein, 1986). t-Haplotypes are recessive sterile and are usually associated with recessive lethals as well. Lethality and sterility prevent the fixation of t-haplotypes.

The association of t-haplotypes with recessive lethals has two explanations. One is based on interdeme selection: because male-sterile t-haplotypes can be fixed, causing the extinction of the entire deme, natural selection may favor those t-haplotypes that have acquired recessive lethal mutations that cannot be fixed (Lewontin, 1962; Silver, 1985). A more plausible explanation is based on kin selection. When a t-bearing female is to produce a sterile t/t son, it might be to the advantage of the $t/+$ sibs if the t/t genotype is embryonic lethal. Such lethality may leave the litter less crowded during and after gestation, hence aiding the survival of the $t/+$ sibs (B. Charlesworth, personal communication).

DNA and chromosome inversion polymorphism

A phylogenetic analysis of four-cutter restriction sites in the α-globin pseudogene ($Hb\alpha$-4ψ) located within the t-complex has been completed (Hammer and

Silver, unpublished results). Nineteen t-haplotypes collected from the ranges of *Mus domesticus* and *M. musculus* were analyzed along with several wild-type forms of chromosome 17 from these species, as well as the outgroups *M. spretus* and *M. cervicolor* (Figure 10). Wild-type forms of *M. domesticus* and *M. musculus* differ by an average of four sites. An average of 12 sites distinguished t-haplotypes from wild-type forms of *M. domesticus* and *M. musculus*, whereas only a single site differed between t-haplotypes from the two species. A parsimony analysis supports the view that t-haplotypes diverged from the lineage leading to the common ancestor of *M. domesticus* and *M. musculus* before the divergence of these two species 1–2 million years ago (Ferris et al., 1983; Bonhomme et al., 1984).

The low level of variation among t-haplotypes collected from the ranges of *M. domesticus* and *M. musculus* gives support to the hypothesis that meiotic drive has facilitated the introgression of t-haplotypes from one species to another. These two kinds of mice meet in central and southern Europe and form a narrow band of hybrid populations. Although mitochondrial DNA has been shown to traverse the hybrid zone, there is little or no flow of nuclear genes, perhaps due to the disruption of coadapted gene complexes (Sage et al., 1986; Vanlerberghe et al., 1986). The genes carried by t-haplotypes may be especially suited to flow between species die to their transmission ratio advantage, as well as their stability as gene complexes. The pattern of variation among all known t-haplotypes contrasts with that for *SD*. The very low levels of differentiation imply either that t-haplotypes have propagated around the world from a common

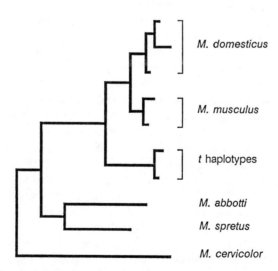

FIGURE 10. A phylogenetic tree of wild-type haplotypes form four species and t-haplotypes from two species of mice. The tree represents the most parsimonious tree based on restriction site data of the *Hbα-4ψ* gene.

ancestor recently, or that there has been a rapid turnover of t-haplotypes in nature.

Finally, the t-complex is associated with many inversions that suppress recombination among loci within the complex. Hammer et al. (1989) performed a phylogenetic survey of the structure of the proximal region of chromosome 17. They concluded that the proximal inversion occurred on the wild-type lineage before the divergence of $M.$ domesticus and $M.$ abbotti. Subsequently additional inversions accumulated on the chromosomal lineage leading to t-haplotypes before the divergence of $M.$ domesticus and $M.$ musculus. This is similar to the SR system in $D.$ persimilis (Wu and Beckenback, 1983), where an inversion associated with meiotic drive arose on the wild-type lineage and was favored as a suppressor of recombination.

SUMMARY

This chapter reviews the evolutionary and molecular genetics of ultraselfish genes, genes that gain an advantage at the expense of others within the same genome. Much of our theory of evolutionary processes depends critically on the assumption that these are rare events. For instance, it was pointed out by Hamilton (1967) that sex-linked meiotic drive would violate Fisher's principle of equal sex ratios. Charlesworth and Hartl (1978) cited a number of studies to show that general principles of population genetics depend decisively on the supposition of Mendelian segregation. With meiotic drive, average fitness is not maximized and the fundamental theorem of natural selection does not hold. The three cases of meiotic drive reviewed above, representing one sex-linked and two autosomal systems, in *Drosophila* and in mice, are interesting precisely because they are unusual at such a fundamental level.

Meiotic drive is indeed an unusual trait but such systems provide a convenient window for studying many other phenomena. We have discussed natural selection of various forms that act to prevent the fixation of all these drive chromosomes. Because of the large effect of meiotic drive, the counterbalancing selection is correspondingly strong and, hence, is within the resolution of our measurement. We have also discussed the various ways in which genomes respond to the presence of ultraselfish elements. Numerous suppressors, enhancers, and even lethals have evolved and delicate linkage relationships among them are preserved. It is as though natural selection has acted to form alliances among elements within the genome to resolve the intragenomic conflict. The studies of meiotic drive also reveal modes of chromosome migration throughout the species range and across the species boundary.

The mechanisms of meiotic drive are likely to be diverse and the molecular nature of meiotic drive is a promising molecular genetic topic. As discussed above, studies of the mechanism of SD may shed light on the functions of heterochromatin, an enigma in the organization of genomes of higher eukaryotes, and may provide much needed clues to such key questions as gene inactivation by heterochromatin, or position effect variegation (Spofford, 1976, Reuter et al., 1990).

Many of the classical questions in evolutionary biology have been difficult to address in molecular terms. These are mostly systems of complex genetic interactions. Recent research on *Segregation Distorter* and the *t*-complex has demonstrated that the current technology, while leaving many questions unanswered, is capable of providing significant new insight. We also believe that many other problems such as the evolution of quantitative traits and the molecular basis of reproductive isolation between closely related species are also ripe for molecular analysis.

CHAPTER 10

DETECTING SELECTION AT

THE LEVEL OF DNA

Martin Kreitman

In the aftermath of a seemingly endless and inconclusive debate between neutralists and selectionists over the evolutionary forces governing protein polymorphism, the possibility that this debate is now descending to the level of DNA must be particularly dismaying to evolutionists. After all, if natural selection and genetic drift cannot be distinguished on the basis of protein allele frequency data (Ewens, 1977), what do the current generation of DNA "gel jocks" hope to accomplish with nucleotide data? Will the problem of distinguishing among the forces governing variation simply be reiterated by the data —albeit at the highest level of genetic resolution? Or does nucleotide data possess features that might allow new approaches to distinguish evolutionary forces. If nucleotide data yield only greater accuracy of estimation, without the possibility for new kinds of comparisons and tests, it is unlikely that the old issues in population genetics will be resolved by the new data.

The evolutionary causes of genetic variation clearly involve both neutrality and selection. Even a superficial comparison of third positions in codons relative to first and second positions indicates that both genetic drift and selective constraints are operating (Kimura, 1983; although see Gillespie, 1986a,b). The purpose here is to emphasize how little we know about the nature and importance of natural selection in shaping nucleotide variation. Without this knowledge, we will not be able to find satisfying answers to broader molecular evolutionary issues.

This chapter has three purposes: (1) To examine our present understanding of nucleotide polymorphism (with a focus on *Drosophila*); (2) to describe a framework, based on the theory of gene genealogies and the method of historical inference, for predicting how variation is patterned, given natural selection and genetic drift; and finally, (3) to show, from an analysis of patterns of variation in a region of DNA encompassing the *alcohol dehydrogenase* locus (*Adh*), how data and theory can be combined to infer the evolutionary history of the regions.

THE PROBLEM OF QUANTIFYING POLYMORPHISM

How much polymorphism is there in a natural population at the level of DNA, and how is this variation distributed within and between genes? To answer this question, another question must first be asked: how is nucleotide polymorphism estimated and how reliable are the estimates? Ignoring for the moment the issue of what part of the genome is being studied—whether the estimate of polymorphism comes from the coding regions of a gene, introns, "silent" positions in codons, noncoding regions, etc.—estimators of polymorphism fall into two basic categories, one based on nucleotide heterozygosity and one based on the number of segregating sites in a sample (see Nei, 1987, pp. 255–265, for a more complete explication). For the purpose of the discussions of polymorphism levels that follow, I will briefly review three of these estimators.

Nucleotide diversity, π, is defined as the probability of a nucleotide site being different in two randomly chosen genes. An unbiased estimator of nucleotide diversity is

$$\hat{\pi} = n/(n-1) \sum \hat{x}_i \hat{x}_j \pi_{ij} \tag{1}$$

where n is the sample size, x_i is the sample frequency of the ith type of DNA sequence, and π_{ij} is the proportion of different nucleotides between the ith and jth types of DNA sequences. In a randomly mating population this is an estimate of the average heterozygosity per nucleotide site. This estimator combines information about both the number of segregating sites and allele frequencies in the population.

The average number of pairwise nucleotide differences between DNA sequences, k, is a related estimator of polymorphism, and it is also based on the number of segregating sites (s_n) in a sample and their site frequencies (Tajima, 1983):

$$\hat{k} = \sum_{i=1}^{s_n} h_i m \tag{2}$$

where h_i is the unbiased estimate of nucleotide diversity of the ith segregating site and m is the number of sites compared. Under the neutral mutation model, the expected value of k is equal to the neutral parameter,

$$E(\bar{k}) = \theta m \tag{3}$$

where $\theta = 4N\mu$ (N is the evolutionary effective population size and μ is the mutation rate per nucleotide per generation). Note that this definition of θ pertains to a single nucleotide site.

The simplest (and possibly the most useful) estimator of polymorphism is the number of segregating sites, s_n, in a sample of size n. s_n depends on sample size (the number of segregating sites increases with sample size) and is therefore not useful for comparing levels of polymorphism among samples of different size. However, under an infinite-site (or allele) selectively neutral mutation model, the expected number of segregating sites is, like the average pairwise difference, related to the neutral parameter θ,

$$E(s_n) = \theta m \ [1 + 1/2 + 1/3 + \cdots + 1/(n - 1)] \tag{4}$$

where n is the number of sequences compared. An unbiased estimator of θ is therefore

$$\hat{u} = s_n \{m[1 + 1/2 + 1/3 + \cdots + 1/(n - 1)]\}^{-1} \tag{5}$$

$\hat{u}$ is a useful estimator of nucleotide polymorphism and it is conveniently estimated from the number of segregating sites in a sample.

The parameter θ, like π, is also related to expected heterozygosity, H, in a straightforward way, and at steady state, the infinite alleles model predicts that

$$H = \theta/(1 + \theta) \tag{6}$$

For small values of θ, which generally hold for per-nucleotide estimates, $H \approx \theta$. Therefore, θ can be considered a rough measure of the probability of a nucleotide site being different in two randomly chosen alleles.

Taking advantage of the ability to estimate θ from s_n and from $\hat{k}$, Tajima (1989) recently developed a test of the neutral theory by comparing the two estimates of θ. An application of this test for the Adh region of $Drosophila$ $melanogaster$ is given later in the chapter.

The variances of the polymorphism estimators, θ, π, and k, are large and provide a sobering reminder of the large stochastic variance associated with selectively neutral evolutionary processes. For nucleotide diversity (π), the variance is

$$V(\pi) = \frac{n + 1}{3(n + 1)m_t} \theta + \frac{2(n^2 + n + 3)}{9n(n - 1)} \theta^2 \tag{7}$$

where m_t is the number of nucleotides in the region being compared. $V(\theta)$ is bounded by the values that apply to free recombination between nucleotide sites and to complete linkage,

$$\theta/S_{n-1} \text{ and } \theta/S_{n-1} + \theta^2[1 + 1/4 + \cdots + (n - 1)^{-2}]/S^2_{n-1} \tag{8}$$

respectively, where $S_{n-1} = 1/1 + 1/2 + \cdots + 1/(n - 1)$. In general, $V(\pi) > V(\theta)$, and for this reason θ is preferred in some instances over π as an estimator of polymorphism.

To illustrate just how large these variances can be, consider the data for 11 Adh alleles in $D.$ $melanogaster$ (see Table 2). Forty-three polymorphic sites were observed in a segment of 2379 base pairs. Forty-two of the 43 polymorphisms occur in the noncoding regions, introns, and synonymous positions within exons. Under the assumption that the polymorphisms are selectively neutral, the estimators of polymorphism and their standard errors (from Nei, 1987) are

$$\pi = 0.0065 \pm 0.0026$$
$$\theta = 0.0062 \pm 0.0028 \text{ (no recombination)}$$
$$\pm 0.0009 \text{ (free recombination)}$$

Because per-nucleotide recombination rates are generally not known (see Hudson and Kaplan, 1988), and because segregating sites may be in linkage disequilibrium (although see Riley et al., 1989), the no recombination model provides a conservative estimate of the variance. Ninety-five percent confidence limits for θ can also be calculated and for the *Adh* data the lower and upper bounds for θ are $\theta_L = 0.003$ and $\theta_U = 0.016$, respectively (see Kreitman and Hudson, 1991, Equation 3). It is clear that only very large differences in levels of polymorphism will prove to be statistically significant. For polymorphism estimates based on the analysis of restriction fragment length polymorphisms (RFLPs), where the effective number of nucleotide sites compared is substantially smaller than for DNA sequence studies, and where there is an additional source of variance —the number of substitutions is imprecisely estimated by RFLPs—the variances are expected to be even larger (Nei and Jin, 1989).

The first question, "how much polymorphism is there in DNA," can now be addressed, but with the important caveat that the expected variances of these estimates are very large and must be considered in interpreting observed differences among regions, loci, or species.

ESTIMATES OF POLYMORPHISM WITHIN SPECIES

DNA sequencing is now the preferred method for estimating levels of polymorphism for two reasons: (1) The recent development of DNA amplification by polymerase chain reaction (Saiki et al., 1988) and direct genomic sequencing methods (Gyllensten et al., 1988; Kreitman and Landweber, 1989; Higuchi and Ochman, 1989) reduces the labor involved in sequencing. (2) RFLP-based estimates of polymorphism are imprecise (as I show below). But at present almost everything we know about the levels and patterns of nucleotide polymorphism in natural populations comes from RFLP studies. In anticipation of the impending shift to DNA sequencing for estimating levels of polymorphism, I will review what we have learned so far from RFLP studies and will highlight what we may additionally expect from sequence studies. The discussion is restricted to *Drosophila*.

Estimates of polymorphism levels in *Drosophila* from RFLP analyses are presented in Table 1 for nine chromosomal regions (hereafter referred to as "loci") and three species, *D. melanogaster*, *D. simulans*, and *D. pseudoobscura*. The first (and most important) question to ask is whether there is any evidence for heterogeneity in levels of polymorphism among genes within species. Het-

TABLE 1. Summary of RFLP polymorphism in *Drosophila*

Species	Locus	Type of study	Sample size	Length of region (kb)	Number of restriction sites scored[a]	Number of segregating sites	û[b]	Reference
melanogaster	*Adh*	6-cutter	18	12	23	4	0.006	Langley et al. (1982)
	Adh	6-cutter	48	13	30	8	0.006	Aquadro et al. (1986)
	Adh	4-cutter	87	2.7	526 bp[c]	17	0.006	Kreitman and Aguadé (1986)
	87A heat shock	6-cutter	29	25	25	2	0.002	Leigh-Brown (1983)
	Amy	6-cutter	85	15	26	3	0.006	Langley et al. (1988a)
	G-6PD	6-cutter	122	13	28	5	0.003	Eanes et al. (1989)
	Rosy	6-cutter	60	40	41	7	0.003	Aguadro et al. (1988)
	Yellow-achaete-scute	6-cutter	64	106	176	9	0.001	Aguadé et al. (1989b)
		6-cutter	109	30.5	63	9	0.003	Eanes et al. (1989)
		6-cutter	49	120	67	10	0.003	Beech and Leigh Brown (1989)
	Zeste-tko	6-cutter	64	20	42	10	0.004	Aguadé et al. (1989a)
	White	6- & 4-cutter	64	45	327	54	0.013	Miyashita and Langley (1988)
	Notch	6-cutter	37	60	58	15	0.005	Schaeffer et al. (1988)
simulans	*Rosy*	6-cutter	30	40	56	28	0.019	Aquadro et al. (1988)
pseudoobscura	*Adh*	6-cutter	20	32	43	27	0.021	Schaeffer et al. (1987)
	Rosy	4-cutter	58	5.2	147	66	0.092	Riley et al. (1989)

[a]The number of nucleotide sites under survey is approximately twice the number of restriction sites times the number of base pairs in the recognition sequence.
[b]Estimates based on methods of Nei and Tajima (1981), Engels (1981), or Hudson (1982). The methods generally give similar values.
[c]This is the number of nucleotide sites under survey.

erogeneity across loci might suggest variation in mutation rates, variation in levels of selective constraint (some regions may contain a greater proportion of nucleotide sites experiencing purifying selection), or varying intensity or kinds of natural selection acting to maintain polymorphism.

Several factors militate against finding significant heterogeneity. As expected, the variances of the estimated θs for the nine loci listed in Table 1, as calculated by Equation (8), are large. The lack of statistical power is compounded by the relatively small number of nucleotide sites being monitored for polymorphism by RFLP analysis (the "effective" number of scrutinized nucleotide sites is approximately two times the number of restriction sites observed times the number of base pairs required for correct recognition by the restriction enzyme). Averaging polymorphism estimates over such large regions (for example, 100 kb for the *yellow-achaete-scute* region) also makes it unlikely that heterogeneities will be revealed.

Given these problems, it should not be surprising if no heterogeneity in polymorphism levels could be detected among the nine regions [*Adh* (Aquadro et al., 1986), 87A heat shock, *Amy*, *rosy*, *yellow-achaete-scute* (Aguadé et al., 1989b), *Zeste-tko*, *white*, *G6pd* and *Notch*]. Three studies of *yellow-achaete-scute* region are presented in Table 1; I used the Aguadé et al. (1989b) study because they report a low level of polymorphism, which will favor obtaining a significant result. The test for heterogeneity of θs I use here, based on a suggestion of Hudson (described in Aguadé et al., 1989a), asks whether the different θ estimates are compatible with one underlying value, which is taken as the unweighted average of the estimated θs. This value is taken to be $\theta = 0.005$ for the nine regions presented in Table 1. If the true θ equals 0.005, the expected number of segregating sites and their variances can be calculated for the given sample sizes and number of nucleotide sites scrutinized (calculated as above) under an infinite site model of mutation with no selection and no recombination (Watterson, 1975). The test proposed by Hudson is

$$X^2 = \Sigma_{1,L}(S_o - S_e)^2/\text{Var}(S_e) \tag{9}$$

where S_o and S_e are the observed and expected number of segregating sites and L is the number of regions (or loci). The test statistic has an approximate χ^2 distribution with $L - 1$ degrees of freedom.

Applying this test to the nine regions, $\chi^2 = 11.7$, $p > 0.1$. Therefore, there is no evidence for heterogeneity in the amount of polymorphism, even though there is a 13-fold difference between the θ estimate for *white* ($\theta = 0.013$) and for *yellow-achaete-scute* ($\theta = 0.001$). This result differs from the conclusion of Aguadé et al. (1989b) that the estimated θ for the *yellow- achaete-scute* region is significantly lower than the assumed parametric value, $\theta = 0.005$ (although see Eanes et al., 1989, who studied the same region and reach the opposite conclusion). The reason for the discrepancy may have to do with the surprising uniformity of θ estimates for the other eight regions. In my view, the problem is not to explain the heterogeneity of θs but rather the uniformity: with the exception of the *yellow-achaete-scute* estimate of Aguadé et al., all the published estimates of polymorphism levels are remarkably similar. One possible explanation—and potentially a strong criticism of this approach—is that the examination of a low density of sites across large regions, characteristic of RFLP studies, obscures heterogeneity in small regions.

One indication that this might be happening is revealed by my study of the *Adh* locus based on a DNA sequence analysis of 11 alleles of *Adh* in *D. melanogaster* (Kreitman, 1983, data given in Table 2). The analysis indicates highly significant heterogeneity in θs, calculated for intron 1, introns 2 and 3, *Adh* coding silent sites, and nontranslated regions [$E(\theta) = 0.006$, $\chi^2 = 24.0$, $p \ll 0.001$] (see also Kreitman and Hudson, 1991). This heterogeneity in the level of polymorphism across regions within the 1858 bp of the *Adh* locus is obscured by RFLP analysis of a larger region encompassing the locus (Aquadro et al., 1986; Table 1).

ESTIMATES OF HETEROGENEITY BETWEEN SPECIES

In contrast to the paucity of evidence for heterogeneity of nucleotide polymorphism levels within species there appears to be substantial differences in the amount of polymorphism across species. A comparison of three *Drosophila* species, shown in Table 1, indicates substantially higher levels of polymorphism in *D. simulans* and *D. pseudoobscura* than in *D. melanogaster*. This observation is also supported, at least for *D. simulans*, from DNA sequence comparisons of a region encompassing *Adh* (Coyne and Kreitman, 1987; Kreitman, unpublished) and from RFLP studies of *per*, *Adh*, and the *Amy* loci (Aquadro, 1989, Table 3).

It is interesting to compare the geographic distributions of the three *Drosophila* species with their levels of polymorphism. Even though *D. melanogaster*

TABLE 2. Summary of silent polymorphism at the *Adh* locus of *D. melanogaster* based on DNA sequences of 11 alleles

Region	Length of region (bp)	Number of silent sites[a]	Number of segregating sites	û
Adh Intron 1	620	616	11	0.006
Adh Intron 2 + 3	135	127	7	0.018
Adh coding silent only	765	192	13	0.231
Adh nontranslated	335	335	3	0.003

Source: From Kreitman (1983).
[a]Four base pairs were subtracted from introns, corresponding to the absolutely conserved junctions—GT and AG. The number of silent sites in coding regions is the "effective" number of silent sites.

and D. *simulans* are sibling species and both have worldwide geographic distributions (they are both human commensals), they appear to have a three- to sixfold average difference in per-nucleotide heterozygosity. Assuming evolutionary steady state, selective neutrality, and equal mutation rates, this also implies a three- to sixfold difference in effective population size. The two species are thought to have originated in Africa (Lemeunier et al., 1986) 2–4 million years ago (Bodmer and Ashburner, 1984; Stephens and Nei, 1985) and expansion out of Africa is postulated to have occurred only recently, possibly in historical time (Lemeunier et al., 1986). Such recent origin suggests that the sizes of the ancestral populations from which the two species expanded must also have been different. This historical difference, if recent enough, would still be reflected in extant levels of polymorphism because a long time is expected between changes in a species abundance and its attainment of a new mutation-genetic drift equilibrium for polymorphism.

The very high level of nucleotide polymorphism in D. *pseudoobscura* supports the idea that the present polymorphism levels in D. *melanogaster* (and possibly D. *simulans*) are not in mutation–selection equilibrium. This species has a level of polymorphism even greater than D. *simulans* but has a geographic range that includes only the western United States, the Pacific Northwest, and part of Mexico. If both D. *melanogaster* and D. *simulans* have expanded their ranges relatively recently, they probably have not yet reached mutational equilibrium for their current large population sizes. But in the absence of direct estimates of effective population size for any of the species, we cannot determine whether any of the species is at mutation–genetic drift equilibrium.

To summarize, RFLP analysis has provided some estimates of polymorphism at the nucleotide level for a number of regions of the *Drosophila* genome, especially for D. *melanogaster*. However, RFLP estimates are based on regions of different lengths; the studies differ in the intensity of surveying for polymorphism, and they cover regions of mostly unknown function. This approach provides almost no information as to whether regions vary in level of polymorphism or on the cause of real differences. To understand the causes of variation, a better theoretical framework for analyzing sequence data is required. Such a framework is being developed, and it is based on the concept of gene genealogy and the method of historical inference.

UNDERSTANDING THE CAUSES OF VARIATION

Although only a few studies of nucleotide polymorphism in natural populations have been published, the data have nevertheless been sufficient to motivate the development of methods for analyzing patterns of polymorphism and distinguishing the effects of natural selection and genetic drift. Some of the recent methods rest on the theory of gene genealogies and the mathematics of coalescent processes. The gene genealogy approach was initially used to study models of selectively neutral mutations .(Kingman, 1980, 1982a,b; Tajima,

1983; Hudson, 1983; Tavaré, 1984; also see Ewens, 1989, and Hudson, 1990, for reviews). More sophisticated models of balancing selection (Kaplan et al., 1988; Hudson et al., 1988), and directional selection (Kaplan et al., 1989), both of which incorporate recombination, population bottlenecks (Tajima, 1989) and certain types of population subdivision and migration (Takahata, 1988; Slatkin, 1989; Slatkin and Maddison, 1989) have been developed.

Coalescent models are based on a property common to all genetic systems: all alleles in a population, no matter how divergent one from another, must derive from a common ancestor. In other words, for any sample of alleles a genealogy must exist with a single node representing the most recent common ancestral allele. Coalescent models are useful in formulating the distribution of branching times back to common ancestors for samples of alleles that have evolved under a given set of evolutionary forces. Under a model in which mutations are Poisson distributed along the branches of the genealogy, all of the moments of the distribution of the number of segregating sites are determined from the distribution of branching times. In this sense time and polymorphism are interchangeable. A model in which only genetic drift is operating can be thought of as a null hypothesis: molecular evolution in the absence of natural selection. Alternative models that incorporate specific kinds of selection can be constructed and the distribution of branch lengths derived. In principle, selection can be distinguished from genetic drift and different kinds of selection resolved if the differences in the distributions of branch lengths or segregating sites mediated by the forces are statistically detectable.

How can the coalescence time for a sample of alleles be estimated? To answer this question, a brief digression into more elementary population genetic theory is necessary. Under models of molecular evolution without selection, the expected number of segregating sites, s_n, in a sample of n genes is a function of both the evolutionary effective population size, N_e, and the mutation rate, μ, as given in Equation (4). For two randomly chosen genes the relationship is simple, $E(s_n \mid n = 2) = \theta$. As described in the previous section, estimates of θ for sample sizes greater than two can be made from polymorphism data in a straightforward way.

It is also possible to relate the number of segregating sites in a sample of n alleles to the total time, T, back to the common ancestor of the sample. For two alleles, if the neutral mutation rate per gene per generation is μ, then the expected number of mutations accumulating per generation is 2μ. For a fixed mutation rate, the expected number of segregating sites depends only on how many generations separate the two alleles from their common ancestor,

$$E(s_n) = 2\mu E(t) \tag{10}$$

where $E(t)$ is the expected time, in number of generations, to the common ancestor of the two alleles. In the case of two alleles, the coalescence time is simply $E(s_n)/2\mu$, or

$$E(t) = 2N_e \tag{11}$$

How can estimates of polymorphism be used to test the neutral model? The simplest way is to compare estimates of θ for two regions of DNA. We may wish to know, for example, whether introns and silent sites of exons exhibit different levels of variation. If both are evolving neutrally with the same mutation rate, then the expected number of segregating sites, adjusted for the lengths of the regions, should also be the same. The neutral hypothesis is rejected if the estimated θs for the two regions are sufficiently different.

If the neutral hypothesis is rejected, does this mean that either the introns and/or sections of the exons are not evolving neutrally? Unfortunately not, because although the absolute mutation rate may be the same among regions, the mutation rate to neutral alleles may vary. Evidence from sequence comparisons clearly indicates that most mutations in coding regions of genes are not selectively neutral but instead must be deleterious. For proteins such as histone H4, where only two amino acid substitutions have accumulated since the divergence of wheat and mouse (Tabata et al., 1983), essentially every mutation leading to amino acid substitutions must have been sufficiently deleterious for natural selection to have prevented their fixation. The mutation rate to effectively neutral alleles must be essentially zero for the amino acid replacement sites of histone H4. Therefore, regions having different proportions of neutral mutable sites will exhibit differences in θ simply by virtue of θ being defined as an average across all sites.

The same is true at the level of polymorphism. Natural selection prevents deleterious mutations from increasing in frequency in the population just as it prevents them from accumulating over evolutionary time as substitutions. Consider the pattern of polymorphism at amino acid replacement sites and silent sites in the 255 codons comprising the coding region of Adh in D. melanogaster. As shown in Table 2, 13 silent polymorphisms were observed in a sample of 11 genes, whereas only one amino acid replacement polymorphism was observed. Of the 765 nucleotide sites only 192 are effectively silent. Therefore, if mutations to neutral alleles occurred at an equal rate between the two classes of sites then only one in four polymorphisms would be expected in the silent sites. Instead, a 13:1 ratio is observed, a result indicating that most mutations in Adh causing amino acid replacements must be deleterious and have been selectively removed from the population.

Getting back to our example comparing introns and silent sites in exons, significantly different levels of polymorphism in introns and exons can be explained by postulating different levels of selective constraint. Introns can contain regulatory sequences, for example, and codon biases are known to vary among genes (Sharp and Li, 1986). Because selective constraint appears to be a ubiquitous feature of genome evolution, a realistic test of selective neutrality must incorporate the possibility of deleterious mutations that may vary across regions.

Taking into account the possibility of selective constraint in a test of neutrality turns out to be straightforward and can be done by using sequence divergence between species to gauge the intensity of constraint. Consider an infinite-sites neutral model for two species that split t generations ago, that are at equilibrium at the time of sampling, and whose population sizes have remained constant. The expected number of nucleotide differences, D, between two genes chosen randomly from each species is,

$$E(D) = 2\mu E(T) \qquad (12)$$

where $T = t + 2N$ (the time since the species split plus the expected time since the divergence of the two alleles in the ancestral population) and μ is the neutral mutation rate. This simple relationship is identical to that for two randomly sampled alleles within a population [Equation (10)], and therefore the ratio of expected polymorphism and divergence, $E(s_n)/E(D)$, is a constant and is independent of the actual value of the mutation rate. Two loci, or two regions of a gene, evolving neutrally but with different proportions of neutral sites, can be thought of as differing only in their neutral mutation rates. Because polymorphism and divergence are equally affected, regions differing in levels of selective constraint can have different levels of polymorphism and divergence, but the ratio of the two is expected to be a constant.

Therefore the neutral model with selective constraint predicts that the ratio of polymorphism to divergence is the same for all genes or regions of the genes considered. This prediction was first formalized into a test for nucleotide data for the *Adh* locus and its 5′ flanking noncoding region in *Drosophila* by Kreitman and Aguadé (1986), although the idea had previously been applied to protein data (Skibinski and Ward, 1982; Ward and Skibinski, 1985). A more sophisticated and conservative test was then formulated by Hudson et al. (1987), and its application to the *Adh* data, like the earlier test, indicated a significant departure from selective neutrality. The patterns of polymorphism and divergence in *D. melanogaster* and *D. simulans Adh* and its 5′ flanking region are incompatible with a model of selective neutrality, and the pattern cannot be explained by differences in selective constraint.

One caveat must be emphasized about the choice of species for between-species comparisons. As formulated, the within- and between-species test assumes that each mutation occurs only at a previously unmutated site. Although the test may not be very sensitive to relaxation of this assumption, (e.g., allowing multiple hits), the two species should have a relatively low level of divergence. Divergence values in the range of 5–10% are certainly appropriate.

How can departures from selective neutrality (with selective constraint already taken into account) be explained? One possibility is that population sizes have not remained constant. This can cause a change in the effectively neutral mutation rate for certain combinations of population sizes and levels of selective constraint. Below a critical population size, slightly deleterious mutations (i.e. those with selective effects close to $1/2N_e$) behave as effectively neutral mutations. Such mutations, however, will be selected against at population sizes above the critical threshold. Hudson et al. (1987) studied the sensitivity of the within- and between-species test to population size fluctuation and concluded that the effect on the test is small.

Another possibility is that some other form of natural selection has been operating—preserving polymorphism in one region (as in the case of balancing selection) or reducing it in another (as in the case of directional selection). Selection does not have to be operating at any of the nucleotide sites under observation: levels of neutral polymorphism are affected by linked sites under selection. For example, a balanced polymorphism is expected to increase the level of polymorphism at linked neutral sites (Strobeck, 1983; Kaplan et al., 1988; Hudson et al., 1988) whereas directional selection is expected to reduce the level of polymorphism at linked neutral sites (Kaplan et al., 1989).

Is directional or balancing selection also expected to similarly influence levels of divergence between species? If so, then the within- and between-species test would not allow these types of selection to be detected. Fortunately, linkage has no effect on the rate of neutral divergence (Birky and Walsh 1988), and, therefore, the effects of directional and balancing selection can be detected by the within- and between-species test.

It is clear why directional selection affects the level of polymorphism at linked neutral sites. Consider a locus, such as *Adh* in *D. melanogaster*, that contains silent polymorphisms. (In a sample of 11 *Adh* genes, 34 polymorphic sites define 10 alleles.) Now consider the consequence of an advantageous substitution somewhere in one of the 10 alleles. As the favored allele sweeps to fixation, all other alleles in the population are eliminated. To the extent that the silent mutations of other alleles do not recombine onto the favored allele, the level of silent polymorphism will decrease. The long-term consequence is a shortening of the coalescent time of alleles in the population. In the extreme case where there is no recombination, as in the mitochondrial genome, a favored substitution will replace all preexisting polymorphism in the whole genome and the coalescent time is reduced to zero. The same principle is at work in the case of periodic selection in bacteria.

Several points should be made about selective sweeps. First, the effect on polymorphism is transient. The level of silent polymorphism around the selected site will have reached a minimum by the time the sweep is complete. Subsequent to the sweep, new silent mutations will begin to accumulate, but only after a very large number of generations will the level of silent polymorphism be restored. Directional selection, therefore, produces a long-lasting reduction in the level of linked neutral polymorphism. Second, the magnitude of the effect on linked neutral polymorphism falls off with recombination distance. Therefore, the size of the region over which polymorphism levels are reduced and the expected magnitude of the reduction depends on many population parameters—the population size, recombination rate, mutation rate, intensity of selection, and the time since the occurrence of the selection event.

Balanced polymorphism is expected to elevate levels of linked neutral polymorphism, but unlike directional selection, the effect is not transient. A stable equilibrium of "excess" linked neutral polymorphism will result and the effect is reduced by recombination. Neutral polymorphisms accumulate around the site of the balanced polymorphism because the selected alleles define two evolutionary lineages that are maintained in the population by balancing selection. The two lineages simply continue to diverge as they get older than they would by genetic drift alone. In coalescent terms, the time back to the common ancestor of the balanced allele lineages is greater than that for neutral alleles.

One problem in detecting the effects of balancing selection by the within- and between-species test is that the size of the region expected to exhibit excess neutral polymorphism may be quite small. The per nucleotide recombination rate is not generally known for higher organisms, although it has been estimated for the *rosy* locus of *D. melanogaster* (Chovnick et al., 1977). With the estimated *rosy* intragenic recombination rate (10^{-8} per base pair per generation) and a mutation rate of approximately 10^{-9} per generation, the "window" of excess neutral polymorphism may have a width of less than 100 bp (Hudson and Kaplan, 1988). Therefore, to detect a balanced polymorphism by the within- and between-species test RFLP data will not suffice. Populational sequence data are required.

An example

The data produced from my study of sequence polymorphism in 11 lines of *D. melanogaster* for the *Adh* locus (Kreitman, 1983) have been analyzed by the within- and between-species approach (Kreitman and Aguadé, 1986; Hudson et al., 1987; Hudson and Kaplan, 1988). I have recently extended these sequences to include two additional regions contiguous to the *Adh* locus, consisting of 1250 bp upstream from *Adh* and 1300 bp downstream from *Adh* (Kreitman and Hudson, 1991). The upstream region contains certain regulatory sequences that are necessary for the correct transcriptional expression of *Adh* (Fisher and Maniatis, 1988), but it is otherwise thought to be nonfunctional.

The downstream region contains the coding region of a recently discovered functional locus (*Adh-dup*), which has been shown to be related to the *Adh* locus as an ancient tandem duplication (Schaeffer and Aquadro, 1987; Cohn 1985). The open reading frames of *Adh-dup* encode a conceptual polypeptide whose amino acid sequence is approximately 50% identical to *Adh*. At the DNA level, however, there is very little remaining similarity. This suggests that the two loci have not engaged in intergenic recombination (at least not for a very long time).

A summary of polymorphism in the 11 lines and divergence between one of them (*Af-s*) and *D. simulans* (Cohn, 1985) is presented in Figure 1. A total of 82 segregating sites was observed, 30 in the 5′ flanking region, 37 in the *Adh* locus, and 15 in *Adh-dup*. There were two amino acid replacement substitutions, a Thr-Lys substitution in exon 4 of *Adh* (which produces the fast and slow allozyme phenotypes) and an Ile-Val substitution in exon 3 of *Adh-dup*.

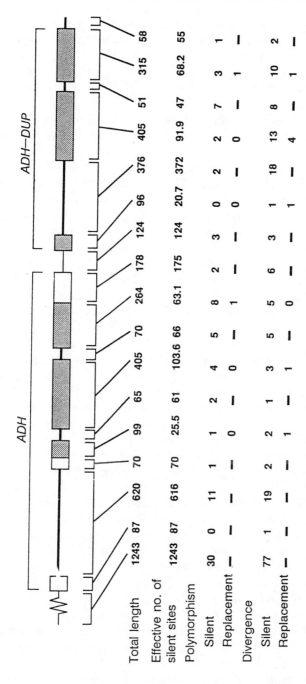

FIGURE 1. Distribution of nucleotide polymorphism in a sample of 11 *D. melanogaster* genes and nucleotide divergence between one *D. melanogaster* and one *D. simulans* gene (data from Kreitman, 1983 and Kreitman and Hudson, 1991).

The pattern of silent polymorphism around the *Adh-f/Adh-s* site is of particular interest because it has been suggested to be a polymorphism under balancing selection (reviewed in Lewontin, 1985; van Delden, 1982; van Delden and Kamping, 1989). In accord with the balancing selection hypothesis, *Adh* exon-4 contains the highest level of silent polymorphism (8 sites in 63).

What does the within- and between-species test reveal? Comparison of the 5′ flanking region with *Adh* (3 coding regions + 2 introns) yields $\chi^2 = 3.7$, with a probability, $p \approx 0.05$. The 5′ flanking region compared to *Adh-dup* yields $\chi^2 = 1.4$ with $p > 0.1$. The comparison of *Adh* with *Adh-dup* produces $\chi^2 = 5.4$ with $p < 0.025$. The tests indicate, therefore, that *Adh* and its two flanking regions are not evolving neutrally. Furthermore, the *Adh* locus deviates from the other two regions in the direction of having higher than expected polymorphism. This is, of course, consistent with a hypothesized balanced polymorphism between *Adh-f* and *Adh-s*.

Observed and expected patterns of polymorphism are shown in a graphic or "window" format in Figure 2. The observed values are the average pairwise number of segregating sites in a window of 100 effectively silent sites (50 on either side of the plotted site). The window is therefore expanded when it contains coding sequence to maintain the desired number of silent sites. If all silent sites were evolving neutrally, the predicted pairwise number of silent sites would be identical at all sites. This prediction clearly is not obtained for the data. A better neutral prediction can be obtained by using the between-species divergence to gauge the relative constraints within each species. This is done in the following manner. The neutral mutation parameter is allowed to differ from site to site and is denoted θ_i for site i. If t is the average time since the most recent common ancestor of a *D. melanogaster* and *D. simulans* sequence (estimated according to Hudson and Kaplan 1988), then for small $\theta_i t$, the probability that a site differs between the two sequences is approximately $\theta_i t$. Therefore, the average value of θ for a region containing l sites is simply the number of sites differing between the two sequences at the l sites divided by the product tl. This predicted value of the pairwise number of differences is also plotted in Figure 2. As the graph indicates, there is a good fit between the observed and expected values for the 5′ flanking region, a large excess of observed polymorphism centered around position 1490 (the site of the *Adh-f/Adh-s* protein polymorphism), and a small deficiency of polymorphism in the *Adh-dup* locus.

The departure from selective neutrality between *Adh* and the 5′ flanking region and between *Adh* and *Adh-dup* can be accounted for by the excess silent polymorphism around the allozyme polymorphism, and the deviation is in the direction predicted for a balanced polymorphism. This deviation may be the evolutionary "footprint" of a balanced polymorphism. Less clear is how to interpret the small "deficiency" of polymorphism in the *Adh-dup* locus revealed in Figure 2. This deficiency may be an indication of a prior selective sweep. Seven of the 14 silent polymorphisms in *Adh-dup* are associated with (or define) two alleles (*Fl-1s* and *Wa-s*, Kreitman, 1983) from the other nine. The re-

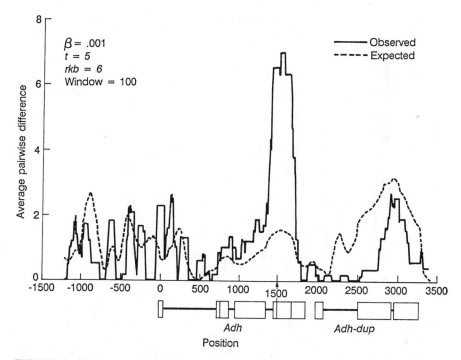

FIGURE 2. Sliding window of observed and expected polymorphism levels in a sample of 11 *D. melanogaster* genes. The average pairwise number of nucleotide differences in a window of 100 "silent" sites is plotted at each nucleotide position. The arrow at position 1490 marks the site of the *Adh-f/Adh-s* amino acid replacement polymorphism. See Kreitman and Hudson (1991) for more detailed information about the sliding window approach.

maining seven polymorphisms each occurs only once in the sample. This frequency distribution is distinctly different from that of *Adh* or the 5′ flanking region, both of which contain many sites with intermediate frequency polymorphisms (Figure 3) (Kreitman and Hudson, 1991).

Tajima (1989) recently developed a test for comparing the observed polymorphism frequency spectrum to that expected under a neutral mutation model. Noting that the neutral parameter can be estimated from the number of segregating sites in a sample, s_n [Equation (5)], or from the average pairwise difference [Equation (3)], Tajima's test determines the probability, assuming neutrality, of obtaining a difference between the two estimates as large as that observed. When the test is applied to the three *Adh* regions, the 5′ flanking region and the *Adh* and *Adh-dup* loci, none exhibits a significant departure from neutrality, but the test statistic becomes significant at the 0.05 level when the *Wa-s* and *Fl-1s* alleles are removed. Additional data will likely resolve whether or not a recent selective substitution has occurred in the *Adh-dup* region.

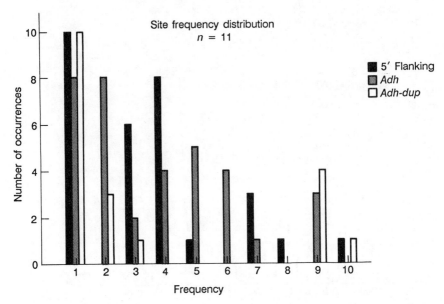

FIGURE 3. Observed and expected frequency distribution of nucleotide polymorphisms in a sample of 11 *D. melanogaster* genes. Expected values are based on an infinite allele neutral model (Tajima, 1989, Equation 51). Sixteen of the 18 sites with frequency two or nine in *Adh* or *Adh-dup* are obtained from the differences between two genes (*Fl-1s* and *Wa-s*) and the remaining nine. The evolutionary cause of this divergence is, at present, unknown.

FUTURE PROSPECTS

Historically, two challenges have faced molecular population geneticists: to develop a methodology for obtaining accurate estimates of genetic polymorphism and to develop the theoretical predictions about how this polymorphism should be distributed under different models of natural selection and genetic drift. Rapid progress has been made in both areas. Sequences are easily obtained from DNA amplified by the polymerase chain reaction, and improvements in methodology will accelerate as the genome project progresses. Additionally, the development of neutral theory based on infinite site (and allele) models, application of coalescent theory to gene evolution, and the incorporation of certain kinds of selection into the models have already resulted in useful tests. Importantly, the gene genealogy approach allows for the possibility of recombination, a necessity for the development of biologically realistic models.

The most serious obstacle to the evaluation of the forces shaping patterns of variation and change may be the way we organize empirical population genetics: many groups working on many organisms. The issues to be resolved at the molecular level do not center around whether a particular phenomenon can occur but how frequently in evolution the phenomenon applies and under what circumstances. Whether balancing selection (indeed, any evolutionary force) is important in regulating polymorphism requires that we know something about the frequency of the various forces acting across the genome. To obtain such estimates we must systematically study many loci in a way that allows us to make the appropriate historical inferences. Our *Adh* data, and the inference of a balanced polymorphism, by themselves tell us little about how evolution works in general and whether balancing selection is important in maintaining protein polymorphism.

To answer these questions, data cannot be collected helter-skelter across organisms. If the within- and between-species method is to be the backbone of the historical inference approach, investigation of population variation at many loci in closely related species will be necessary. Evolutionists who are interested in this problem must be willing to follow the example of genetics (or genome sequencing projects), where many groups work on a small number of model species, in contrast to a time-honored population biology approach, where work is spread across many different and unrelated organisms. It seems likely that if we can solve the problem of how to divide up the work, we may finally be able to understand the role of natural selection in maintaining genetic variation and put to rest some issues raised by the old neutralist–selectionist debate.

CHAPTER 11

POLYMORPHISM AND EVOLUTION OF THE MAJOR HISTOCOMPATIBILITY COMPLEX LOCI IN MAMMALS

Masatoshi Nei and Austin L. Hughes

The major histocompatibility complex (MHC) is a large multigene family that plays an important role in the vertebrate immune system. Glycoproteins (MHC molecules) encoded by genes of this complex bind foreign antigens (peptides) and present them to T cells, thereby triggering appropriate immune responses. Some MHC loci are extraordinarily polymorphic and others are virtually monomorphic. Curiously, the polymorphic loci are highly expressed, whereas the monomorphic loci are often unexpressed or only marginally expressed.

The mechanism of maintenance of polymorphism at the MHC loci has been debated for more than two decades, yet no consensus has been reached. Before the function of the MHC became known, wild speculations were made concerning the mechanism of maintenance of the polymorphism. One of the most popular early hypotheses was frequency-dependent selection (Snell, 1968; Bodmer, 1972). In the late 1960s and early 1970s, there were data suggesting that some alleles at the MHC (HLA) loci in humans are associated with particular chronic diseases, even in heterozygous condition. For this reason, the possibility of overdominant selection (heterozygote advantage) was rejected, and frequency dependence was favored as an alternative type of diversity-enhancing selection.

In 1974, the function of MHC molecules was finally deciphered by Zinkernagel and Doherty (1974), who showed that they bind (restrict) foreign antigens and present them to the T lymphocytes. Soon after this discovery, Doherty and

Zinkernagel (1975) suggested that MHC polymorphism could be maintained by heterozygote advantage. However, this view was not accepted by many population geneticists, because it was difficult to obtain clear-cut supporting evidence.

In the past 10 years, the application of new biochemical techniques, such as gene cloning and DNA sequencing, to the study of the genomic organization of the MHC has contributed greatly to an understanding of the mechanism of maintenance of polymorphism at the molecular level (Klein, 1986). Using X-ray crystallography, Bjorkman et al. (1987a,b) identified the antigen recognition site (ARS) of the class I MHC molecule. This identification led Hughes and Nei (1988) to reason that if there is any positive Darwinian selection at the MHC loci, it must occur at the ARS. Examining the pattern of nucleotide substitution at the ARS and the remaining regions of the genes, they then showed that positive selection is indeed operating at the ARS and argued that the most likely type is overdominant selection.

In this chapter, we first review essential aspects of the genomic organization and function of the MHC genes and then discuss their polymorphism and evolution.

ORGANIZATION AND FUNCTION OF THE MHC GENES

MHC genes are classified into two groups, class I and class II, which are evolutionarily related, although distantly. Class I and class II genes are located on the same chromosome but form separate clusters, which are separated by another cluster of unrelated genes (Figure 1). (The genes belonging to the third cluster are often called class III genes.) In the mouse, two class I genes (K2 and K) are separated from the other class I genes, being located on the centromeric (left) side of the class II genes. Note that each of the gene maps in Figure 1 refers to a particular haplotype, and that the maps are not the same for all haplotypes in either human or mouse. For example, the class I locus D in the mouse is represented in only a small number of haplotypes. Note also that the exact number of MHC gene loci is not known for the human.

Class I genes

Class I MHC loci can be divided into classical and nonclassical I groups. Classical genes encode glycoproteins that are expressed on the surface of all nucleated cells and present peptides derived from intracellular pathogens to cytotoxic T lymphocytes (CTLs) (Figure 2; Klein, 1986). The classical class I molecule has three extracellular domains (α_1, α_2, and α_3), a transmembrane portion, and a cytoplasmic tail (Figure 3). The α_3 domain associates noncovalently with β_2-microglobulin, a polypeptide encoded by a gene that lies outside the MHC but is distantly related to the MHC genes. Bjorkman et al. (1987a,b) showed that the ARS of the classical class I MHC molecule is located in the

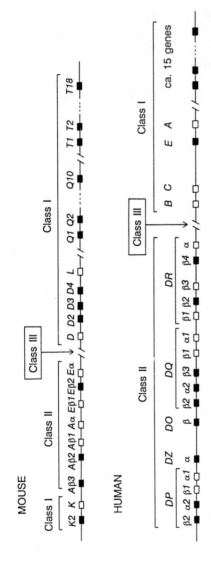

FIGURE 1. The arrangement of class I and class II MHC genes in the mouse and the human genome. Open boxes represent functional genes (classical genes), whereas black boxes stand for genes that are not expressed or expressed at a low level (nonclassical genes and pseudogenes). The gene arrangement for the mouse is that for the *d* haplotype; other haplotypes may have different numbers of genes. In the human genome, there are five class II gene regions (*DP, DZ, DO, DQ, DR*). The *DR* region shown here is that of the *DR3* haplotype; other haplotypes may have different numbers of genes in this region. Different authors use slightly different nomenclatures for human class II genes. The α and β chain genes are often designated by *A* and *B*, respectively (e.g. *DPA1* instead of *DPα1*, etc.). *DZα* is also known as *DNα*

or *DNA*. In the *DQ* region, the genes β₂ and α₂ are sometimes called *DXβ* and *DXα* (or *DXB* and *DXA*). The *DQ* region gene β₃ is also known as *DVβ*. The gene β4 in the *DR* region is a pseudogene consisting only of an isolated β₁ domain exon. In the human class I, the exact number of genes is not known for any haplotype. Orthologous relationships exist between mouse and human class II genes or gene regions but not for class I genes. The orthologous relationships of class II genes are as follows: Aβ3 is orthologous to the *DP* region β chain genes; Aβ2 to *DOβ*; Aβ1 and Aα to *DQ* region genes; *E* region genes to *DR* region genes. The gene maps presented here are based on information from Klein and Figueroa (1986) and Bodmer (1989).

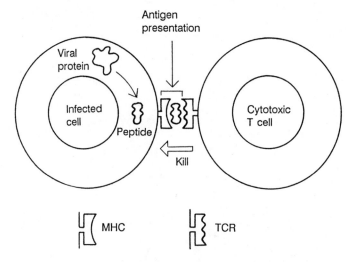

FIGURE 2. Function of the class I MHC molecule. Foreign (viral) proteins enter the cell through an infectious process. These proteins are degraded into small peptides (15–20 amino acids long) inside the cell, and the peptides bound by MHC molecules are transported to the cell surface and presented to cytotoxic T cells (CTLs). T cell receptors (TCR) then recognize and bind the foreign peptides displayed by the MHC molecules. Once this recognition occurs, CTLs kill the infected cell.

α_1 and α_2 domains and consists of 57 amino acid residues (Figure 4). Examples of classical class I loci are *HLA-A, -B,* and *-C* in the human and *H2-K, -D,* and *-L* in the mouse. Most mammals apparently have one to three classical class I loci (Klein, 1986).

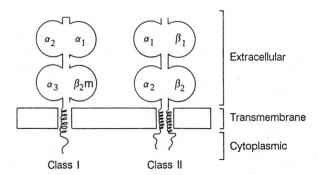

FIGURE 3. The molecular structure of class I and class II MHC molecules. The class I molecule consists of three extracellular domains (α_1, α_2, and α_3), transmembrane region, and cytoplasmic region. The antigen recognition site is located in the α_1 and α_2 domains. β_2-microglobulin (β_2m) associates noncovalently with the α chain. The class II MHC molecule is a heterodimer consisting of noncovalently associated α and β chains. The α and β chains consist of two extracellular domains (α_1, α_2, and β_1, β_2, respectively), transmembrane region, and cytoplasmic region.

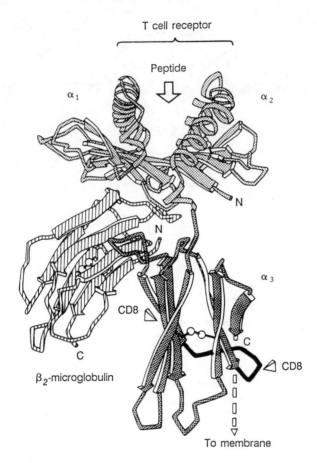

FIGURE 4. Structure of the α_1, α_2, and α_3 domains of a class I MHC molecule with a β_2-microglobulin chain. The β strands are shown as thick arrows in the amino to carboxyl direction and the α helices as helical ribbons. A foreign peptide is bound by the groove (antigen recognition site) produced by the α_1 and α_2 domains, and the combined entity of the peptide and the antigen recognition site is believed to be recognized by the T cell receptor. From Lawlor et al. (1990).

The molecular structure of a class I gene consists of eight exons and seven introns, as shown in Figure 5. Exon 1 codes for the 5′ untranslated region and the amino acids in the leader peptide, and exons 2, 3, 4, and 5 encode the α_1, α_2, and α_3 extracellular domains and the transmembrane portion of the MHC molecule, respectively. Exons 6 and 7 and part of exon 8 encode the amino acid sequence of the cytoplasmic portion of the molecule.

As mentioned above, the function of the classical class I molecule is to present foreign antigens (peptides) to CTLs, which recognize foreign antigens (nonself antigens) only when they are presented together with the MHC mol-

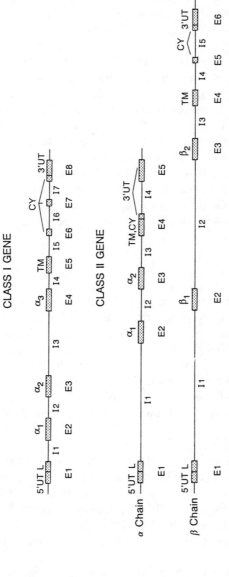

FIGURE 5. Exon–intron organization of class I and class II genes. E, exon; I, intron; L, leader peptide; α_1, α_2, α_3, β_1, β_2, external domains; TM, transmembrane region; CY, cytoplasmic region; 5' UT and 3' UT, 5' and 3' untranslated regions.

ecules (self antigens) (Figure 2). CTLs then kill infected host cells. Skin grafts are rejected in humans and mice, unless the recipient and the donor have the same genotype, because the antigens of the donor individuals are perceived as foreign by the recipient.

The products of nonclassical class I loci lack both the wide tissue distribution and the high degree of polymorphism characteristic of classical loci and ordinarily do not function to present foreign antigens to T cells (Tewarson et al., 1983; Howard, 1987). Examples of nonclassical class I genes are the *Q* (*Q1*, *Q2*, . . .) and *TL* (*T1*, *T2*, ...) loci in the mouse and the *E* locus in the human. (The *Q* and *TL* loci are often called the *Qa* and *Tla* loci, respectively.) The class I MHC also contains a number of pseudogenes in the human, mouse, and some other mammalian species (Klein and Figueroa, 1986), and in most species the nonclassical loci and pseudogenes greatly outnumber the classical loci. For example, in the mouse there are 2 or 3 classical loci (depending on the haplotype) but 24—30 nonclassical and pseudogene loci in the class I MHC.

Class II genes

Class II MHC molecules are expressed on antigen-presenting cells of the immune system and consist of noncovalently associated α and β chains, which are encoded by separate genes (Klein, 1986). Each chain is composed of two extracellular domains (designated as α_1 and α_2 in the α chain and β_1 and β_2 in the β chain), a transmembrane portion, and a cytoplasmic tail (Figure 3). The molecular structure of a class II gene is shown in Figure 5. By analogy with the class I molecule, an antigen recognition site for the class II molecule has been provisionally identified (Brown et al., 1988); it consists of 19–20 amino acid residues of the first domain of the α chain and 15–16 residues of the first domain of the β chain.

The class II molecule has essentially the same function as that of the class I molecule, except that it presents a foreign peptide to the helper T cell (Th cell). Here too, the foreign antigen (peptide) is bound to a host MHC antigen (MHC-restricted), and the combination of the peptide and the MHC molecule is recognized by the Th cell. The Th cell then interacts with the B cell and sends differentiation signals that cause the B cell to transform into an antibody-secreting plasma cell.

Most mammals apparently have two or three pairs of functional class II α and β chain genes. In each of these pairs, at least the β chain locus is highly polymorphic, and sometimes the α chain is as well. Examples of such pairs are the *Aα* and *Aβ1* loci in the mouse and the *DQα1* and *DQβ1* loci in the human (Figure 1). In addition to these polymorphic loci, there are certain class II loci that are poorly expressed and monomorphic; examples are the *Aβ2* locus in the mouse and the *DOβ* locus in the human. As with class I loci, there are also some pseudogenes.

EXTENT OF MHC POLYMORPHISM

Heterozygosity and the number of alleles

MHC loci show an extraordinarily high degree of polymorphism, and no other protein-coding loci in the mammalian genome match them in this regard. For example, heterozygosity at the human *HLA-A* and *HLA-B* loci is about 90% in many populations, and the number of alleles detected in a sample of 200 individuals ranges approximately from 15 to 30 (Roychoudhury and Nei, 1988). Similarly, the mouse *H2-K* and *H2-D* loci are highly polymorphic; the numbers of major alleles detected at these loci are 54 and 55, respectively (Klein, 1986). The extent of allelic polymorphism in other organisms has not been well studied, but skin grafting experiments, serological tests, and restriction fragment analyses indicate that many vertebrate species are highly polymorphic for both class I and class II loci (Table 1).

There are, however, some species in which all the MHC loci are virtually

TABLE 1. MHC polymorphism in vertebrate species

Species	Method[a]	MHC polymorphism[b]		References
		Class I	Class II	
Chicken	SE	High	High	Simonsen et al. (1982)
	RFLP	N.A.	High	Warner et al. (1989)
Cheetah	SG	Absent	N.A.	O'Brien et al. (1985)
Cottontop tamarin	IEF	Low	Moderate	Watkins et al. (1988)
Balkan mole rat	RFLP	Low	Absent	Nizetic et al. (1988)
Israeli mole rat	RFLP	High	N.A.	Nizetic et al. (1985)
Fin whale	RFLP	Low	Absent	Trowsdale et al. (1989)
Sei whale	RFLP	Low	Absent	Trowsdale et al. (1989)
Syrian hamster	RFLP	Low	Moderate	McGuire et al. (1985)
	SE	Low	Moderate	Streilein (1987)
Rat	RFLP	High	N.A.	Palmer et al. (1983)
Domestic cat	RFLP	Moderate	Moderate	Yuhki et al. (1989)
	SE	Moderate	Moderate	Winkler et al. (1989)
Pig	RFLP	High	High	Chardon et al. (1985)
Mouse	All methods	High	High	Klein (1986)
Human	All methods	High	High	Klein (1986)

[a]SG, Skin grafting; IEF, isoelectric focusing electrophoresis; RFLP, restriction fragment length polymorphism; SE, serology.
[b]N.A., data not available.

monomorphic. An example is the cheetah, in which skin grafting can easily be made between unrelated individuals (O'Brien et al., 1985). In the Syrian hamster (McGuire et al., 1985), relatively little polymorphism has been detected at the class I and class II loci (Streilein, 1987). The fin whale also shows a low level of MHC polymorphism, as do mouse populations living on small islands in the North Sea (Figueroa et al., 1986).

Variation at the DNA level

In recent years, DNA sequences have been determined for many different alleles at the polymorphic MHC loci in humans and mice. One can therefore examine the extent of DNA variation by calculating nucleotide diversity (π), which is the average number of nucleotide differences per site for all pairwise comparisons of DNA sequences studied (Nei, 1987). Values of π estimated for the *HLA-A* and *H2-K* loci are presented in Table 2, together with those for other loci (or gene regions) in higher organisms. The π values for other loci or DNA regions range from 0.0002 to 0.011. For the MHC loci, however, π is about 10 times higher than that for other nuclear or mitochondrial genes. This indicates that at the MHC loci the number of polymorphic alleles is large and the extent of nucleotide divergence between alleles is high. Note that the $\pi = 0.04-0.07$ is

TABLE 2. Estimates of nucleotide diversity (π; heterozygosity at the nucleotide level)[a]

DNA or gene region	Organism	Method	n	bp	π
mtDNA	Human	R	100	16,500	0.004
mtDNA	*D. melanogaster*	R	10	11,000	0.008
β-Globin	Human	R	50	35,000	0.002
Growth hormone	Human	R	52	50,000	0.002
Notch gene region	*D. melanogaster* (1)	R	37	60,000	0.005
White locus region	*D. melanogaster* (2)	R	38	45,000	0.011
Factor IX	Human (3)	S	22	2460	0.0002
Adh locus (C)	*D. melanogaster*	S	11	765	0.006
Prochymosin (C)	Bovine (4)	S	8	1146	0.004
Growth hormone (C)	Pig (4)	S	6	651	0.007
Class I MHC (*HLA-A*)	Human (5)	S	5	274	0.043
Class I MHC (*H2-K*)	Mouse (5)	S	4	273	0.077

Sources: (1) Schaeffer et al. (1988); (2) Langley and Aquadro (1987); (3) Koeberl et al. (1989); (4) D. J. McConnell and P. M. Sharp (unpublished); (5) Hughes and Nei (unpublished); see Nei (1987) for others.
[a]*n*, Number of DNA sequences examined; bp, base pairs; R, restriction enzyme technique; S, DNA sequencing; C, coding region.

much higher than the average number of nucleotide differences per site (0.016) for the η pseudogene region between the human and the chimpanzee (Miyamoto et al., 1987).

Another interesting contrast in DNA polymorphism between the MHC and other loci is that the nucleotide differences between alleles at the other loci largely involve synonymous substitution, whereas at the MHC loci there are many nonsynonymous (amino acid altering) substitutions. Nonsynonymous substitutions are particularly prevalent in the antigen recognition site (ARS) region of the gene (Figure 6).

PATTERN OF NUCLEOTIDE SUBSTITUTION

One way of studying the nature and pattern of natural selection operating at the DNA level is to examine the number of substitutions per synonymous site (d_S) and the number of substitutions per nonsynonymous site (d_N). In the absence of selection on a DNA region, d_S and d_N should be more or less the same. If there is purifying selection, d_S is expected to be higher than d_N, because selection would occur at the amino acid level. By contrast, if there is positive Darwinian selection, d_N is expected to be higher than d_S. As mentioned earlier, the antigen specificity of the MHC molecule is determined by the amino acids in the ARS. Therefore, if there is positive selection at the MHC loci, it should occur at the ARS region of the gene. We therefore examined the d_S and d_N values for the ARS region and the remaining region separately (Hughes and Nei, 1988, 1989a). Values of d_S and d_N were computed for each pair of DNA sequences available at each polymorphic locus in humans and mice by Nei and Gojobori's (1986) method I.

Table 3 shows the mean d_S and d_N values for all sequence comparisons at each class I polymorphic locus. In the case of the human HLA-A locus, d_N is much higher than d_S in the ARS, the difference being significant at the 0.1% level when the standard errors computed by Nei and Jin's (1989) method are used. In the other regions, d_S is higher than d_N. The same pattern of nucleotide substitution is also observed for the HLA-B and HLA-C loci. It is interesting to see that d_S is more or less the same for all three regions examined, whereas d_N in the ARS is significantly higher than d_S in any region. This clearly indicates that amino acid substitutions in the ARS are enhanced by positive Darwinian selection. By contrast, d_N is much smaller than d_S in the non-ARS regions, suggesting that amino acid substitutions are generally subject to purifying selection in these regions.

Table 3 also includes the results of allelic comparisons of the mouse H2-K and -L loci. The d_N in the ARS is again substantially higher than the d_S in any region. Therefore, the pattern of nucleotide substitution for the mouse MHC loci is essentially the same as that for the human MHC loci. (The unusually low value of d_S for α_3 may have been caused by interlocus genetic exchange; Hughes and Nei, 1988.)

TABLE 3. Mean numbers of nucleotide substitutions per synonymous site (d_S) and per nonsynonymous site (d_N) expressed as percentages between alleles from the same class I MHC polymorphic loci in humans (HLA-A, -B, and -C) and mice (H2-K and D)[a]

Locus (Number of sequences)	Antigen recognition site (ABS) ($N = 57$)		Remaining codons in α_1 and α_2 ($N = 124, 125$)[b]		Domain α_3 ($N = 92$)	
	d_S	d_N	d_S	d_N	d_S	d_N
Human						
A (5)	3.5	13.3***	2.5	1.6	9.5	1.6**
B (4)	7.1	18.1**	6.9	2.4*	1.5	0.5
C (3)	3.8	8.8	10.4	4.8	2.1	1.0
Overall means	4.7	14.1***	5.1	2.4	5.8	1.1**
Mouse						
K (4)	15.0	22.9	8.7	5.8	2.3	4.0
L (4)	11.4	19.5	8.8	6.8	0.0	2.5**
Overall means	13.2	21.2*	8.8	6.3	1.2	3.6**

[a]The difference between d_S and d_N is significant at the 5% level (*), the 1% level (**), or the 0.1% level (***) (see Hughes and Nei, 1988, for details). N, Number of codons compared.
[b]N is 124 for the mouse and 125 for the human.

We conducted a similar statistical analysis for class II MHC loci. In the mouse, there are seven class II loci in the H-2^d haplotype (eight in some other haplotypes) (Klein and Figueroa, 1986). However, only three of them (Aα, Aβ1, and Eβ1) are polymorphic, the others being either monomorphic or unexpressed. The number of class II loci in the human is considerably larger than that in the mouse, but only about five loci are polymorphic (DPβ1, DQβ1, DQα1, DRβ1, and DRβ3). (There are two polymorphic loci for DRβ, but they are treated here as a single locus, because they are sometimes difficult to distinguish.) Other loci are again either monomorphic or unexpressed. For some reasons, the β chain loci are more polymorphic than the α chain loci. In the present work, we studied all polymorphic loci.

The d_S and d_N values obtained for the class II MHC loci are presented in Table 4. In both mouse and human polymorphic loci, d_N is again much higher

FIGURE 6. Nucleotide sequences for domains α_1, α_2, and α_3 of the five alleles from ▶ the human HLA-A locus. The amino acid residues involved in the antigen recognition site (ARS) are marked with +. "−" indicates that the nucleotide is identical with that of allele A2.

α_1

```
    GGC TCT CAC TCC ATC AGG TAT TTC TTC ACA TCC GTG TCC CGG CCC CGC GGC GGC TTC GTG GCA GTG CCC TTC ATC GCA GTG GGC TTC GTG CGG TTC GTG GAC AGC GAC GCC
A2                                                                                                                        +              +              +
A3  --- --C --- --G ... ... -A- --C --- ... ... ... ... ... ... ... ... ... ... ... ... --C ... ... ... ... ... ... ... --G -GC ... ... ...
A11 --- --G ... ... ... ... -C- --C --- ... ... ... ... ... ... ... ... ... ... ... ... --C ... ... ... ... ... ... ... --G ... ... ... ...
Aw24 --- --G ... ... ... ... -A- --C --- ... ... ... ... ... ... ... ... ... ... ... ... --C ... ... ... ... ... ... ... ... ... ... ... ...
Aw68 --- --G ... ... ... ... -A- --C --- --C ... ... ... ... ... ... ... ... ... ... ... ... ... ... ... ... ... ... ... ... ... ... ... ...
```

TABLE 4. Mean numbers of nucleotide substitutions per synonymous site (d_S) and per nonsynonymous site (d_N) expressed as percentages between alleles at mouse and human class II MHC loci[a]

Locus (Number of sequences)	Putative recognition site (ABS) (N=15–20)		Domain 1 excluding ABS (N=64–78)		Domain 2 (N=94)	
	d_S	d_N	d_S	d_N	d_S	d_N
β Chain loci						
Mouse						
Aβ1 (7)	0.0	30.0***	4.0	6.7	7.3	1.3*
Eβ1 (4)	11.6	41.5*	1.8	5.2	0.9	0.6
Human						
DPβ1 (3)	3.9	19.0	2.4	2.8	5.3	0.6
DQβ1 (8)	13.7	26.5	8.5	6.7	5.6	1.6
DRβ (14)[b]	15.0	45.7**	8.0	4.5	8.3	3.3**
α Chain loci						
Mouse						
Aα (6)	3.2	23.7***	2.8	2.6	7.2	0.7**
Human						
DQα1 (4)	21.7	27.0	8.0	4.3	4.0	2.4

[a]The difference between mean d_S and d_N is significant at the 5% level (*), the 1% level (**), or the 0.1% level (***) (see Hughes and Nei 1989a for details). N, Number of codons compared.
[b]DRβ1 and DRβ3 are treated as a single locus.

than d_S in the ARS, although the difference is not necessarily significant because of the small number of codons involved. In the present case, even the entire domain 1, including both ARS and non-ARS regions, generally shows a higher d_N value than d_S. Some amino acid residues outside the putative ARS may be involved in antigen recognition. In the conserved (α_2 or β_2) domain region, however, d_N is always smaller than d_S. Therefore, class II polymorphic loci show the same pattern of nucleotide substitution as that of class I loci.

TRANSSPECIFIC POLYMORPHISM

A pair of neutral alleles (not two randomly chosen genes) at a locus may persist for 2N generations in a population but not much more, where N is the effective population size (Takahata and Nei, 1990). In *Drosophila melanogaster*, the

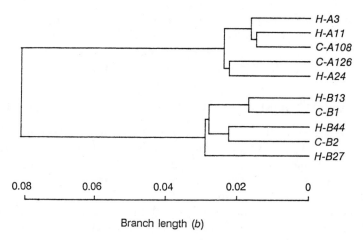

FIGURE 7. Phylogenetic tree for alleles from the class I MHC loci A and B in the human (*HLA-A* and *-B*) and the chimpanzee (*CHA-A* and *-B*). This tree was obtained by UPGMA, and *b* represents the branch length measured in the number of nucleotide substitutions per site. H, human; C, chimpanzee.

long-term effective population size seems to be about 2×10^6 (Kreitman, 1983). Since there are about six generations a year in natural populations of *D. melanogaster*, $2N$ generations correspond to about 700,000 years. Indeed, there is evidence that the *F* and *S* alleles at the alcohol dehydrogenase locus have coexisted for 610,000–1,000,000 years (Stephens and Nei, 1985). In general, however, polymorphic alleles are relatively short-lived, and the same pair of polymorphic alleles is rarely shared by related species (Coyne, 1976). Exceptions to this rule are the MHC loci. McConnell et al. (1988) have shown that two polymorphic allelic lineages at the Aβ1 locus are shared by three species of mice (*Mus*) that apparently diverged about three million years ago. Figueroa et al. (1988) showed that two polymorphic allelic lineages marked with insertions/deletions at the Aβ1 locus are shared by mice and rats, which diverged at least 10 million years ago. Shared polymorphisms have also been reported for the class I and class II MHC loci in humans and chimpanzees, which diverged 5–7 million years ago (Lawlor et al., 1988; Mayer et al., 1988; Fan et al., 1989).

Figure 7 shows the phylogenetic tree of alleles at the human *HLA-A*, *-B* and *-C* loci and the chimpanzee *ChA-A* and *-B* loci. The human allele *H-A11* is closer to the chimpanzee allele *C-A108* than to any other human alleles at the A locus, whereas the human allele *H-A24* is closer to the chimpanzee allele *C-A126* than to any other human alleles. Here we can clearly see that a pair of polymorphic allelic lineages is shared by humans and chimpanzees. The B locus also shows a similar shared polymorphism. These polymorphic lineages apparently have coexisted in both human and chimpanzee populations for a long time. For example, the allelic pair *H-A11* and *H-A24* in humans or *C-A108* and *C-A126* in chimpanzees are estimated to have been polymorphic for more than 20 million years if we use the rate of synonymous substitution estimated by Li and Tanimura (1987) for hominoids. This indicates that polymorphic allelic lineages persist in the population through speciation events. Klein (1987) called this the transspecific polymorphism.

MECHANISMS OF MAINTENANCE OF MHC POLYMORPHISM

It is now clear that MHC polymorphism is characterized by several distinctive features and that any hypothesis for the mechanism of maintenance of MHC polymorphism must be able to explain all of the following observations: (1) The extent of polymorphism is extraordinarily high. (2) The number of nucleotide differences between alleles is unusually high. (3) In the antigen recognition site of the gene, the rate of nonsynonymous substitution is higher than the rate of synonymous substitution. (4) Polymorphic alleles (allelic lineages) may persist in the population for tens of millions of years.

As mentioned earlier, many different hypotheses explaining MHC polymorphism have been proposed in the past. The major ones are (1) a high mutation rate (Bailey and Kohn, 1965), (2) neutral mutations (Klein, 1987), (3) gene conversion (Lopez de Castro et al., 1982; Ohta, 1982), (4) maternal–fetal incompatibility (Clarke and Kirby, 1966; Hedrick and Thomson, 1988), (5) mating preference (Yamazaki et al., 1976), (6) overdominant selection (Doherty and Zinkernagel 1975; Hughes and Nei, 1988), and (7) frequency-dependent selection (Snell, 1968; Bodmer, 1972).

Let us now examine each of these hypotheses in the light of the present knowledge about the MHC.

High mutation rate, neutral mutation, and gene conversion

The hypothesis of a high mutation rate is certainly capable of explaining the higher degree of polymorphism, but it cannot explain the observed relationship of $d_N > d_S$ in the ARS region of the gene. Furthermore, it is now known that the rate of synonymous nucleotide substitution in the MHC genes is not unusually high compared with that of other nuclear genes (Klein, 1986). For example, the number of synonymous substitutions between the human and mouse class I genes is about 0.32, which is less than half the average value of d_S for other genes (Hayashida and Miyata, 1983).

Klein (1987) proposed the hypothesis of neutral mutations to explain the transspecific polymorphism. (He actually proposed that positive selection may occur after a drastic change in environment but that MHC alleles are neutral through most of evolutionary time.) However, this hypothesis is incapable of explaining the long persistence of polymorphic alleles in rodents and hominoids, since, as mentioned above, a pair of neutral alleles rarely persists in a population for more than $2N$ generations. The extent of protein polymorphism in rodents suggests that the long-term effective population size (N) of a species is of the order of 10^5 (Nei and Graur, 1984). If we assume that there are two generations in a year in nature, $2N$ generations correspond to 10^5 years, a period clearly too short compared with the observed persistence time of MHC polymorphic alleles. Note also that the neutral mutation hypothesis cannot explain the observed relationship of $d_N > d_S$ in the ARS.

Another hypothesis that does not invoke selection is that of gene conversion. This hypothesis obviously cannot explain the observation that $d_N > d_S$ in the ARS. Furthermore, it has been very difficult to establish clear-cut evidence for gene conversion, and, hence, this hypothesis has been controversial (Klein and Figueroa, 1986). Nevertheless, there is indirect evidence that unequal crossover, exon shuffling, or gene conversion generates genetic exchange between different alleles at the same MHC locus (Holmes and Parham, 1985) or at different loci (e.g., Weiss et al., 1983; Hughes and Nei, 1989b). Therefore, intragenic or intergenic recombination seems to be a source of genetic variation. However, the frequency of occurrence of these events seems to be quite low (Klein et al., 1989; Lawlor et al., 1990), and they cannot be a major factor of MHC polymorphism.

Maternal–fetal incompatibility

This hypothesis was first proposed by Clarke and Kirby (1966) to explain James' (1965) report that the fetal growth rate in mice is increased when mother and father are of different MHC haplotypes. The idea behind the hypothesis is that a histoincompatible fetus stimulates the mother's immune response, which in turn makes the fetus more vigorous. This postulated maternal–fetal interaction is expected to produce a net heterozygous advantage (Hedrick and Thomson, 1988). However, this hypothesis is not based on a solid body of experimental data. Subsequent studies of the mouse (Hamilton et al., 1985) and the rat (Palm, 1974) failed to confirm James' (1965) original report. In James' work, the role of the MHC was emphasized because the growth-enhancing effect was most marked when females were immunized against the male's MHC prior to mating (James, 1967). However, Clarke (1971) was unable to replicate the effect of preimmunization to the male's MHC, and subsequent studies have produced inconsistent results depending on the MHC haplotypes used (reviewed by Wegmann, 1984).

Certain human data have been interpreted as support for the maternal–fetal incompatibility hypothesis. For example, Gill (1983) observed a high rate of spontaneous abortion when mother and father shared the same MHC alleles. But this result can be explained equally well by the occurrence of deleterious recessive alleles at one or more loci closely linked to the MHC (Schacter et al., 1984). Perhaps the most powerful argument against this hypothesis is the fact that the chicken has a high degree of MHC polymorphism even though there is no maternal–fetal interaction.

Mating preference

It is often claimed that mice mate disassortatively with respect to MHC genotype and that such a mating pattern promotes MHC polymorphism (Partridge, 1988), but the evidence for such disassortative mating is quite meager. Certainly there is evidence that mice and rats can discriminate by olfaction between individuals that apparently differ only at MHC loci (Yamazaki et al., 1979; Singh et al., 1987). Yamazaki et al. (1983b) have shown that mice can discriminate by urine odor between congenic individuals homozygous for the K^b allele and those homozygous for the K^{bm1} allele. (The MHC molecules encoded by these alleles are known to differ only in three amino acid positions.) However, showing that mice have the ability to make such discriminations is not the same as showing that they actually use this ability in nature, much less that they use it as a basis for mate choice.

The claim that mice mate disassortatively with respect to MHC is based on the experiments of Yamazaki et al. (1976), in which males from inbred strains were allowed to mate with two congenic estrous females differing only in MHC haplotype. Males of some strains showed a preference for females whose MHC genotype differed from their own, but other strains showed no preference or even preferred females of the same MHC genotype. In some cases, the direction of the preference depended on the alternative females offered. Thus, for example, males of the *bb* strain preferred *aa* or *kk* females to *bb* females but preferred *bb* females to *dd* females. Furthermore, there are several reasons for questioning the relevance of such experiments to natural populations. First, in most species of animals, female choice of males is much more important in determining mating patterns than is male choice of females (Partridge and Halliday, 1984). Second, Yamazaki et al. (1976) did not use the mate-choice experimental design that is standard in behavior science. Since the two females could interfere with each other in mating, it is possible that behavioral interactions between the females determined the result of the tests. Egid and Brown (1989) recently conducted some experiments using a better design in which they demonstrated a preference by female mice for males of different MHC genotype. However, since they used only two strains, their results should be regarded as preliminary. Third, individual odors of mice in natural conditions are likely to be influenced by many factors—including protein products of many non-MHC loci, diet, and so forth—in addition to MHC products in the urine. These considerations cast strong doubt on the importance of mating preference in the maintenance of MHC polymorphism. Note also that mating pattern in man is hardly affected by MHC genotypes (Rosenberg et al., 1983).

Overdominant selection

The overdominance hypothesis has a solid biological basis. Showing that different MHC molecules bind different foreign antigens, Doherty and Zinkernagel

(1975) argued that a heterozygote having two different MHC molecules would be more resistant to infectious diseases than a homozygote with only one type of MHC molecule. Since then, many authors have demonstrated for both class I and class II loci that allelic products differ in their ability to bind and present foreign peptides to T cells (Berzofsky et al., 1979; Buus et al., 1986; Enssle et al., 1987; Sette et al., 1989; Celis and Karr, 1989). Probably because resistance to viral disease is a complex phenomenon, involving not only the class I and class II MHC but also the antibody system, immunologists have so far identified only a few cases in which resistance to a natural viral pathogen is correlated with the MHC genotype of the host. The best documented case is that of the resistance to Marek's disease virus, a pathogen of the domestic fowl (Longenecker et al., 1976). In this case, different MHC haplotypes show several gradations of resistance to this virus; the B^{21} haplotype is strongly resistant, several other haplotypes show moderate resistance, and still others are highly susceptible (Longenecker and Mosmann, 1981). No such clear-cut examples have been found in mammals, but there are some suggestive data. In humans, for example, statistical associations have been reported between the presence of certain MHC alleles and resistance or susceptibility to a wide variety of infectious agents (Tiwari and Terasaki, 1985, chapters 18 and 23). The cheetah, which is virtually monomorphic for the MHC loci, is highly susceptible to certain pathogens (O'Brien et al., 1985), as is the cottontop tamarin, which also has a low level of MHC polymorphism (Watkins et al., 1988).

As noted earlier, MHC polymorphism is mainly caused by variation in the amino acid sequence of the ARS of the MHC molecule, which is located in the α_1 and α_2 domains of the molecule in the class I MHC and in the first domain of the α and β chains in the class II MHC. As shown in Tables 3 and 4, the rate of nonsynonymous nucleotide substitution is enhanced in the ARS, apparently because of positive Darwinian selection ($d_N > d_S$). This is exactly what is expected with overdominant selection (Maruyama and Nei, 1981), because a new mutant allele is almost always in heterozygous condition and thus enjoys a selective advantage over more common alleles, which occur in both heterozygous and homozygous condition.

The overdominance hypothesis also accounts for the other features of MHC polymorphism mentioned above. That is, the high heterozygosity (H) and the large number of alleles (n_a) observed can easily be explained by this hypothesis, since these values increase as Nv and Ns increase, where N, v, and s are the effective population size, mutation rate per locus, and selective advantage of heterozygotes over homozygotes, respectively (Table 5; Kimura and Crow, 1964; Wright, 1966; Maruyama and Nei, 1981). For example, when $Nv = 0.01$ and $Ns = 1000$, the expected values ($\overline{H}$ and $\overline{n}_a$) of H and n_a become 0.95 and 20, respectively, for a sample of 1000 individuals. When $Nv = 0.01$ and $Ns = 10,000$, $\overline{H}$ and $\overline{n}_a$ become 0.98 and 56, respectively (Wright, 1966). If a species is subdivided into subpopulations as in man and mice, $\overline{n}_a$ becomes even higher. Therefore, the speculation of Lawlor et al. (1990) that overdominant selection

TABLE 5. Means of heterozygosity ($\bar{H}$), number of alleles ($\bar{n}_a$), coalescence time ($\bar{T}_c$), and rate of amino acid substitution ($\bar{a}$) for neutral mutations (Ns = 0) and overdominant selection (Ns > 0)[a,b]

Nv	Ns	$\bar{H}$	$\bar{n}_a$	$\bar{T}_c$	$\bar{a}$
0.001	0	(0.004)	(1.0)	—	(1.0)
	33	0.698	3.6	387.0	10.8
	100	0.812	5.9	468.0	10.2
	10,000	(0.977)	(49)	—	—
0.01	0	0.056	1.3	1.5	0.8
	33	0.755	4.6	50.4	6.2
	100	0.842	7.2	80.8	7.7
	10,000	(0.980)	(56)	—	—
0.1	0	0.216	2.5	2.4	0.9
	33	0.814	7.3	13.1	3.3
	100	0.877	10.9	19.3	4.3
	10,000	(0.984)	(73)	—	—

[a]These results were obtained by computer simulation (Takahata and Nei, 1990).
[b]N, Effective population size; v, mutation rate per locus; s, selective advantage of heterozygotes over homozygotes. $\bar{T}_c$ is given in terms of N generations. $\bar{a}$ is given relative to the expected value for neutral mutations, i.e., $\bar{a} = v$. All values in this table are the means for 20 replications. It is clear that overdominant selection enhances all of $\bar{H}$, $\bar{n}_a$, $\bar{T}_c$, and $\bar{a}$ compared with the case of neutral mutations. The values in parentheses were obtained by analytical formulas; in these cases the simulation was not conducted because of the large computer time required. Sample size was assumed to be 100 individuals except for the cases where $\bar{H}$ and $\bar{n}_a$ were determined analytically; in these cases sample size was assumed to be 1000.

alone would not be able to maintain a large number of alleles such as 40 or 50 is not supported by population genetics theory.

The long-term persistence of polymorphic alleles can also be explained by overdominant selection, as can be seen from the results of a computer simulation presented in Figure 8. This figure shows the phylogeny of different alleles sampled at an evolutionary time (present). All alleles eventually go back to the common ancestral allele. The time at which all the alleles converge to this ancestral allele is called the coalescence time for polymorphic alleles, which can be used to obtain an idea of how long polymorphic alleles persist in the population. Figure 8 shows that the coalescence time is about 900N generations. As mentioned above, the effective population size for a rodent species is about 10^5. Therefore, 900N generations are about 9×10^7 generations or about 45 million years. Thus, there is no problem in accounting for the long persistence of MHC polymorphic alleles. Table 5 shows that the long persistence of poly-morphic alleles occurs whenever Ns is large and the mutation rate is relatively low.

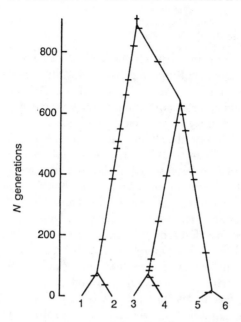

FIGURE 8. Allelic genealogy for overdominant alleles. This genealogy was obtained by computer simulation (Takahata and Nei, 1990). The "−" symbol denotes the occurrence of a mutation. $Nv = 0.001$ and $Ns = 100$ were assumed.

To study the pattern of selection involved at the MHC loci, Hedrick and Thomson (1983) and Klitz et al. (1986) applied Watterson's (1977) test of neutrality to polymorphic alleles at the class I and class II loci in humans (*HLA* loci). Finding an excess heterozygosity relative to the number of alleles, they concluded that the polymorphism at the MHC loci is maintained by some kind of balancing selection. Since overdominant selection is one form of balancing selection, their results are consistent with the overdominance hypothesis.

Earlier we mentioned that in some organisms or populations the level of MHC polymorphism is very low. In these organisms, the long-term effective population size seems to be quite low. For example, the population size of the cheetah has been estimated to be 1500 to 25,000 individuals and seems to have experienced several bottlenecks in the past (O'Brien et al., 1985). O'Brien et al. (1983) examined 52 protein loci for 55 individuals but found no variability. Therefore, the low level of MHC polymorphism in this organism is most probably caused by the small effective population size. Similarly, the low level of MHC polymorphism in the mouse populations in the North Sea can be explained by the small effective population size. These observations suggest that the selective advantage of heterozygotes over homozygotes is quite small, probably of the order of 0.01–0.1. If the selection coefficient is small, overdominant selection is effective in increasing heterozygosity only in large populations (Table 5).

Frequency-dependent selection

The hypothesis of frequency-dependent selection has been advocated by many authors in the past 20 years. Surprisingly, however, this hypothesis is ill-defined, and the supporting arguments are highly speculative. The most popular model of frequency-dependent selection assumes that host individuals carrying a recently arisen mutant allele have a selective advantage because pathogens will not have had time to evolve the ability to infect host cells carrying a new mutant antigen (Snell, 1968; Bodmer, 1972). This will result in a constant turnover of alleles in the population because old alleles lose resistance to pathogens. Although this model generates a higher value of d_N than d_S, it can explain neither the high degree of polymorphism nor the long-persisting polymorphism at the MHC loci (Hughes and Nei, 1988; Takahata and Nei, 1990).

There are some other forms of frequency-dependent selection that potentially maintain a high degree of polymorphism. Mathematically, one can show that selection due to minority advantage can enhance the level of polymorphism as much as overdominant selection (Takahata and Nei, 1990). But there is no evidence that this type of selection actually occurs. Damian (1964) proposed a model of molecular mimicry in which a parasite gains resistance to the host immune system by producing a peptide that mimics the host MHC molecule, and he suggested that this mechanism generates polymorphism in immune system loci in the host. Takahata and Nei (1990) showed mathematically that this model may indeed generate polymorphism under certain conditions; but in practice, the conditions required do not appear to be generally satisfied in natural populations. Furthermore, there is no convincing evidence that molecular mimicry of MHC molecules actually occurs (Damian, 1987).

Nevertheless, it is presently difficult to reject the hypothesis of frequency-dependent selection, particularly minority advantage, because the experimental data are too few and overdominant selection and minority advantage often produce similar population dynamics of genes. It is hoped that more experimental or statistical studies will be conducted to distinguish between the hypotheses of overdominant selection and frequency-dependent selection. One such study would be to examine the fitness of a homozygote for a rare allele at a locus. If the frequency-dependent selection hypothesis is correct, this genotype should have a higher fitness than a heterozygote for a pair of common alleles. In contrast, if the overdominance hypothesis is correct, the reverse would be true.

EVOLUTION OF MHC GENES

As mentioned earlier, class I MHC loci are of two types, classical and nonclassical, the latter being apparently largely nonfunctional. The number of classical class I loci is one to three in all mammalian species thus far studied. The human and chimpanzee genomes each has three classical loci. In the mouse, most haplotypes have only two classical loci, whereas some (e.g., H-2^d) have three. In the miniature swine there are only two classical loci (Singer et al., 1987), whereas in the rat (Howard, 1987) and in some rabbit strains (Tykocinski et al., 1984), there is only one classical class I locus.

At first sight, this limited number of classical class I loci appears to be contradictory to the overdominant selection hypothesis, because under this hypothesis individuals with many heterozygous loci are expected to be at an advantage over those with fewer heterozygous loci. In practice, however, there is natural selection that acts to limit the number of expressed MHC loci. That is, for each self MHC antigen that exists a certain proportion of the T-cell population must be eliminated as self-reactive. If the proportion of the T-cell repertoire that is eliminated becomes very large, this would impose a prohibitive burden on the organism (Howard, 1987; Hughes and Nei, 1989a; Lawlor et al., 1990). Consequently, the number of polymorphic loci probably cannot increase very much. It should also be noted that the number of alleles at the MHC loci need not be as large as the number of different T-cell receptors (> 10^7) because an MHC molecule recognizes a group of foreign peptides sharing a common structural motif (Sette et al., 1988).

Although the above argument accounts for the limited number of classical class I genes, it does not explain why there are so many nonclassical genes in the genomes of humans, mice, and, probably, other mammalian organisms. How did they evolve? To answer this question, we studied the phylogenetic relationships of both classical and nonclassical genes from the human (H), mouse (M), rabbit (R), pig (P), bovine (B), and chicken (C), using almost all available DNA sequences (Hughes and Nei, 1989c). Figure 9 shows a phylogenetic tree for the 45 genes examined. In this figure the genes or alleles marked with an asterisk are nonclassical or pseudogenes, whereas the others are classical. In all organisms, a nonclassical gene is more closely related to classical genes of its own species than it is to any other genes of any other species. Mouse nonclassical genes such as Q and TL are more closely related to mouse non-classical or classical genes than they are to any other genes, including human and pig nonclassical genes. Thus in each mammalian order classical and non-classical loci diverged relatively recently.

This finding supports the hypothesis that nonclassical genes have arisen independently in separate lineages as a result of unequal crossover events (Rogers, 1985; Klein and Figueroa, 1986; Howard, 1987). Under this hypothesis, one copy of a duplicated classical locus may become nonclassical as a result of mutations that reduce its expression. Once nonclassical genes are produced, they may again be duplicated. Over long periods of evolutionary time, however, nonclassical genes would eventually degenerate into pseudogenes because of deleterious mutations or would be removed from the genome by unequal crossing over (Figure 10). This would explain why distantly related mammalian species do not share nonclassical loci, whereas closely related species such as humans and chimpanzees may share them. This hypothesis is also supported

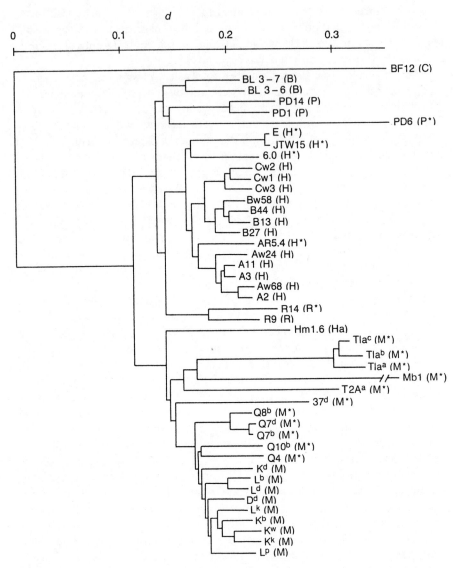

FIGURE 9. Phylogenetic tree for mammalian classical and nonclassical class I MHC genes. This tree was constructed by the neighbor-joining method (Saitou and Nei, 1987). The chicken class I gene *B-F12* was used as an outgroup. M, mouse; Ha, hamster; R, rabbit; H, human; P, pig; B, bovine; C, chicken; *, nonclassical gene or pseudogene. Nonclassical and pseudogenes usually have a long branch indicating acceleration of nucleotide substitution. Branch lengths are measured in the number of nucleotide substitutions per site.

Time 1

Time 2

Time 3

Time 4

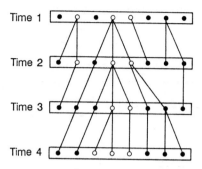

FIGURE 10. Schematic diagram illustrating relationships among classical (represented by open circles) and nonclassical (solid circles) class I loci in a hypothetical MHC. The figure shows the composition of the MHC at four different evolutionary times, which are separated by long periods (about 20 million years).

by the fact that the nonclassical Q genes in the mouse have DNA segments homologous to regulatory sequences involved in the universal expression of classical class I genes but that they have accumulated numerous nucleotide substitutions in these segments (Hughes and Nei, 1989c). However, it should be mentioned that some nonclassical genes are expressed at a low level in some tissues and that molecular biologists (e.g., Strominger, 1989) are still debating the function of these genes. It is possible that some of these genes maintain the original function or have acquired a new function.

Evolution of class II MHC genes is somewhat different from that of class I MHC genes. In this class of genes there are not as many unexpressed genes as in the case of the class I MHC, and it is possible to identify orthologous genes between humans and mice. The α chain and β chain genes in the class II MHC are quite different in both humans and mice and apparently diverged much earlier (probably more than 370 million years ago, Klein, 1986) than the time of mammalian radiation (about 75 million years ago). The divergence of different groups of the α and β chain genes also occurred before the mammalian radiation. Thus, DQ sequences in the human are more similar to A sequences in the mouse than they are to human DR sequences. Figure 11 shows the evolutionary relationships of genes from different loci of the human and mouse. In both α and β chains there are pairs of orthologous genes between the two organisms. Even the $A\beta2$ gene in the mouse and the $DO\beta$ gene in the human, of which the function is unknown, show an orthologous relationship. (This evolutionary pattern is quite different from that of class I genes.) At present, it is not clear why this difference exists between the two classes of genes. In the class II MHC, α and β chain genes are arranged in alternating fashion on the chromosome, and the mature glycoprotein is a heterodimer consisting of non-covalently associated α and β chains. Perhaps the need for synchronous transcription of paired α and β chain genes generates selective pressure against

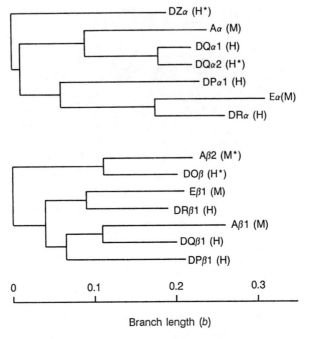

FIGURE 11. Phylogenetic trees for the class II MHC genes from humans and mice. H, human; M, mouse. Branch lengths are measured in the number of nucleotide substitutions per site.

duplication of class II loci, whereas this constraint is absent in the case of class I MHC.

It should, however, be noted that there are pseudogenes or apparently nonfunctional genes that have recently diverged from the functional genes. For example, genes DQβ2 and DRβ2 in Figure 1 diverged from DQβ1 and DRβ1, respectively, much later than the time of mammalian radiation. These genes apparently correspond to nonclassical class I genes and have essentially the same evolutionary pattern as that of the latter. Unexpressed or largely unexpressed class II loci may be called nonclassical class II loci, whereas the remaining loci may be called classical class II loci.

SUMMARY AND CONCLUSIONS

Classical MHC loci are extremely polymorphic, and there are several features that are unique to MHC polymorphism. (1) The heterozygosity is as high as 90% in many human and mouse populations. (2) The number of nucleotide differences between alleles is unusually high. (3) In the antigen-binding site of the gene, the rate of nonsynonymous substitution is much higher than that of synonymous substitution. (4) Polymorphic alleles (allelic lineages) persist in the population for tens of millions of years.

These observations cannot be explained by the neutral mutation hypothesis, even if a high mutation rate is assumed. They are also incompatible with the gene conversion hypothesis, though gene conversion itself may occur occasionally. To explain the above observations, some form of balancing selection is necessary. One of the most reasonable forms of balancing selection to explain them is overdominant selection or heterozygote advantage. The hypothesis of overdominant selection is well supported by the function of MHC molecules and the pattern of nucleotide substitution in MHC genes. From the mathematical point of view, the above features of the MHC polymorphism can also be explained by frequency-dependent selection such as minority advantage. However, it is not clear how realistic this model is for the case of MHC polymorphism. It seems that more careful experimental studies are necessary to distinguish between the overdominance and frequency-dependent selection hypotheses.

The function of monomorphic nonclassical genes is currently debated. Our study of the evolutionary pattern of nonclassical genes suggests that they have arisen from classical genes and are largely nonfunctional or are in the process of degeneration into nonfunctional genes. The low level of polymorphism of nonclassical genes seems to reflect the absence of balancing selection at these loci.

CHAPTER 12

POPULATION GENETICS OF HLA

Philip W. Hedrick, William Klitz, Wendy P. Robinson, Mary K. Kuhner, and Glenys Thomson

The major histocompatibility complex (MHC) in humans, known as the HLA (human leukocyte antigen) system, consists of at least 50 closely linked genes whose products control a variety of functions concerned with the regulation of immune responses. The great variety and complexity of molecular, genetic, and physiological phenomena of the HLA region make it an extraordinary resource and exemplary system for studying population genetics. However, the evolutionary dynamics of HLA are extremely complex and understanding them provides a great challenge to population geneticists.

GENERAL BACKGROUND

The HLA system was discovered as a blood-group-like system detected on white blood cells (for reviews see Terasaki, 1980; Albert et al., 1984; Dupont, 1989). The major impetus for the early investigation of the HLA system was the need to match organ donors and recipients for the antigens important in tissue transplantation. Although this has turned out to be much more difficult than was at one time hoped, from this effort has grown our knowledge of the HLA system and homologous systems in other species. It is now known that these histocompatibility systems contain many genes that control a variety of functions (see, e.g., Klein 1986) including determination of cell surface molecules, immune response differences, components of the complement system, and possibly other related functions connected with cell–cell regulation, hormone receptors (Edidin, 1986), and maternal–fetal interaction (Gill, 1983).

The HLA class I and II alleles were originally defined using the ability of selected polyclonal antisera to identify and discriminate specific antigenic (allelic) variants. This included both so-called public and private alleles. Public alleles include, for example, *Bw4* and *Bw6*, which divide *HLA-B*-locus products into two discrete sets. These are in turn further subdivided into 40 serologically defined *HLA-B* alleles described in the Tenth HLA Workshop (Dupont, 1989). The functional significance of these allelic definitions remains open to question, but at present HLA serology is the primary means for rapidly describing HLA variation, and the method by which virtually all available population data have been accumulated. The Tenth HLA Workshop (Dupont, 1989) launched an ambitious effort to define additional HLA variation by utilizing a battery of methods, including the analysis of restriction fragment length polymorphisms (RFLPs), monoclonal antibodies, biochemistry, T-cell clones, and DNA sequence information. A great deal of additional diversity has been revealed.

What has been missing is a means to cast this wealth of information on genetic variation into a biologically meaningful, functional context. The results of X-ray crystallography on *HLA-A2* by Bjorkman et al. (1987b) and *HLA-Aw68* by Garrett et al. (1989) make it possible to assign the variation (in the form of DNA base pairs, amino acid sequence, or antigenic sites) to its relationship to the antigen-presenting function of the MHC product. Now the critical question can be asked of genetic variation at the histocompatibility loci revealed by any technique: does the observed variation play a role in antigen recognition and T-cell interaction? This assignment is now possible for all variation observed from DNA and protein sequence data. The genetic basis of serologically defined variation is rapidly being determined so that the functional relevance of the serological definitions will soon be known.

GENETICS

By a combination of family and somatic cell hybrid studies, the HLA system was mapped to the short arm of human chromosome 6. The HLA region consists of 3500 kilobases (kb) of DNA covering about two map units and containing at least 50 genes (Dupont, 1989). It includes the class I genes (*A*, *B* and *C*), the class II genes (*DR*, *DQ*, and *DP*), and the class III or complement region (Figure 1). The complement region contains several genes involved in the classical (C2, C4A, C4B) and alternate (factor B) complement pathways as well as the genes for cytochrome *P-450* 21-hydroxylase (21-OHA and 21-OHB). Also indicated in Figure 1 is the location of two genes determining tumor necrosis factors, *TNF-A* and *TNF-B*.

The "classical" class I genes *HLA-A*, *-B*, and *-C* encode a 43-kD transmembrane heavy chain that is associated with β_2-microglobulin (encoded by chromosome 15) and is expressed on most cells (Figure 2). These cell surface antigens present processed foreign antigen for recognition by killer and effector T cells. There are also a large number of "nonclassical" loci that are similar to the classical loci at the sequence level; however, only one (*HLA-E*) has been detected serologically and it appears that many of the others are pseudogenes. The mouse MHC contains a family of nonclassical sequences, the *Qa* loci, some of which are known to be expressed. (The existence of *Qa* homologues

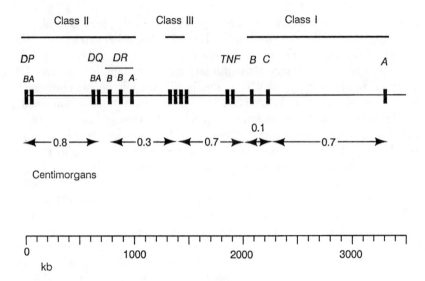

FIGURE 1. Map of the HLA region showing both the level of recombination (in centimorgans) and physical map distances (in kilobases).

in humans is controversial.) The three-dimensional structure of a human class I antigen (Figure 2) shows that the antigen recognition site (or pocket) consists of a groove formed by two α-helices sitting on top of a β-pleated sheet (Bjorkman et al., 1987). The antigen-binding properties of the MHC molecule are determined by the amino acids lining the antigen pocket. Two subunits, α_1 and α_2, which are 90 and 92 amino acids long, respectively, form this pocket (30 and 27 amino acids for α_1 and α_2 are actually part of the antigen recognition site). α_1 and α_2 are encoded by exons 2 and 3 of the class I genes, an example of exact exon–domain correspondence.

Class II antigens comprise two noncovalently associated transmembrane polypeptides, a 31-kD α chain and a 28-kD β chain, that function in presentation of foreign antigen to helper T cells and are expressed on B lymphocytes, macrophages, and monocytes. Expressed loci in the class II region each consist of at least one A and one B gene (Trowsdale, 1987); chromosomes can have one or more *DRB* genes. The class I and class II genes share structural homologies and are both members of the immunoglobulin supergene family.

A distinctive feature of the HLA system is the high level of polymorphism exhibited by the class I and class II loci. By use of serological and cellular typing methods, more than 20 antigens have been defined for the *HLA-A* locus, 40 for the *B* locus, 10 for *C*, 14 for *DRB1*, 6 for *DP*, 7 for *DQ*, and 24 for the mixed lymphocyte culture response *Dw* (Dupont, 1989), making this region far more polymorphic than any other known human system. Use of RFLPs, monoclonal antibodies, isoelectric focusing, and T-cell typing has allowed further subdivision of many specificities. The class III loci *C2*, factor B, *C4A*, and *C4B*

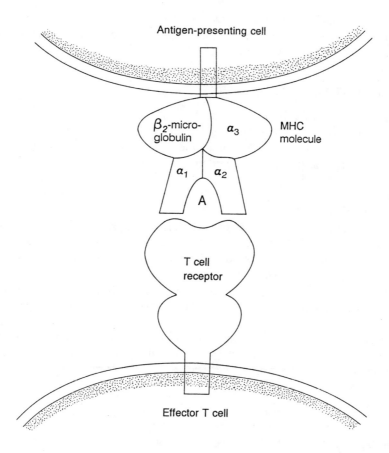

FIGURE 2. A class I MHC molecule presenting an antigen to a T cell. The antigen A is bound in the antigen recognition site of the MHC molecule. The antigen recognition site is composed of amino acids from subunits α_1 and α_2; subunits α_3 and β_2-microglobulin form most of the remainder of the molecule. The T-cell receptor, the variability of which is generated somatically to produce unique clones, makes specific contact with both the MHC molecule and the antigen.

exhibit polymorphism (although to a lesser extent than for the class I and II loci) with 4, 5, 9, and 11 alleles respectively (Albert et al., 1984).

Nonrandom association (linkage or gametic disequilibrium) between certain combinations of antigens at different loci (termed haplotypes) is another predominant feature of the HLA loci. There are characteristic groups of antigens exhibiting elevated frequencies and gametic disequilibrium in different ethnic and racial groups. For example, in Caucasians the antigens A1, B8, and DR3 show significant nonrandom associations. A number of other combinations of alleles exhibiting significant gametic disequilibrium also exist in these popula-

tions. Although a number of cases of strong linkage disequilibrium are found
in each population group, many combinations of antigens do not exhibit significant gametic disequilibrium.

ASSOCIATION WITH DISEASES

Many diseases have been shown to be associated with HLA antigens (e.g.,
Tiwari and Terasaki, 1985); these diseases affect all organ systems, although an
autoimmune etiology is a common theme. The most commonly used statistical
measure of an association between an HLA marker and a disease is the odds
ratio (OR) which is the cross-product ratio in a 2 × 2 contingency table containing the frequency of the antigen among patients versus its frequency in
controls (Woolf, 1955).

Some of the diseases known to be associated with specific HLA antigens are
listed in Table 1. The discovery of HLA associations with insulin-dependent
diabetes mellitus (IDDM) helped distinguish this early onset diabetes from non-
insulin-dependent diabetes, which is now known to have a distinctive etiology.
The case of ankylosing spondylitis is one of the most striking examples of HLA–
disease association. The frequency of the antigen B27 in Caucasian patients is

TABLE 1. HLA disease associations as shown by the frequency of HLA antigens in
patients with various diseases and in control populations and the odds ratio[a]

Disease	Antigen	Race[b]	Frequency Patients	Frequency Controls	Odds ratio
Narcolepsy	DR2	C	1.0	0.22	129.8
		O	1.0	0.34	358.1
Ankylosing spondylitis	B27	C	0.89	0.09	69.1
		O	0.81	0.01	354.4
		N	0.58	0.04	54.4
Reiter's disease	B27	C	0.47	0.10	8.2
Insulin-dependent diabetes mellitis	B8	C	0.40	0.21	2.5
	B15	C	0.22	0.14	2.1
	DR3	C	0.52	0.22	3.8
	DR4	C	0.74	0.24	9.0
	DR2	C	0.04	0.29	0.1
Rheumatoid arthritis	DR4	C	0.68	0.25	3.8
		O	0.66	0.39	2.8
		N	0.44	0.10	5.4

[a]Data adapted from Tiwari and Terasaki (1988) and Thomson et al. (1988)
[b]C, Caucasian; O, Oriental; N, Negro.

0.89, compared to only 0.09 in controls (OR = 69.1). Furthermore, a high frequency of B27 in ankylosing spondylitis patients is also found in Orientals and Negroes. DR2 is strongly associated with narcolepsy in Caucasian, Black, and Asian populations. Initial studies showed that all narcoleptic individuals had DR2, whereas among healthy Caucasians the frequency of DR2 is 0.22 (OR = 129.8). Subsequent cases of DR2-negative narcoleptic patients have been reported.

One explanation for these associations is that disease susceptibility is a direct result of the presence of a particular HLA antigen. This may, in particular, be the case for ankylosing spondylitis and narcolepsy. In the case of ankylosing spondylitis, most patients have the B27 antigen, this association appears to be in all racial groups, and the B27 haplotypes in patients are indistinguishable from control haplotypes. For narcolepsy and DR2, the same observations, except the last, hold. DR2 haplotypes in patients can be differentiated by A and B alleles from control haplotypes. This observation does not exclude a direct role of the antigen in disease: a particular (as yet undefined) subset of DR2 haplotypes may be the predisposing agent. A second explanation for an HLA–disease association is that an increased frequency of a particular antigen(s) with a disease is the result of gametic or linkage disequilibrium between the antigen locus and the alleles at a nearby locus that confer susceptibility to the disease. It is difficult to differentiate between these two alternative possibilities.

For some diseases, different racial groups exhibit different HLA associations. Insulin-dependent diabetes mellitus is strongly associated with DR3 and DR4 in Caucasians, while in the Chinese the association is with DR3 and DRw9, and in the Japanese, the association is with DR4, DRw8, and DRw9 (DR3 is very rare in the Japanese) (see Thomson, 1988 for references). The value of ethnic comparison of disease associations in terms of delineating disease predisposing components and the evolution of disease predisposing genes cannot be overemphasized.

It is now clear that many of the associations of HLA and disease may involve heterogeneity in both HLA components and non-HLA genetic components (Thomson, 1988). In fact, the pattern of inheritance of these diseases is often complex because of incomplete penetrance, variable disease expressivity, disease heterogeneity, and interaction among multiple predisposing loci.

Insulin-dependent diabetes mellitus and DQB position 57

The role of HLA in predisposition to IDDM provides a good demonstration of the multiple ways in which HLA genes interact singly and in combination to produce the same disease. Because of the polymorphism and gametic disequilibrium among the several class I and II loci, it has been difficult to define the genetic element(s) responsible for this disease. The basic approach has been to find new serological or molecular variants and then look for increased frequencies of these variants in the affected population compared to controls.

It has been known for many years that the serologically defined alleles DR3 and DR4 are associated with IDDM (see Table 1). Molecular and serological subdivisions of DR4 based on variation at the nearby DQB locus may localize susceptibility factors further. The enthusiasm for the reductionist approach reached its apogee with the announcement that variation at a single amino acid position in the DQB chain (position 57) explained much of the HLA component of IDDM susceptibility (Todd et al., 1987). Position 57 is in fact a reasonable candidate for such a role because it forms part of the lining of the antigen pocket of the MHC molecule, which determines the antigen-binding characteristics. Several lines of evidence, however, have been pointed out that make the two-allele disease model (one "disease" allele and one "healthy" allele) for IDDM untenable (Segall, 1988; Klitz, 1988; Thomson et al., 1988, as well as several earlier papers). For one thing, the "susceptible and protective" amino acids at position 57 are not consistently present and absent in affected and unaffected individuals, respectively. Other findings not explained by the two-allele model are that the allele DR2 confers an additional degree of protection, that the DR3/DR4 heterozygotes are more susceptible than other DR3 or DR4 genotypes, and further that the DR3 and DR4 predisposing haplotypes show different modes of inheritance (see e.g. Thomson, 1988). In addition, more than one locus of a single haplotype may be necessary to produce susceptibility. For example, the most susceptible DR4 haplotypes are those that contain both the appropriate DQB gene (defined at position 57) and a particular DRB1 gene (Sheehy et al., 1989).

When all HLA–DR haplotypes are examined for susceptibility to IDDM, many distinguishable effects are uncovered (Klitz et al., 1990). These include susceptibility conditional on parental sex, differences within a susceptibility class (e.g., DR4) and genotypic interactions dependent on a particular haplotype, for example, DR3-B18 and DR4. The overall picture is that several genetic elements can all lead to the same end point: in this case, the autoimmune destruction of the insulin-producing cells of the pancreas.

SINGLE-LOCUS VARIATION

Although the relationship of HLA variants to particular diseases suggests that selective forces have had substantial effects on the alleles in the HLA region, it is important that we objectively examine the variation in this region and not make a priori assumptions about selective factors. The neutral mutation theory gives a useful starting point for predicting distributions of allele frequencies if selection were not acting. Recent reviews (Klein, 1987; Serjeantson, 1989) have, in fact, suggested that HLA variation is essentially neutral, a conclusion that we hope to dispel in the ensuing discussion (see also the recent studies of Hughes and Nei, 1988, 1989a). In the neutral model, the expected equilibrium heterozygosity (or homozygosity) in a population is a function of the combined effects of genetic drift and mutation, thereby providing theoretical predictions against which the observed genetic variation in a population or a sample may be compared.

Neutrality test for allelic variation

Ewens (1972) developed a sampling theory to predict the distribution of alleles observed in a sample of size n taken from a population at equilibrium under neutrality. Watterson (1978a,b) extended this approach and developed a test that allows the comparison of the estimated homozygosity expected in a sample of size n containing k alleles to the homozygosity expected under neutrality. This conditional homozygosity F is defined as

$$F = \Sigma p_i^2 \tag{1}$$

where p_i is the frequency of the ith allele in the sample of size n. (Note that this test does not examine deviation from Hardy-Weinberg proportions but examines the allelic frequency array expected under neutrality conditional on the sample size and the number of alleles observed.)

Using this approach, Hedrick and Thomson (1983) compared the conditional homozygosity in 22 samples at both the A and B loci of the HLA region to neutral expectations. In all cases, the estimated homozygosity was less (estimated heterozygosity was greater) than expected, indicating a more even allelic frequency distribution than under neutrality. The homozygosity was significantly less ($p < 0.05$) than neutral expectations for 25 of the 44 cases (see Table 2 for some of these data). A variety of evolutionary factors that potentially influence the level of homozygosity conditional on n and k, such as gene flow, population bottlenecks, unidentified alleles, and balancing selection that could possibly decrease conditional homozygosity (increase heterozygosity) relative to neutral expectations, were examined. Some form of balancing selection is the explanation most consistent with the level of conditional homozygosity at the A and B loci in the populations studied. The relatively high rate of gene conversion in the HLA region (see below) is probably not a candidate for explaining the low conditional homozygosity because this statistic is generally independent of the mutation rate and therefore also of gene conversion rate (Watterson, 1989).

Using population data from the Ninth International Histocompatibility Workshop (Albert et al., 1984), Klitz et al. (1986) applied this approach to samples for nine loci of the HLA region. As in the previous samples, loci A and B had homozygosities statistically significantly less than neutral expectations (both $p < 0.001$). In addition, the other loci that encode membrane glycoproteins, C, DQ, and DR, as well as locus Glo-I (an enzyme locus about five map units to the left of the class II region), had estimated homozygosities significantly less than neutral expectations (all $p < 0.001$). However, the four complement loci had estimated homozygosities either consistent with (BF, $C4B$, and $C4A$)

TABLE 2. Number of alleles (*k*), sample size (*n*), expected and observed homozygosity, and the significance level for different populations at the *HLA-A* and *B* loci

Population	*k*	*n*	Homozygosity[a]		*p*
			Expected	Observed	
HLA-A					
Caucasian					
American	18	1734	0.215	0.134	<0.1
French	17	874	0.233	0.139	<0.05
Italian	17	1044	0.233	0.113	<0.025
African blacks	17	286	0.204	0.100	<0.01
Japanese	17	1878	0.233	0.217	—
HLA-B					
Caucasian					
American	31	1734	0.121	0.065	<0.01
French	31	874	0.121	0.068	<0.01
Italian	29	1049	0.131	0.073	<0.025
African blacks	25	286	0.135	0.089	—
Japanese	29	1900	0.130	0.075	<0.025

Source: From Hedrick and Thomson (1983).
[a]Hardy–Weinberg homozygosity in a sample drawn from a population at equilibrium under neutrality (expected) and the Hardy–Weinberg homozygosity found in different populations (observed).

or exceeding (C2) ($p < 0.01$) neutral expectations. These results are particularly interesting because they suggest that the class III complement loci, which are embedded in the HLA region, display quite different evolutionary histories from the class I and II HLA loci despite their close linkage and the background of extensive disequilibrium in the region.

Amino acid heterozygosity

Amino acid and DNA sequence data have been obtained for a number of HLA alleles (Parham et al., 1988 and references therein; Kato et al., 1989). In general, the alleles at a locus are quite different from one another, with only approximately 95% amino acid identity among alleles at either *HLA-A* or *HLA-B*. This variation is concentrated in the exons encoding the α_1 and α_2 domains; it is especially striking in the amino acid residues of the antigen recognition site (Parham et al., 1988).

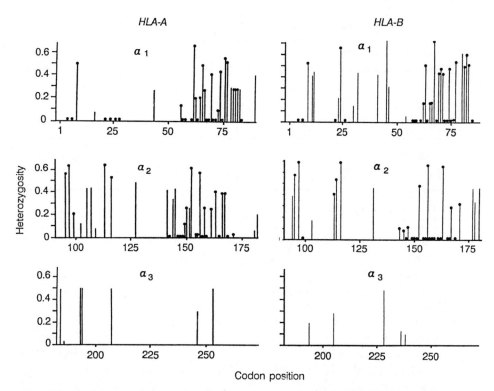

FIGURE 3. The amino acid heterozygosity for loci *HLA-A* and *HLA-B* divided into the α_1, α_2, and α_3 domains. Closed circles indicate codons in the antigen recognition site (ARS) (all are indicated) and open circles indicate non-ARS codons with a nonzero heterozygosity.

To illustrate the extent of this variability, we have calculated heterozygosity for individual amino acids in exons 2, 3, and 4 (corresponding to the three extracellular domains α_1, α_2, and α_3 domains respectively). Figure 3 shows amino acid heterozygosities based on 12 *HLA-A* alleles (representing 93.1% of the population) and 18 *HLA-B* alleles (73.8% of the population) (data from Parham et al., 1988; Kato et al., 1989; Ennis et al., 1989). Allele frequencies in the Caucasian population (from Albert et al., 1984) have been normalized to sum to unity. In cases where several sequences were available for one serotype, the lowest-numbered sequence (i.e., A2.1 to represent all A2 sequences) was chosen. There is relatively little variation among sequences from the same serotype (<1%; Hedrick et al., unpublished).

Amino acid heterozygosity was calculated assuming Hardy-Weinberg proportions as

$$H = 1 - \Sigma p_i^2 \qquad (2)$$

TABLE 3. The average individual amino acid heterozygosity for exon codons for subunits α_1, α_2, and α_3 broken down for loci A and B, and codons in the antigen-recognition site (ARS) and those that are non-ARS[a]

Locus	α_1		α_2		α_3
	ARS (30)	Non-ARS (60)	ARS (27)	Non-ARS (65)	(92)
HLA-A	0.172	0.024	0.244	0.040	0.031
HLA-B	0.228	0.057	0.191	0.041	0.005

[a]The numbers of sites in each category are given in parentheses.

where p_i is the population frequency of the given amino acid at the position under consideration. For example, if seven alleles comprising 70% of the population have arginine at a given position, while the remaining five alleles, comprising 30% of the population, have valine, $H = 0.42$; 42% of individuals will be heterozygous for this position.

The positions of highest heterozygosity are concentrated in the antigen recognition site (ARS) as shown by the closed circles in Figure 3. Invariant positions do exist in the ARS and some highly variable sites elsewhere. Exon 1, which encodes the transcribed leader, also contains a few highly variable positions; the remainder of the molecule is less variable (data not shown). Across exons 2–4, the average heterozygosity at ARS positions is 20.8%; at other positions, 3.1% (Table 3). This high heterozygosity at positions critical to antigen recognition suggests positive selection in favor of diversity. This pattern was also demonstrated by the analysis of Hughes and Nei (1988) and Nei and Hughes (this volume) showing that in the ARS nonsynonymous codon changes are proportionally more common than silent ones. The patterns of variability in HLA-A and HLA-B differ in detail, suggesting some specialization of each locus for antigen recognition.

MULTILOCUS VARIATION

Under neutrality, a population at equilibrium has an expected nonrandom association, gametic disequilibrium, between alleles at different loci. Even if there is no selection among different alleles at a locus, the combined effects of genetic drift and mutation result in an interlocus association that is highest when there is limited recombination (e.g., Ohta and Kimura, 1969; Hill, 1975). Hedrick and Thomson (1986) used the computer simulation approach of Hudson (1983) to determine the extent of gametic or linkage disequilibrium expected in a sample of size n with k and l different alleles at two loci. These disequilibrium values also depend upon the amount of recombination as measured by the quantity $4Nc$ where c is the rate of recombination between the loci. The gametic or linkage disequilibrium for a given gamete (or haplotype) is defined as

$$D_{ij} = x_{ij} - p_i q_j \qquad (3)$$

where x_{ij} is the observed frequency of gamete $A_i B_j$, while p_i and q_j are the frequencies of alleles A_i and B_j at loci A and B. The expected frequency of gamete $A_i B_j$ is $p_i q_j$, assuming no association of the alleles (see Hedrick et al., 1978, for a review). The range of this measure of gametic disequilibrium is a function of the allelic frequencies and although no ideal measure of association exists (Hedrick, 1987; Lewontin, 1988) it is often useful to consider D'_{ij}, the value of D_{ij} normalized by the maximum value it could have for this combination of allele frequencies and sign (Lewontin, 1964).

Haplotype $A1$-$B8$ has the highest positive disequilibrium in many Caucasian populations. In a large (5202 individuals) and relatively homogeneous sample from Denmark (Hansen et al., 1979), $D = 0.077$ and $D' = 0.728$. In other words, the disequilibrium between $A1$ and $B8$ is 72.8% of the maximum possible. Although there are cases, such as $A1$-$B8$, of exceptionally large disequilibrium values, the values for other haplotypes show little or no statistical association.

Disequilibrium and physical map distance

A number of different approaches, some of which are strongly correlated, have been suggested to measure the overall gametic disequilibrium when there are multiple alleles at two loci (see, e.g., Hedrick and Thomson, 1986; Hedrick, 1987). For example, one measure (Maruyama, 1982) is

$$D^* = \frac{D^2}{H_A H_B} \qquad (4)$$

where

$$D^2 = \Sigma\Sigma D_{ij}^2$$
$$H_A = 1 - \Sigma p_i^2$$
$$H_B = 1 - \Sigma q_j^2$$

When there are two alleles at a locus, then the expectation of D^* in a finite population is approximately linear with Nc, i.e., D^* decreases linearly with increasing Nc (Hill and Robertson, 1968; Ohta and Kimura, 1969; Sved, 1968). Furthermore, Sved (unpublished) has shown that with multiple alleles, D^* is also approximately linear with Nc.

Baur et al. (1984) estimated the haplotype frequencies in Caucasians for all pairwise combinations of eight loci (A, B, C, DQ, DR, Bf, C4A, and C4B)

TABLE 4. The kilobase distance, estimated recombination fraction, and disequilibrium as measured by D^* between eight HLA loci

	Locus pair	kb[a]	c	D^*
Class I	B–C	150	0.001	0.033
	A–C	1150	0.007	0.005
	A–B	1200	0.008	0.009
Class II	DQ–DR	225	0.0	0.076
Class III	C4A–C4B	30	0.0	0.030
	C4A–Bf	45	0.0	0.010
	C4B–Bf	75	0.0	0.009
Class I–II	A–DQ	2650	0.018	0.003
	A–DR	2425	0.018	0.003
	B–DQ	1450	0.010	0.007
	B–DR	1225	0.010	0.010
	C–DQ	1600	0.011	0.009
	C–DR	1375	0.011	0.008
Class I–III	A–C4A	1900	0.015	0.006
	A–C4B	1930	0.015	0.003
	A–Bf	1855	0.015	0.003
	B–C4A	700	0.007	0.015
	B–C4B	730	0.007	0.008
	B–Bf	655	0.007	0.013
	C–C4A	850	0.008	0.008
	C–C4B	880	0.008	0.016
	C–Bf	805	0.008	0.007
Class II–III	DQ–C4A	750	0.003	0.009
	DQ–C4B	720	0.003	0.009
	DQ–Bf	795	0.003	0.007
	DR–C4A	495	0.003	0.010
	DR–C4B	525	0.003	0.012
	DR–Bf	570	0.003	0.016

[a]Data adopted from Carroll et al. (1984), Trowsdale and Campbell (1988), Dunham et al. (1987), and Caroll (1987) (detailed in Robinson, 1989).

with sample sizes ranging from 333 to 2130. Table 4 gives the D^* values for these 28 locus pairs as well as the kilobase distance and estimated amount of recombination between these loci (Robinson, 1989). Figure 4 shows a plot of the extent of disequilibrium found as a function of kilobase distance for these locus pairs. Obviously, there is a general monotonic trend of decreasing D^* with greater physical distance as expected theoretically. (Rare alleles, those with frequencies less than 0.005, were combined together in one allelic class.) The correlation of kilobase distance and D^* is 0.84 ($r^2 = 0.71$), and that between

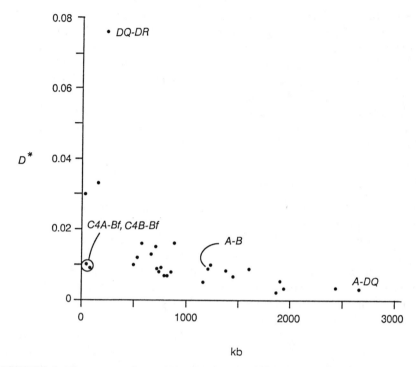

FIGURE 4. The pairwise disequilibrium values using D^* between eight HLA loci as a function of kilobase distance.

the recombination estimate and D^* is 0.81 ($r^2 = 0.66$) (Robinson, 1989). For kilobase and recombination distance, $D^* = 0.96$ and $r^2 = 0.92$.

The disequilibrium values observed can be tested against neutral expectations. Confidence limits have been determined for the D^* estimates for values of $4Nc = 0$, 1, 10, 100, and 500, for sample sizes of 100, 200, and 400 haplotypes, and for up to eight alleles per locus (Hedrick and Thomson, 1986). Using these estimates, we find that the data best fits values for N, the effective population size, of 5,000–10,000 (Robinson, 1989). Although the DR–DQ disequilibrium seems quite large, this value is consistent with any value of $4Nc \leq 10$. As no recombination has been observed between these two loci, this is not an unreasonable upper limit. On the other hand, the C4A-Bf and C4B-Bf disequilibrium values seem rather small, and no recombination has ever been observed between these loci either. Given that neither of the C4 loci has more than six alleles (rare alleles are lumped), these D^* values are consistent with any value of $4Nc$ between 0 and 100.

Although there is general concordance between the physical map distance and the recombination fraction in the HLA region, there are suggestions of greater recombination rates in certain regions, so-called recombinational hotspots. There is convincing evidence for a recombinational hotspot in the mouse *H2-I* region where the majority of recombinants has been localized to a 1.7-kb interval within a region of 150 to 200 kb (Steinmetz et al., 1982; Kobori et al., 1984). Bodmer and Bodmer (1989) state they have identified a recombinational hotspot in the *HLA-DQ* region based on the absence of gametic disequilibrium between restriction fragment length polymorphisms at the *DQ* and the closely linked *DX* loci. However, because most measures of gametic disequilibrium are dependent on allelic frequencies, care should be taken in inferring recombinational hotspots from disequilibrium data (Hedrick, 1988a).

Disequilibrium pattern analysis

Selection may affect the distribution of particular disequilibrium values when there are multiple alleles without strongly affecting the overall average magnitude of disequilibrium between two loci. We have developed an approach, termed disequilibrium pattern analysis, designed to examine the distribution of multiallelic disequilibrium values and identify patterns that are consistent with past selective events (Thomson and Klitz, 1987; Klitz and Thomson, 1987). A selected haplotype in a two-locus multiallelic system produces a distinct pattern of linkage disequilibrium values for all generations while the selection is acting. The pattern is also maintained for many generations after the selection event, until the disequilibrium pattern is eventually broken down by genetic drift and recombination. Related haplotypes, sharing an allele with a selected haplotype, assume an expected value of gametic disequilibrium proportional to the frequency of the unshared allele and have a single negative value of the normalized gametic disequilibrium. The disequilibrium pattern predicted under selection is robust with respect to the influence of gene flow and random genetic drift.

As an example of the disequilibrium pattern approach, Figure 5 gives the observed distribution of D_{ij} and D'_{ij} values for allele $A1$ and alleles at the B locus from a Caucasian population (Klitz and Thomson, 1987). In Figure 5A, $A1$-$B8$ is the main haplotype in the positive space with two low-frequency haplotypes having low positive disequilibrium (D), while the rest of the related haplotypes (non-$B8$ and $A1$) have disequilibrium values proportional to the frequency of the unshared B allele. Using normalized disequilibrium values (D') as shown in Figure 5B for $A1$ haplotypes, we find an approximate alignment of the negative values with the more common haplotypes. The values fall between -0.6 and -0.8. Rare haplotypes, for example, $A1$-$B47$, $A1$-$B38$, and $A1$-$B13$, depart furthest from this alignment, apparently due to sampling effects. It is likely that the misalignment of $A1$-BX occurs because the "blank" allele BX is an mixture of unidentified B locus alleles.

This pattern is most easily explained by selection favoring the $A1$-$B8$ haplotype (Thomson and Klitz, 1987). There are six haplotypes that have disequilibrium patterns consistent with a past selective event, $A1$-$B8$, $A3$-$B7$, $A25$-$B18$, $A29$-$Bw44$, $A2$-$B15$, and $A1$-$B17$, the first two listed being the most striking

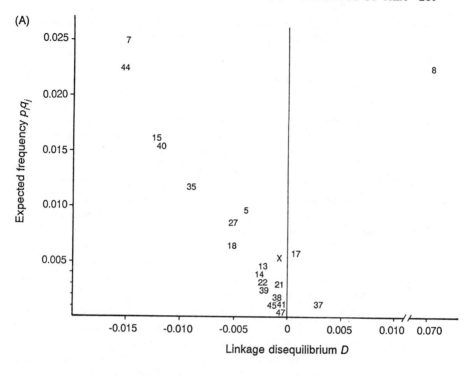

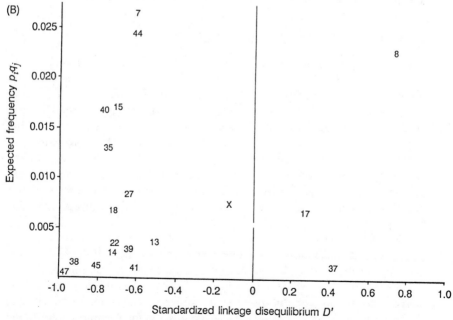

FIGURE 5. All haplotypes containing the allele *A1* where the numbers indicate the antigen designation at the *B* gene for (A) the disequilibrium measure *D* and (B) the normalized measure *D'*.

(Klitz and Thomson, 1987). Many of the other haplotypes have disequilibrium distributions completely different from these expectations.

Constrained disequilibrium values

Robinson (1989) developed a further method of analysis that uses information from the relationships of pairwise D values between three alleles at different loci that may be used to distinguish which of these alleles has most likely been selected. For example, for the HLA haplotype $A1$-$B8$-$DR3$, the observed D' values for the pairs $A1$-$B8$, $A1$-$DR3$, and $B8$-$DR3$ are 0.65, 0.32, and 0.61, respectively. These values, combined with the values of a normalized measure of disequilibrium taking account of constraints on pairwise disequilibrium values imposed by a three-locus model (0.90, 0.09, and 0.91, respectively), indicate that $B8$ is the most likely of these three alleles to have been selected. Examining disequilibrium values among the HLA loci, Robinson (1989) identified five HLA haplotypes with normalized and constrained pairwise disequilibrium values indicative of selection. Three of them implicated selection at a B locus allele (or a very closely linked locus): $A1$-$B8$-$DR3$ (described above), $A1$-$B57$-$DR7$ ($B57$ is a split of $B17$), and $A3$-$B35$-$DR1$. [Four of the A–B haplotypes identified by Klitz and Thomson (1987) as selected did not show up in this analysis. This is not unexpected as the two methods test different features of the data.] Additionally, Robinson identified selection acting on or near two rare Bf alleles ($F1$ and $S07$): haplotypes $A30$-$B18$-$BfF1$-$DR3$ and $C6$-$B50$-$BfS07$-$DR7$.

GENE CONVERSION AND RECOMBINATION

The complex patterns of similarity between MHC sequences have led most researchers to conclude that information is being exchanged between alleles and loci by gene conversion or reciprocal recombination (e.g., Lopez de Castro, 1989). Bosch et al. (1989) identified sequences in the second exon of MHC genes that correspond to sequences associated with gene conversion in other organisms. They suggest this as a general explanation for the apparently high incidence of exchange events in this region. Until recently, our understanding of the role of gene conversion in the evolution of higher organisms has been inhibited by lack of appropriate data to observe and sample the process. This is changing with the accumulation of sequence data from the highly polymorphic multigene families of the MHC (Parham et al., 1988).

Various approaches have been taken to identifying genetic exchange events. Nathenson et al. (1986) took a direct approach to the detection of genetic exchange in the mouse MHC. They selected for spontaneous mutants of K^b, a mouse class 1 allele, by assaying skin transplant rejection in inbred mice. The MHC DNA of almost all of the mutants formed contained clusters of changes spanning a small area. Potential DNA donors for these clusters were found elsewhere in the MHC, some in classical and others in nonclassical loci.

Within-locus exchange could not be measured, since the mice were K^b homozygotes.

Comparison of overall sequence similarity can reveal large-scale recombination events. Holmes and Parham (1985) show that Aw69 is identical to Aw68 in exon 2 and to A2 in exons 3 through 5, and suggest that Aw69 is the result of a single recombination between these sequences. A similar analysis by Hughes and Nei (1989b) and Nei and Hughes (this volume) of the class I loci A, B, C, and E suggests that exons 4 and 5 of A are derived from E.

A different approach looks for clustering of shared bases among otherwise dissimilar sequences. Table 5 shows the best-known example of this: A24 and A32 share 5 substitutions with a group of B alleles over a 9-bp region. The similarities between A24, A32, and the B alleles extend only a short distance beyond the region shown. Elsewhere, A24 and A32 are more similar to the other A alleles than to any B alleles. Parham et al. (1989) list 16 apparent instances of such between-locus exchanges based on comparison of 40 sequences. The proportion of exchange events that occur between loci rather than within loci in the HLA region remains an open question. Within-locus events are much harder to recognize because similarities between alleles of the same locus may be due to common ancestry rather than genetic exchange. Work by Stephens (1985) and Sawyer (1989) on the statistical detection of gene conversion events suggest ways in which the two may be distinguished, although the uneven distribution of variability across the MHC genes complicates such analysis.

Through sequence comparisons, it has been possible to reconstruct the evolutionary history of class 1 MHC genes in the mouse (Kuhner et al., 1991). Using a sample of 23 available mouse class 1 DNA sequences, 28 putative conversion and recombination events were identified. Based on this analysis it appears that gene conversion occurs throughout the length of the gene without preference for site variability or coding function. In the mouse MHC, this process is apparently so widespread that almost no features differentiate the various classical class I loci. In humans, however, there remain many nucleotide positions differentiating HLA-A, -B, and -C (Parham et al., 1989).

TABLE 5. The DNA sequence over a 9-bp region in exon 3 showing the similarity of alleles A24 and A32 to B alleles

Alleles	Positions 311–319
(1) All A except A24, A32	ccct gcgcg
(2) B8, B18, Bw41, Bw42, Bw46, Bw60, Bw65	a--------
(3) A24, A32, B27.2, B51, Bw52, Bw58	t--gc-t-c
(4) B13, B44	---gc-t-c
(5) B27.1, Bw47	------t-c

Some human alleles have very close analogs in other primates (Lawlor et al., 1988; Mayer et al., 1988; Gylleston and Ehrlich, 1989), leading to speculation that most human alleles are very old (Mayer et al., 1988) and implying that neither gene conversion nor mutation occurs at a high rate in these loci. However, the A2 group of HLA-A molecules has not been detected either serologically (Balmer et al., 1978) or by sequencing in the chimpanzee, although it is an abundant and diverse allele class in humans. HLA evolution may involve a simultaneous preservation of old alleles and rapid elaboration of new alleles. Both point mutations generating new variation and genetic exchange reassorting it into new combinations have clearly played key roles in the creation of the extraordinary HLA diversity.

SEGREGATION DISTORTION

Segregation distortion occurs when heterozygous individuals produce unequal proportions of their constituent gametes. For example, alleles at the t locus in the house mouse, *Mus musculus*, which maps close to the mouse MHC system, H-2, are favored in males by segregation distortion but also are usually recessive lethals. There was some suggestion that humans may have a locus near the HLA region that causes segregation distortion, a putative t homologue (e.g., Awdeh et al., 1983). A detailed population study by Klitz et al. (1987) found no evidence for segregation distortion in the HLA region. In addition, the gene homologous to that central to the segregation distorter activity of the mouse t haplotypes has been found in humans on the same chromosome as the HLA region, but genetically far removed, located on the long arm of chromosome 6 (Willison et al., 1987).

Segregation distortion has been suggested as an important factor generating or maintaining gametic disequilibrium (e.g., Alper et al., 1985), begging the following questions. How are the observations of gametic disequilibrium in the MHC region related to the phenomenon of segregation distortion? Can segregation distortion generate gametic disequilibrium or can it influence the rate of decay of gametic disequilibrium?

In a theoretical study, Hedrick (1988b) demonstrated that segregation distortion cannot de novo generate gametic disequilibrium between a segregation distortion allele and alleles at another locus. Furthermore, segregation distortion results in a faster rate of decay of standing disequilibrium between the segregation distortion locus and a neutral locus than if there is no segregation distortion. The increased rate of decay occurs because the difference in allelic frequencies between male and female gametes generated by single sex segregation distortion results in an excess of double heterozygotes and thereby causes a faster rate of decay than if there were Hardy-Weinberg proportions. Overall then, contrary to the contention of Alper et al. (1985) that male transmission bias can maintain disequilibrium, it appears that segregation distortion will increase the rate of decay of gametic disequilibrium.

Klein and his coworkers (e.g., Figueroa et al., 1985) suggest that t haplotypes in mice are associated with particular H-2 alleles (see also Nadeau, 1983, 1986). This is probably not due to segregation distortion, but to the suppression of recombination in t-locus heterozygotes by the inversions normally found associated with t-locus haplotypes (e.g., Silver, 1985).

MODES OF SELECTION

A distinctive feature of the HLA data at the single-locus level is the high degree of polymorphism of the loci in combination with a relatively even distribution of allele frequencies for the class I and II loci. In addition, many of the alleles appear to be very old. Multiple locus comparisons show high levels of linkage disequilibrium for some antigen combinations. These observations are consistent with the notion that the high levels of variation at these HLA loci are maintained by a selective mechanism(s), and that possibly all the HLA alleles have been subject to some degree of selection.

A variety of viability selection models have been suggested to be important for HLA genes including heterozygote advantage, frequency-dependent selection models based on host–pathogen interactions, selection for particular haplotype combinations, and genetic hitchhiking models. In addition to viability selection, other modes of selection have been proposed in the MHC region, including maternal–fetal effects and nonrandom mating.

Viability selection

The most easily identified source of selection on the MHC is that associated with the autoimmune diseases discussed earlier. It has been suggested because most of these diseases are fairly uncommon and generally affect individuals after reproduction, that their impact on genetic variation at the HLA loci may not be great. Some evidence indicates that the frequency of the disease predisposing variants in the population may be high, indicating the possibility of complex selective effects.

On the other hand, a major function of MHC histocompatibility molecules is in presenting foreign antigen to stimulate an immune response to invading pathogens (Zinkernagel, 1979). Histocompatibility alleles have been shown to differ in their ability to create an immune response to a variety of infectious agents (van Eden et al., 1983), suggesting that the epidemic diseases of the human past may have played a central role in determining the HLA haplotype and allelic frequencies observed in human populations today.

The result of such immune differences may, in the simplest model, be represented by a symmetrical constant fitness model in which all heterozygous haplotypes have a higher fitness than all homozygous haplotypes (Black and Salzano, 1981). Black and Salzano suggest that such an advantage may result because "disease resistance by specific antigens seems to require the presentation to the responding cells of one of their own histocompatibility antigens as a part of the total antigenic pattern that the responding cell recognizes (Blanden, 1980). Responding cells that carry two antigens for each locus would have more opportunities to respond than would homozygous cells." A frequency-dependent selection model in which genotypes with a rare allele have an advantage would work in much the same way. Bodmer (1972) suggested that a new allele would allow greater protection against pathogens than more common alleles to which pathogens may have evolved resistance.

Black and Salzano (1981) present both family and population data as supportive evidence for strong selection in South American Indians. First, in a group of families in which the parental genotypes were known only 9 two-locus homozygotes were observed while 20.5 were expected. Second, in a population survey only 45 two-locus homozygotes were observed out of 459 individuals when 73.4 were expected. For these data, Hedrick (1990) has estimated that to account for this homozygous deficiency, the selection coefficient against homozygotes must be around 0.42.

However, the overdominance hypothesis is open to the objection that multiple HLA loci, each fixed for a particular variant, would be more advantageous than a few loci with high heterozygosity. Gene duplication has certainly happened many times in the class I region; yet humans have only three classical class I loci. A competing hypothesis is frequency-dependent selection. A person whose HLA type is different from his neighbors' may be less likely to contact diseases from them. Such a mode of selection is difficult to distinguish from overdominance, and, of course, more than one type of viability selection may well be operating in the MHC region.

Maternal–fetal interactions

Approximately 30% of the couples having two or more spontaneous abortions do not have a demonstrable basis, such as a chromosomal or anatomical abnormality, for this fetal loss (Thomas et al., 1985). A number of studies indicate that such couples often share antigens for HLA loci (see Table 6 for a summary of 14 studies). Note that the frequency of shared antigens is higher for aborting couples in all comparisons but this difference is most striking when two or three loci are examined simultaneously. There are, however, reports not consistent with this trend (e.g., Oksenberg, 1984).

The immunological hypothesis suggests that the presence of an immune response occurring when the mother and fetus differ at the HLA loci is necessary for proper implantation and fetal growth (e.g., Gill, 1983, and references therein). In other words, sharing of HLA antigens in a parental couple results in a fetus similar to the mother and consequently an immune response by the mother to the fetus that is abnormal in some way. Another hypothesis based on linked lethal alleles does not adequately explain these data in several respects (Hedrick, 1988c).

TABLE 6. The prevalence of shared antigens at different HLA loci for normal couples having a history of recurrent spontaneous abortions (sample size in parentheses)

Locus (loci)	Normal couples	Aborting couples
A^a	0.422 (408)	0.505 (325)
B^a	0.243 (408)	0.314 (325)
C^a	0.219 (114)	0.504 (115)
A, B^b	0.072 (83)	0.230 (152)
A, B, DR^b	0.220 (150)	0.505 (109)

Source: After Thomas et al. (1985).
[a]Share one or two antigens.
[b]Share two or more antigens.

Both single-locus and two-locus theoretical models have been developed to examine this mode of selection (Hedrick and Thomson, 1988). This immunologically based model appears to have the potential to maintain many alleles at a single locus and to result in substantial gametic disequilibrium when recombination is comparable to that observed between HLA loci A, B, and DR. Overall, it appears that this mode of selection has the potential to strongly affect genetic variation in the HLA region.

Nonrandom mating

Although there is no evidence for nonrandom mating with respect to HLA types, experimental studies in mice have demonstrated mating preferences with regard to H-2 types (e.g., Yamazaki et al., 1976, 1978). In these studies males generally mate preferentially with females of a different MHC type (Table 7).

TABLE 7. The type of female selected by males when presented with two females of the same and different genotypes (sample size in parentheses)

	Type of female selected		Preference (same:different)
	Same as male	Different from male	
Not fostered (82)	0.305	0.695	0.431
Allogeneic foster nursing (74)	0.689	0.311	2.26

Source: From Yamazaki et al. (1988).

For example, the first row of Table 7 gives a summary of the type of female selected by male mice when they are presented with a choice of females of their own *H-2* genotype and those different at *H-2* (the genetic background is identical). The preference value (after Haldane, 1956) is quite strong against females of the same genotype as the male.

Recently, Yamazaki et al. (1988) reported experiments that suggest that this preference is the result of familial imprinting. The second row of Table 7 gives the male preference when the male is raised by foster parents that differ at their MHC type. In this case, males prefer females genetically like themselves, a reverse of the previous experiments, illustrating that male mating preference is acquired through parental imprinting.

There is substantial evidence that mice (and other organisms) can distinguish MHC types, an ability that could easily result in nonrandom mating. For example, Yamazaki et al. (1983) mated females with males of a given MHC type. These females were then exposed (in an adjacent compartment) to the same male (stud male), another male of the same genotype as the first (a syngeneic male), or another of the same genotype except for the H-2 region (a congeneic male). The proportion of females that returned to estrus was low (around 10%) when the second male was the stud or a syngeneic male but was fivefold higher when the second male differed in the H-2 region.

CONCLUSIONS

Our understanding of the structure and function of the HLA genes, their disease associations, and the evolutionary features of this multigene family has benefited from recent advances in molecular biology, immunology, disease modeling, and population genetics. Considerable progress has been made in developing methods to analyze what at first seemed an intractable challenge: determination of the evolutionary events operating in the HLA region, and their relationship to the function of the HLA genes.

The rate at which MHC diversity evolves remains controversial. On the one hand, certain human alleles are extremely similar to chimpanzee alleles (Mayer et al., 1988) suggesting a gradual evolutionary rate similar to that seen in enzyme loci. On the other hand, the high linkage disequilibrium seen among many HLA alleles is indicative of more recent events. There are species like the Syrian hamster with little or no MHC polymorphism. Efforts have been made to measure the MHC mutation rate directly. Nathenson et al. (1986) reported a mutation rate of approximately 10^{-4} in the K^b haplotype of the mouse locus *H-2K*, but other studies (Klein, 1978) have suggested that this value is unusually high.

The detection of selection in the HLA region is particularly important to population, disease, and evolutionary studies. Some haplotypes that have been subject to recent selection have been identified. Although the selection acting on the region is presumably strong, only a small fraction of actual selection events has been, or probably can be, identified. Although various mechanistic possibilities can be proposed involving immune function and response, the specifics of the selective process elude us at present. There is, however, substantial evidence that various types of viability selection, maternal–fetal selection, and nonrandom mating can influence MHC variation.

The patterns of haplotype disequilibrium can be informative in determining the progenitor of a new allele. Such information is invaluable for determining the relative age of alleles, and in combination with DNA sequence analysis, the mechanism by which a new allele arose (i.e., point mutation, gene conversion, or recombination) as well as understanding cross-reactivity of antigens. Observations from such studies are intimately related to the origin and evolution of genetic variation including disease predisposing genes. Ethnic comparisons in this regard are invaluable.

As we have illustrated, population genetics provides an appropriate context in which to integrate the evolutionary factors influencing HLA variation and to test competing hypotheses suggested to explain the molecular variation and function of this multigene family. Far from replacing the need for population genetic analysis, the new information obtained from the application of molecular techniques has created an even greater challenge for population geneticists (Bodmer and Bodmer, 1989).

CHAPTER 13

EVOLUTION OF

GLOBIN GENE FAMILIES

Ross C. Hardison

The globin genes encode the polypeptides of heme-binding proteins such as hemoglobins, myoglobins, and leghemoglobins. All these hemoproteins bind and release oxygen in a regulated manner, although they do this in widely different tissues (red blood cells, muscle, and plant root nodules, respectively). Every organism that uses hemoglobin to transport oxygen produces several types of hemoglobin at progressive stages of development.

The molecular analysis of globin genes has provided many insights regarding gene regulation and evolution. The globin genes are expressed in specific tissues in a developmentally regulated manner. Several inherited hemoglobinopathies are widespread in the human population, including structurally abnormal hemoglobins, such as sickle-cell hemoglobin, and deficiencies in production of hemoglobins, such as the thalassemias. Possible therapies for these diseases include (1) replacement of the defective gene by a normal gene, and (2) administration of drugs designed to modulate the expression of the genes. For example, by allowing fetal hemoglobin to continue to be expressed in adult life, there is considerable amelioration of the symptoms of sickle-cell anemia. The mechanisms of normal gene regulation must be understood to achieve success in either of these approaches, and, consequently, considerable effort is being expended toward that end.

Globin genes are also widely studied as exemplary models of molecular evolution. The ready availability of hemoglobins from a wide variety of species allowed many different amino acid sequences to be determined, thus providing one of the more extensive data sets available for studies of molecular phylogenies (Dayhoff et al., 1972). Recombinant DNA techniques were initially applied in the 1970s to genes with abundant products, such as globins and immunoglobulins, and, again, the database for evolutionary studies was greatly increased by the extensive sequence analysis of the isolated cDNAs and genes. Globin gene sequences permit a much deeper analysis of the mechanism and rate of evolution and refine the phylogenetic analysis of this small segment of chromosomal DNA. The linked clusters of globin genes have been a particularly fertile ground for these studies and have also provided considerable new information for the studies of regulation of these genes. Many sequences are observed to be well conserved in the flanking and internal regions of the globin genes. In virtually all cases of evolutionary conservation of a particular sequence, further analysis has identified an important role for that sequence in regulation. It is the synergism of evolutionary and regulatory approaches that has made the globin gene system particularly informative.

In this chapter, I will review several of the general features of globin gene structure and the organization of the gene families, and briefly summarize the analysis of the complete sequence determination of the β-like globin gene cluster in representatives of three different mammalian orders. Many important topics that have recently been reviewed elsewhere are not covered here in detail; these include globin gene regulation (Collins and Weissman, 1984; Karlsson and Neinhuis, 1985), a phylogenetic analysis of globin gene sequences (Goodman et al., 1987), evolution of nonvertebrate hemoglobins (Riggs, 1990), and adaptive changes in hemoglobins from different species (Perutz, 1983).

OVERVIEW OF GLOBINS IN PHYLOGENY

Globins are widely distributed in animals and plants. All vertebrates carry oxygen on tetrameric hemoglobin in their red blood cells and store oxygen on monomeric myoglobin in their muscles (Dickerson and Geis, 1983). Invertebrates such as the arthropod *Chironomus thummi* produce extracellular hemoglobins (some monomeric and some dimeric) in their hemolymph (Antoine and Niessing, 1984). The best-characterized globin in plants is the leghemoglobin encoded in the soybean genome that is produced in root nodules following symbiotic infection by rhizobia to fix nitrogen. The monomeric leghemoglobin also binds oxygen, sequestering it from the oxygen-sensitive nitrogen-fixing apparatus and providing oxygen to the energy-generating electron transport system (Hyldig-Nielsen et al., 1982). Dimeric hemoglobins are also found in nitrogen-fixing, nonleguminous plants (Appleby et al., 1983). Hemoglobins are also found in fungi, such as yeast (Keilin, 1953) and in bacteria, such as *Vitreoscilla* (Wakabayashi et al., 1986).

This wide distribution of globins indicates that their genes are very old; and, in fact, they are members of a large superfamily of proteins that includes all the heme-binding proteins. The heme-binding cleft of hemoglobins is structurally related to that of the cytochromes (Argos and Rossmann, 1979). Thus although significant primary sequence homology is not detectable between globins and cytochromes, they may have descended from an ancient gene encoding a heme-binding domain (Blake, 1981).

By aligning the amino acid sequences of the globin polypeptides (or the

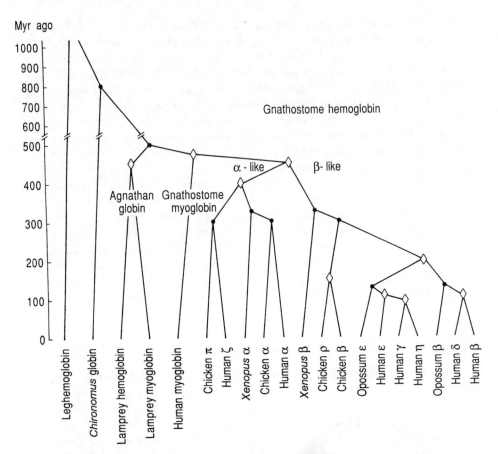

FIGURE 1. Phylogenetic tree of representative globin genes. The most parsimonious phylogenetic tree for globins illustrating speciation events (filled dots) and gene duplications (open diamonds) is plotted along an axis with the paleontological dates estimated for those events (Goodman, 1981). This diagram was selected from the extensive analyses in Czelusniak et al. (1982), Goodman et al. (1987), and Koop and Goodman (1988).

coding regions of the genes), one can build phylogenetic trees based on the minimum number of events required to generate the diversity among them (maximum parsimony approach). The tree in Figure 1 illustrates many of the major events in the evolution of globin genes (Czelusniak et al., 1982; Goodman et al., 1987; Koop and Goodman, 1988; Nei, 1987). The plant leghemoglobins diverged from other globins possibly as early as the animal/plant divergence a billion or more years ago. Speculations that leghemoglobin genes were horizontally transmitted into legumes, possibly via insect vectors (Jeffreys, 1982), have not received support from studies of insect globins (Antoine and Niessing, 1984); and the discovery of hemoglobin in the nonleguminous plant *Parasponia*

(Appleby et al., 1983) suggests an early origin for this gene. Arthropod globins form a separate branch, illustrated by the *Chironomus* globin in Figure 1. Other invertebrates, such as annelids and molluscs, each form a separate branch, diverging from the vertebrates more than 600 Myr ago (Goodman et al., 1988). The branch to globins in jawless vertebrates (agnathans), such as the lampreys, diverged about 500 Myr ago. Both hemoglobins and myoglobins are found in the agnathan branch.

The pattern of gene duplication and speciation has been most extensively studied in the jawed vertebrates (gnathostomes). The gene duplication separating the gnathostome myoglobins from the hemoglobins occurred about 500 Myr ago. Although the timing of this event relative to the divergence of agnathan globins is not certain (Goodman et al., 1987), it is clear that myoglobin evolved at least twice, once in the agnathan lineage and again in the gnathostome lineage. The gene duplication leading to the α-like and β-like globins occurred about 450 Myr ago. At that time the globin gene probably encoded a monomeric protein, as is the condition of myoglobin and several of the other early globins. But after this gene duplication, the genes were free to evolve into the current tetrameric form that is finely regulated by allosteric effects (Dickerson and Geis, 1983). Both the α-like and β-like genes duplicated several more times in their subsequent evolution, leading to families of genes expressed at different developmental stages. (The term ζα-like refers to the π, α, and θ globins, as distinct from the β-like globins, which include ε, ρ, γ, η, δ, and β.) Surprisingly, these duplications were not fixed in the population coordinately. The duplication that generated embryonic (avian π and mammalian ζ) and adult (α) α-like globins greatly preceded the analogous duplications in the β-like globin gene cluster (Czelusniak et al., 1982; Proudfoot et al., 1982). Hence the embryonic π and ζ genes are probably orthologous (separated by speciation events, albeit long ago), and are of equivalent age in birds and mammals, whereas the embryonic ρ gene separated from the adult β gene in chickens long after birds and mammals had diverged. Thus, the avian ρ gene and the mammaliam ε gene are not orthologous, even though both are β-like globin genes expressed in early embryonic development.

GLOBIN GENE STRUCTURE

One of the surprises from the early investigations of globin gene structure was the discovery that DNA that encodes mRNA is divided into segments (exons) by intervening sequences or introns (Jeffreys and Flavell, 1977, Tilghman et al., 1978a). Both exons and introns are transcribed by RNA polymerase into a continuous pre-mRNA, and the introns are subsequently removed by splicing (Tilghman et al., 1978b). Introns were initially discovered in mammalian β-globin genes, but they were soon found in α-globin genes (Leder et al., 1978) and eventually in virtually all globin genes (Figure 2). The two introns are found in equivalent positions in all vertebrate α-like and β-like globin genes,

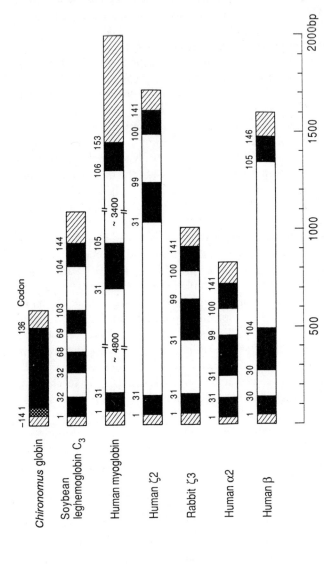

FIGURE 2. Structure of representative globin genes. The structures of several globin genes are shown with exons divided into protein coding regions (filled boxes) and untranslated regions (hatched boxes) while introns are shown as open boxes. The stippled box is the region coding for a signal peptide in the *Chironomus* globin. Codon numbers are above the genes, which are drawn to the scale below except for the introns in the myo-globin gene. References are as follows: *Chironomus* globin IV (Antoine and Niessing, 1984), soybean leghemoglobin c3 (Brisson and Verma, 1982), human myoglobin (Jeffreys et al., 1984), human ζ2 (Proudfoot et al., 1982), rabbit ζ3 (Cheng et al., 1988), human α2 (Liebhaber et al., 1980), and human β (Lawn et al., 1980).

as well as myoglobin (i.e., codons 30, 104, and 105 in the β-globin gene are homologous to codons 31, 99, and 100 in the α-globin gene and to codons 31, 105, and 106 in the myoglobin gene; Figure 2). Both the wide distribution of introns and, in particular, their equivalent positions in the different globin genes provide compelling evidence that the ancestral globin gene contained introns in these positions. The alternative model requires that introns have been inserted into equivalent positions three different times, a very unlikely scenario.

Despite their constancy in position, the globin gene introns vary considerably in size, from 77 bp in α-globin genes (Cheng et al., 1986) to 4800 bp in myoglobin (Jeffreys et al., 1984). Even closely related genes, such as human ζ2 and rabbit ζ3, show considerable differences in sizes of introns (Figure 2), largely accounted for by multiple tandem repeats of a short sequence in the human intron. In contrast, most β-like globin gene introns are of similar size, as illustrated by the human β-globin gene in Figure 2.

The functional significance of the exon–intron structure of the globin genes was not at all clear initially. Exons were proposed to encode discrete domains of protein structure, providing the opportunity for rapid generation of new protein structures by exon shuffling (Gilbert, 1978). The central exon of the vertebrate globin genes encodes the heme-binding pocket (Eaton, 1980; Craik et al., 1980), but the analysis of folded domains in globins showed that this heme-binding region is separated into two discrete compact structures (Gō, 1981). This apparent discrepancy was resolved by the analysis of the leghemoglobin genes of soybeans (Jensen et al., 1981; Brisson and Verma, 1982; Hyldig-Nielsen et al., 1982). These genes have three introns, with the proximal and distal introns corresponding to those found in vertebrate globin genes and the middle intron separating the homologue to the vertebrate central exon into two exons (Figure 2). This is precisely the arrangement that would be expected if each domain of the globin polypeptide was encoded by a separate exon. Thus the simplest model is that the ancestral globin gene was divided into four exons, as is the case in the leghemoglobin genes. In the ancestor of the vertebrate globin genes, the middle intron was lost, fusing the two exons to produce one exon encoding the heme-binding region. Thus the exon–intron structure of globin genes probably has its origins in the shuffling of exons to produce novel proteins, as predicted by Gilbert (1978) and Blake (1978).

A curious but informative exception to the general structure of globin genes is found in the midge, *Chironomus thummi*. During its aquatic larval stage, *C. thummi* produces 12 different hemoglobins; they are synthesized in the fat body and are secreted into the hemolymph. Isolation and sequencing of the genes for hemoglobins III and IV yielded a remarkable result—these genes have no introns (Figure 2). The promoter and RNA processing signals are intact in these genes, and a 14-amino acid signal peptide is encoded at the 5′ end, as expected for secreted proteins (Antoine and Niessing, 1984). Hence these are intronless, functional globin genes. Given the evidence that the ancestral globin gene had three introns, it is most likely that the introns were removed, but in a process that leaves a functional gene. Processed pseudogenes have been observed for α-globins and other genes (Leder et al., 1981), in which the mRNA was apparently copied by a reverse transcriptase activity into cDNA that inserted back into the genome, in all cases at a site far away from the parental gene. Antoine and Niessing (1984) suggest that in the case of the C. *thummi* globin genes, the cDNA replaced the parental gene, perhaps by a cDNA-mediated gene conversion event. A similar situation has been described for the removal of an intron from one of the rat preproinsulin genes (Soares et al., 1985), although in this case the partially processed gene has transposed to a new site. It is important to note that this intron-less gene structure is not typical of all invertebrates. For example, the gene for chain *c* of the extracellular hemoglobin from the earthworm *Lumbricus terrestris* has the two-intron three-exon structure characteristic of vertebrate globin genes (Jhiang et al., 1988).

ARRANGEMENTS OF GLOBIN GENE CLUSTERS

Most functional globin genes occur in clusters; this has been demonstrated in plants, arthropods, amphibians, birds, and mammals (Figure 3). One obvious exception is the gene for myoglobin; it is present in a single copy in both seals (Blanchetot et al., 1983) and humans (Jeffreys et al., 1984). In virtually all gene clusters, the genes are expressed at different times of development; the *Xenopus* genes (Hosbach et al., 1983) are expressed either in the larval stage (denoted by L) or the adult stage (A), and each of the genes in the clusters from birds and mammals is expressed in embryonic [ρ, ε, γ (usually), π, and ζ], fetal (human γ and goat $β^F$), or adult (β, δ, and α) life (Collins and Weissman, 1984). Even the soybean leghemoglobin genes are expressed differentially during root nodule development after infection with rhizobia (Lee et al., 1983). These multiple genes doubtless arose from gene duplications, but the apparent necessity for differential expression during development presumably has provided selective pressures to maintain the multiple copies. Although multiple genes are the rule, the numbers, sizes, and distances between the genes vary widely (Figure 3), from the very compact gene cluster in *Chironomus* (5 kb) to the extended β-like gene cluster in goats (140 kb).

The duplication that led to the evolution of vertebrate α-like and β-like genes about 450 Myr ago (Figure 1) would likely have left the two genes adjacent, as is the case for most gene duplications. The α-like and β-like globin genes are still tightly linked in amphibians (Jeffreys et al., 1980), as is illustrated by cluster I in *Xenopus laevis* (Figure 3). However, in the lineage to birds and mammals, the α-like and β-like gene clusters became separated on different chromosomes (Collins and Weissman, 1984). Additionally, the single myoglobin gene in humans is located on a separate chromosome from the α-like and β-like globin gene clusters (Jeffreys et al., 1984). Once separated, the α-like and β-like globin gene clusters in mammals became very different in sequence. The α-like gene clusters are very G + C-rich and have characteristics of HTF

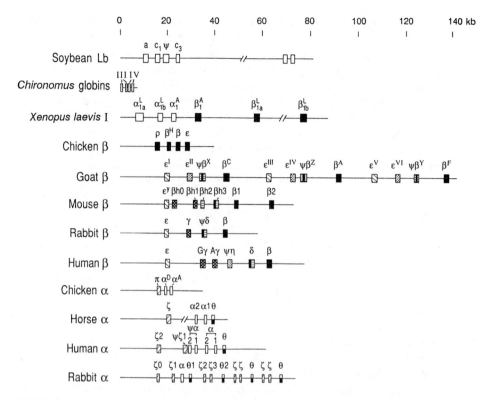

FIGURE 3. Arrangements of selected globin gene clusters. The organizations of several globin gene clusters are mapped. Each gene (exons plus introns) is shown as a single box, coded to show the orthologous relationships for the mammalian genes, as follows: descending diagonal, ϵ; dense stippling, γ; light stippling, η; horizontal lines, δ; filled boxes, β; ascending diagonal, ζ or π; open box, α; and half-filled box, θ. For *Xenopus* and chicken, the α-like genes are shown as open boxes (except π) and the β-like genes are shown as filled boxes, denoting the duplications that are distinct from those in mammals (Figure 1). All genes are arrayed left to right from the 5' end to the 3' end, except for the *Chironomus* genes, which have diverging transcription units for each pair of genes (III and IV). For *Xenopus laevis*, only one of the two gene clusters found in the tetraploid genome is shown. Breaks in the lines signify that the linkage has not been established. References are as follows: soybean leghemoglobins (Hyldig-Nielsen et al., 1982; Lee et al., 1983), *Chironomus* globins (Antoine and Niessing, 1984), *Xenopus* cluster I (Hosbach et al., 1983), chicken β (Dolan et al., 1981), goat β (Townes et al., 1984), mouse β (Jahn et al., 1980; Hardies et al., 1984; Hill et al., 1984; Hutchison et al., 1984), rabbit β (Lacy et al., 1979; Hardison 1984), human β (Fritsch et al., 1980; Efstratiadis et al., 1980), chicken α (Dodgson et al., 1981; Engel et al., 1983), horse α (Clegg et al., 1984), human α (Lauer et al., 1980; Proudfoot et al., 1982; Nicholls et al., 1987; Hsu et al., 1988), and rabbit α (Cheng et al., 1988; Cheng and Hardison, 1988).

islands (Fischel-Ghodsian et al., 1987; Cheng et al., 1986), whereas the β-like gene clusters are more A + T-rich (Collins and Weissman, 1984; Margot et al., 1989). Indeed, each gene cluster is contained within a very long chromosomal segment (much larger than 200 kb) of homogeneous base composition, or isochore (Bernardi, 1989). Mammalian α-like globin gene clusters are in the most G + C-rich isochore fraction, whereas the β-like globin gene clusters are in an A + T-rich isochore (Bernardi et al., 1985). The separated α-like and β-like globin gene clusters contain genes that are coordinately expressed; equal amounts of α-like and β-like globin must be synthesized to produce the hemoglobin heterotetramer. One would have expected that this coordinated expression would be achieved by utilizing identical regulatory schemes; but such does not appear to be the case. The regulation of the human α-globin gene is substantially different from that of the β-globin gene in transfected cells (Humphries et al., 1982; Charnay et al., 1984; Wright et al., 1984), the basis for which may lie in the radically different sequences surrounding the genes (G + C-rich for α-like genes, and A + T rich for β-like genes).

Several examples of block duplications of globin genes have been described. The first example was the β-like globin gene cluster in goats (Townes et al., 1984); it is a triplication of a set of four genes ϵ-ϵ-$\psi\beta$-β (Figures 3 and 4). Similar block duplications also occurred in the β-like gene clusters of two other artiodactyls, sheep (Garner and Lingrel, 1988, 1989) and cows (Schimenti and Duncan, 1985). The β gene at the 3' end of each gene set is expressed at a different time of development in goats; β^F is fetal, β^C is juvenile, and β^A is adult (Lingrel et al., 1985). Thus the recruitment to expression at different developmental stages occurred after the block triplication. The block duplications are not restricted to artiodactyls; the α-like gene cluster in rabbits was also generated by multiple duplications of a set of genes (Cheng et al., 1987, 1988). In this case the basic gene set was ζ-ζ-α-θ; the α-globin gene was lost from one gene set after the first duplication, and, subsequently, the ζ-ζ-θ gene set duplicated (Figure 3). Although the genes in the rabbit α-like globin cluster do not appear to have been recruited for differential expression after the duplications, this example does share several features with the β-like gene cluster in goats. In both cases, several of the genes resulting from the block duplications are pseudogenes, i.e., they do not encode a functional globin. In goats, ϵ^I and ϵ^{II} both encode embryonic globins, and each of the β genes is active; but the remaining genes (ϵ^{III} through ϵ^{VI} and the $\psi\beta$ genes) are probably pseudogenes (Lingrel et al., 1985). Likewise in rabbits, $\zeta1$, $\zeta2$, $\zeta3$ and the θ genes cannot encode functional globins, despite the fact that $\zeta1$ and $\zeta3$ are competent for transcription in transfected cells (Cheng et al., 1986, 1988). These block duplications not only generate an abundance of pseudogenes but also produce novel polymorphisms. Both the β-like gene cluster in sheep and the α-like gene cluster in rabbits show extensive polymorphism in the number of duplicated gene sets (Garner and Lingrel, 1988, 1989; Cheng and Hardison, 1988). Not surprisingly, little selective pressure seems to be exerted on these duplicated

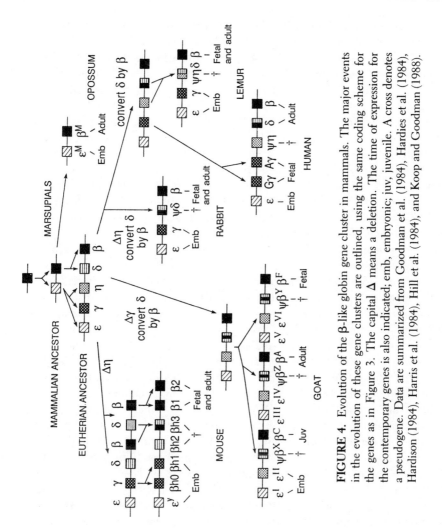

FIGURE 4. Evolution of the β-like globin gene cluster in mammals. The major events in the evolution of these gene clusters are outlined, using the same coding scheme for the genes as in Figure 3. The capital Δ means a deletion. The time of expression for the contemporary genes is also indicated; emb, embryonic; juv, juvenile. A cross denotes a pseudogene. Data are summarized from Goodman et al. (1984), Hardies et al. (1984), Hardison (1984), Harris et al. (1984), Hill et al. (1984), and Koop and Goodman (1988).

gene sets, and perhaps the duplications reflect an underlying propensity for these gene clusters to recombine.

Many, but not all, vertebrate globin gene clusters are arranged in the order of their temporal expression. This is readily apparent in the mammalian α-like and β-like clusters illustrated in Figure 3. The embryonic ε, γ (both are β-like) and ζ (α-like) genes are at the 5′ ends of the gene clusters, whereas the adult δ and β (β-like) genes are at the 3′ ends of the gene cluster. For the α-like gene clusters, the θ gene (Marks et al., 1986), located at the 3′ end of the cluster, has an uncertain status; it is a pseudogene in horses (Clegg, 1987) and rabbits (Cheng et al., 1986), but low levels of transcription have been observed for the θ gene in humans (Leung et al., 1987; Hsu et al., 1988). The functional α-globin genes are 3' to the ζ gene; hence, this gene cluster also largely fits with the correlation between temporal order of expression and order in the gene cluster. The chicken α-like cluster also fits this correlation; π is embryonic and α^A and α^D are adult (Dodgson et al., 1981; Engel et al., 1983). The chicken β-like cluster is a clear exception; the two embryonic genes, ρ and ε, are located at the two ends of the gene cluster (Dolan et al., 1981). However, ρ and ε underwent a conversion event at their 5' ends that may have placed ε under embryonic control (Roninson and Ingram, 1982; Dodgson et al., 1983). The other clear exception is the Xenopus gene cluster; like the chicken β-like cluster, it is arrayed symmetrically: α larval–α adult–β adult–β larval (Hosbach et al., 1983). There is no obvious need or advantage to having genes arrayed in their temporal order of expression, and, as will be seen below, this general arrangement for mammalian β-like globin gene clusters probably is a consequence of common ancestry.

Another common feature of vertebrate hemoglobin gene clusters is the duplication of the adult α-globin gene. Two α-globin genes are found in chickens, horses, humans (Figure 3), and several other birds and mammals. These result from at least three separate duplications. The duplication to produce α^D and α^A in birds occurred about 400 Myr ago (Czelusniak et al., 1982); no homologue to α^D is found in mammals. The duplication that generated α2 and α1 in the horse probably occurred early in eutherian evolution because genes orthologous to both these α-globin genes are found in mammals of several orders (Hardison and Gelinas, 1986; Sawada and Schmid, 1986). A third gene duplication in the stem-simians generated the pair of α-globin genes in humans, apes and monkeys; the pseudogene ψα1 in humans is orthologous to the active gene α2 in goats and horses, and the ancestor to α1 was duplicated to produce the pair of α-globin genes seen in higher primates (Sawada and Schmid, 1986). These separate duplications keep the diploid number of α-globin genes at four in these several species, suggesting that gene number is important. However, rabbits have only one α-globin gene (Cheng et al., 1986). This is true even for a population of rabbits that had substantial polymorphism in other parts of the gene cluster (Cheng and Hardison, 1988). Hence, in rabbits two genes in the diploid genome can equal the output of hemoglobin from four genes in other mammals, suggesting that the rabbit α-globin gene may be regulated rather differently from that in other mammals.

EVOLUTION OF β-LIKE GLOBIN GENE CLUSTERS IN MAMMALS

The evolution of globin gene clusters has been studied in greatest detail for the β-like genes in mammals. As indicated in the parsimony tree in Figure 1, all the β-like globin genes in mammals are descended from a common gene that duplicated about 200 Myr ago to produce the ancestors of the ε-γ-η and the δ-β groups of genes. This preceded the time of the divergence of stem-marsupials from stem-eutherians about 100 to 135 Myr ago (Cifelli and Eaton, 1987). One would expect that the descendants of these two genes are present in marsupials (as well as eutherians) and, in fact, Koop and Goodman (1988) demonstrated that the opossum has only two β-like globin genes, ϵ^M and β^M (Figure 4). This arrangement in opossums is consistent with an initial duplication of β-like genes in the mammalian ancestor. The promoter region of the opossum ϵ^M gene is very similar to that of eutherian ε genes, indicating that the gene ancestral to ε-γ-η was already expressed in embryonic life; presumably the ancestral δ-β gene was already an adult gene (Koop and Goodman, 1988).

An extensive analysis of β-like globin genes in various orders of eutherian mammals led several groups of investigators to conclude that the β-like globin gene cluster had expanded to include ε-γ-η-δ-β prior to the eutherian radiation about 80 Myr ago (Hardies et al., 1984; Hardison, 1984; Harris et al., 1984; Goodman et al., 1984). One of the important problems in elucidating these proposed evolutionary pathways for multigene families was to establish adequate criteria for distinguishing between paralogous relationships (those between genes separated by gene duplications) and orthologous relationships (those between genes separated by speciation events). For these gene families, the orthologous genes have substantial matching sequences in the noncoding regions (far flanking and intronic sequences) that are readily apparent by dot-plot analysis. The paralogous genes generated by the duplications prior to the eutherian radiation (top of Figure 4) retained matching sequences only in the protein-coding regions of the genes. Of course, more recently duplicated genes, such as the γ genes in catarrhine primates (illustrated by the human gene cluster in Figure 4), exhibit more sequence similarity than the orthologous comparisons. As shown in Figure 4, descendants of each of the genes proposed for the eutherian ancestral cluster are found in representatives of different orders. However, the orthologous genes are not all expressed at the same time of development. The γ-globin gene is normally expressed during embryonic life in rodents [βh1, (Farace et al., 1984)], rabbits (Rohrbaugh and Hardison, 1983), and probably lemurs (Harris et al., 1986). In higher primates, including New World and Old World monkeys, apes and humans, the γ-globin genes are expressed in fetal life (Collins and Weissman, 1984). This almost certainly reflects a recruitment of the γ gene for expression in fetal life in the higher primates, possibly to accommodate the longer gestational period. Evidence of this proposed recruitment is provided by the observation that when a human γ-globin gene is placed in transgenic mice, it is expressed during embryonic but not fetal development, thus mimicking the pattern of expression of the orthologous mouse gene rather than its natural pattern in humans (Chada et al., 1986). A similar recent recruitment of a β gene for fetal expression has already been mentioned in the context of the block duplications in the goat β-like globin gene cluster.

Some of the recombination events in the evolution of each of the contemporary gene clusters are illustrated in Figure 4. The η gene was deleted separately in rodents (represented by mouse) and in lagomorphs (rabbit). The η gene fused with δ in a Lepore-like event in lemurs (Jeffreys et al., 1982), and it is an inactive pseudogene in the higher primates (Harris et al., 1984). In fact, only in artiodactyls (goat) is it known to be active; in this mammal it encodes the embryonic ϵ^{II} globin (Lingrel et al., 1985; Goodman et al., 1984). The δ gene has been involved in multiple gene conversions [lagomorphs (Hardison and Margot, 1984), catarrhine primates (Martin et al., 1983), and artiodactyls (Hardies et al., 1984)] and gene fusions [rodents (Hardies et al., 1984) and prosimian primates (Jeffreys et al., 1982)]. Additionally, the γ gene has duplicated two different times, once in rodents (Hill et al., 1984) and again in the catarrhine primates (Harris et al., 1984). The duplicated γ-globin genes have also undergone multiple gene conversions in different lineages (Slightom et al., 1980; Goodman et al., 1987).

All three of the genes that are involved most frequently in exchanges (γ, η, and δ) are located in the center of the gene cluster. Hotspots for recombination have been mapped to regions within the human γ-globin gene (Slightom et al., 1980) and 3′ to the human δ-globin gene (Orkin et al., 1982; Treco et al., 1985). Pseudogenes, which may arise from exchange events, tend to be located in the center of the gene cluster as well. In contrast, the genes on the outside of the gene cluster, especially ε, are not involved in recombinations as frequently (Figure 4).

Not only are different frequencies of recombination among globin genes observed, but the rates of nucleotide substitution vary among genes. The frequency of nonsynonymous substitutions (leading to amino acid replacements) for the ε-globin gene is significantly lower than for other β-like globin genes in comparisons among human, rabbit, mouse, and goat (Shapiro et al., 1983; Hardison, 1984). In comparisons among lemur, human, and rabbit genes, the noncoding regions of ε show a somewhat slower rate than comparable regions in γ and β, but, in general, the rate of synonymous (silent) substitution and noncoding sequence divergence was uniform over most of the primate β-like globin gene cluster (Harris et al., 1986). However, significant differences were seen in the rate of nonsynonymous substitutions. For example, the rate of substitution was faster for the γ gene in the branch leading to New World monkeys and humans. This increased rate occurred when this gene was being recruited for expression in fetal (as opposed to embryonic) life, and it is possible that it resulted from adaptive evolution as this globin adopted a new role in oxygen transport in fetal life. [A similar conclusion was reached by Goodman et al. (1987) using a different data set.] However, as pointed out by Harris et al. (1986), this apparent acceleration in rate can also be explained by a period of relaxed selection on γ as it was being recruited for fetal expression; perhaps it was a pseudogene for a short period prior to its recruitment. A similar increase in the frequency of nonsynonymous substitutions occurred for the β gene in the brown lemur lineage, and a substantial decrease in the synonymous substitution rate is seen for the β gene in the rabbit lineage.

COMPARISON OF THE β-LIKE GLOBIN GENE CLUSTERS OF HUMANS, RABBITS, AND MICE

Comparison of the nucleotide sequences of the β-like gene clusters of human (compiled by Collins and Weissman, 1984), rabbit (Margot et al., 1989), and mouse (Shehee et al., 1989) provides an unprecedented opportunity to investigate the evolution of a segment of the genome that contains several genes under a variety of modes of regulation. These gene clusters are shown in more detail in Figure 5, along with their associated repetitive DNA sequences. These repeats fall into two classes, short interspersed repeated elements (or SINEs) and long interspersed repeated elements [or LINEs (Singer, 1982)]. Both classes are proposed to spread through the genome by an RNA-mediated transposition mechanism involving transcription into RNA (by RNA polymerase III for SINEs and by RNA polymerase II for LINEs), copying into cDNA by reverse transcriptase, and reinsertion into the genome (Weiner et al., 1986). The SINES are typically 300 bp or less and are present in several hundred thousand copies in the haploid genome (reviewed in Schmid and Shen, 1985). Each mammalian order has a distinctive SINE; humans have Alu repeats, mice have B1 repeats, and rabbits have C repeats (Cheng et al., 1984), all of which are represented as open triangles in Figure 5.

The most frequently occurring LINE is called an L1 repeat. Individual copies can be up to 7500 bp long; they are frequently truncated at the 5′ end and are present in 50,000 to 100,000 copies per haploid genome (reviewed by Singer and Skowronski, 1985; Weiner et al., 1986). The L1 repeats contain two long open reading frames; the 3′ end of the first reading frame and the second reading frame are well conserved in L1 repeats in different mammalian orders (Singer and Skowronski, 1985; Loeb et al., 1986; Demers et al., 1989). (The L1 repeats are shown as filled triangles and arrows in Figure 5.) These DNA repeats in mammals share some of the general characteristics of the transposable elements in *Drosophila* discussed by Charlesworth and Langley (this volume).

As summarized in Figure 5, long stretches of sequence homology occur around each of the globin genes and extend through the DNA between them. Examples of this are the regions between δ and β in the rabbit and human and around the ε gene in all three species. These patterns of intergenic sequence similarity provide strong support for the interpretation that these genes are orthologous. The most likely explanation for the extensive sequence similarity far from the coding regions is that the regions are descended from the same gene cluster in the last common ancestor. In contrast, as noted earlier, paralogous genes show matches only in the coding region (Margot et al., 1989; Shehee

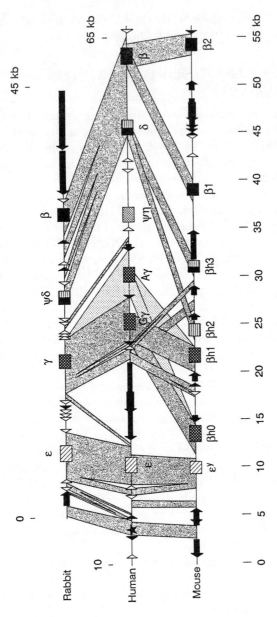

FIGURE 5. Summary of matching sequences among the β-like globin gene clusters of rabbit, human and mouse. The gene clusters are shown with genes (coded as in Figure 3) and repeats (open triangles for SINES—C repeats in rabbits, Alu repeats in human, and B1 repeats in mouse—and filled triangles and arrows for L1 repeats). The matching sequences are shown as shaded regions between the gene clusters. Data are summarized from the dot-plot analyses in Margot et al. (1989) and Shehee et al. (1989) and sequence alignments in Huang et al. (1990).

et al., 1989), indicating the expected conservation of sequences under purifying selection (removal of mutations that disrupt globin function).

The rabbit and human β globin sequences are more similar than human and mouse sequences (Figure 5). This result is consistent with the finding of Wu and Li (1985) that the rate of molecular evolution is more rapid in rodents than in other mammals. Comparisons between mammalian α-like globin gene clusters do not show the extensive matches in the intergenic regions that are observed between the β-like globin genes (Hardison and Gelinas, 1986). Thus there are indications of different rates not only in different lineages (rodents evolving faster than lagomorphs or primates) but also in different regions of the same genome; noncoding sequences seem to be evolving faster in the α-like globin gene cluster than in the β-like gene cluster.

Virtually all the single-copy DNA in the rabbit β-like globin gene cluster show some degree of similarity with sequences in the human gene cluster (Figure 5). The matches extend to the end of the single-copy DNA isolated from the rabbit gene cluster, but it is not known how far along this cluster these matching sequences extend. The human η gene does not match with the rabbit sequence, as expected because of the proposed deletion of η from rabbits (Figure 4). The breakpoints for the deletion cannot be precisely defined, but they must lie between ψδ and the C repeats 5′ to it. The η gene was also deleted from mouse, probably from the region between βh1 and βh2.

The long segments of matching sequences are frequently interrupted by an interspersed repeat in one or another of the species (Figure 5), and none of the repeats is located at the same position in any two gene clusters. Any repeat that was present in the ancestral gene cluster and has remained in the contemporary gene clusters should be readily apparent in these comparisons (because the single-copy matches are readily detected). Consequently one can conclude that insertion of all the repeats present in the contemporary gene clusters occurred after the eutherian radiation—there are no conserved repeats. Several other lines of evidence indicate that many of the repeats are recent transpositions into the gene cluster. Some of the L1 repeats in the BALB/c mouse gene cluster (shown in Figure 5) are not present in the gene clusters of other strains of mouse (Burton et al., 1985; Shyman and Weaver, 1985). The L1 insertion between ε and γ in the human gene cluster is apparently present in all catarrhine primates (Old World monkeys, apes, and humans) but is absent in New World monkeys and prosimians (Harris et al., 1984). In fact, the distances between the β-like globin genes are uniform among the catarrhine primates, indicating that few large rearrangements of DNA (insertions and deletions) have occurred in the past 20 Myr (Harris et al., 1984). These data set upper and lower bounds on the time of transposition of most of the repeats in this gene cluster. Few if any of the present repeats were in the eutherian ancestor about 85 Myr ago, but the bulk of the repeats had appeared in the stem-catarrhine about 20 Myr ago. However, the transposition of repeats did not stop at that point. Repeats were inserted into the globin gene clusters of primates as recently as 16 Myr ago, as shown by the insertion of an L1 repeat into the spider monkey ψη gene after its divergence from owl monkey (Fitch et al., 1988). Also the insertion of an Alu repeat between δ and β in gorilla occurred after its split from other great apes about 8 Myr ago (Trabuchet et al., 1987). In fact, some repeats are still actively transposing, as shown by the appearance of new L1 repeats within a single human generation (Kazazian et al., 1988).

The L1 repeats flanking the human γ-globin genes are found precisely at the breakpoints of the duplication units. Shen et al. (1981) discovered a short sequence present at all three junctions of the duplication unit, and Rogan et al. (1987) recognized this as a truncated L1 repeat. Thus in a model of gene duplication by unequal crossing over, these L1s could be sequences that allowed the initial out-of-register pairing to generate the duplication (Shen et al., 1981). Alternatively, the L1 repeats could be inserting at the chromosome breaks that accompany the duplications, as evidenced, for example, by the insertion of L1 repeats at the deletion junction of a (γδβ) thalassemia (Mager et al., 1985).

The juxtaposition of several repeats in the gene clusters reveals several hotspots for insertions. Examples of L1 repeats inserting repeatedly into the same site are apparent in all three species (Figure 5), and the clusters of C repeats in rabbits are actually examples of repeats inserting into one another (Krane et al., 1991). The region between ε and γ is particularly susceptible to insertions (in rabbit and human) and duplications (to produce βh0 and βh1 in mouse and $^G\gamma$ and $^A\gamma$ in humans). Also, the 5' flank of γ shows a very complex pattern in the comparisons between mouse and human. The DNA matches along the 5' flank of mouse βh1 and human $^G\gamma$, but sequences from this same region in humans matches with the 5' flank of βh3 (orthologous to δ) (Figure 5; Shehee et al., 1989). These latter matches are out of order as one proceeds along the gene cluster, and perhaps they reflect a different recombination event involving the 5' flank of γ in mouse—for example, an exchange between the flanks of δ and γ. Certainly the region between ε and γ has experienced a variety of recombination events, as has the region between γ and δ (deletion of η, insertions of repeats, conversions between δ and β). Clearly the gene cluster has hotspots for recombination.

In contrast, the most stable region of the gene cluster is the ε globin gene and its 5' flank. As pointed out above, this gene has a slower rate of evolution than other β-like globin genes both in nonsynonymous substitutions and in substitutions in the noncoding DNA. Despite the multiple insertions of repeats 5' to the coding region, the gene and its 5' flank also show the most extensive matches in the dot-plot comparisons summarized in Figure 5; this is true for comparisons of both rabbit and mouse with human. One might surmise that this highly conserved region contains some important regulatory element, and this is indeed the case. Sequences located from 6 to 14 kb 5' to the human ε gene strongly enhance expression of globin genes linked in *cis*, and expression in transgenic mice depends on the copy number but not the position of a globin gene linked to these sequences (Grosveld et al., 1987; Forrester et al., 1987).

These strong positive regulatory sequences form DNase hypersensitive sites in the chromatin of erythroid cells (Tuan et al., 1985) and are called dominant control regions or locus activating regions (Grosveld et al., 1987; Forrester et al., 1987). The hypersensitive site closest to the human ε-globin gene is indicated by the star in Figure 5; it is located in one of the few extragenic regions that is highly conserved in all three species. Thus, it is reasonable to propose that the conservation of sequences around the ε gene may reflect the functional importance of this dominant control region, and it will be useful to examine the possible role of this and other conserved regions in regulation, especially those that are marked by DNase hypersensitive sites in chromatin.

A way to identify functionally important sequences is to search for sequences that are evolving at relatively slow rates. The sequences flanking the ε-globin gene in rabbits and humans are diverging at about 4 to 5×10^{-9} substitutions per site per year (Margot et al., 1989), a value that is consistent with the neutral rate based on the divergence rates for synonymous substitutions and for pseudogenes in mammals (reviewed in Nei, 1987). A slower rate of divergence is observed in the intergenic sequences of humans and several apes and monkeys (Maeda et al., 1988), which is consistent with a deceleration in the divergence rate in higher primates. Thus most of the sequences in the flanking regions are evolving at the neutral rate, although ε seems to be changing slightly slower than the other β-like globin genes (Harris et al., 1986). Consequently, any sequences changing more slowly than this neutral rate are prime candidates for regulatory or other important sequences. These sequences must be analyzed at a high level of resolution, however, because most of the recognition sequences for promoter and enhancer binding proteins are very short, from 5 to 17 bp long (reviewed in Maniatis et al., 1987).

CONCLUDING REMARKS

The analysis of globin gene structure and sequence continues to be a fertile field for investigations of phylogeny, function and history of introns, mechanisms of gene duplication and transposition, and the evolution and regulation of multigene families. A growing body of sequence data covering all the sequences in a gene cluster, including intergenic and repeated sequences, is providing new insights. As these data are extended and their analysis carried out with greater precision and resolution, testable hypotheses will be generated. One limitation at this point is the inability of currently available programs to align extremely long sequences. Progress in this area is being made (Huang et al., 1990), and it is hoped that these new alignments will provide guides for the design of site-directed mutagenesis experiments to test the proposed functions of the candidate sequences revealed by conservation of the alignments.

Acknowledgments

Chapter 1   Carl R. Woese
   The author's studies in prokaryote phylogeny are supported by grants from the National Science Foundation (Systematics) and the National Aeronautics and Space Administration.

Chapter 2   Robert K. Selander, Pilar Beltran, and Noel H. Smith
   The senior author thanks Charlotte and Emily Selander for companionship in the course of preparation of the manuscript. The authors' research was supported by USPHS Grants AI22144 and AI24332, and N. H. S. held a postdoctoral fellowship in molecular studies of evolution from the Alfred P. Sloan Foundation.

Chapter 3   Robert F. DuBose and Daniel L. Hartl
   The authors wish to thank Daniel Dykhuizen, Louis Green, Jeff Lawrence, Kyoko Maruyama, Howard Ochman, and Stanley Sawyer for helpful discussions on technical and theoretical aspects of the work described here, as well as for comments on previous versions of the manuscript. This work was supported by NIH grants GM30201 and GM40322.

Chapter 5   Shozo Yokoyama
   I would like to thank Ruth Yokoyama for initially drawing my attention to the molecular evolution of oncogenes and Takashi Gojobori for his collaboration on the present project. This work was partially supported by grants from the National Institutes of Health and National Science Foundation. This work was also supported by DMB890002N and utilized the CRAY X-MP/48 system at the National Center for Supercomputing Applications at the University of Illinois at Urbana-Champaign.

Chapter 7   Michael T. Clegg, Gerald H. Learn, and Edward M. Golenberg
   We thank Dr. W. Dennis Clark for helpful comments on the manuscript, and we thank Dr. M. Sugiura for providing the 1989 update of the tobacco cpDNA sequence. Supported in part by NSF Grant BSR-8500206 and NIH Grant GM11825.

Chapter 9   Chung-I Wu and Michael F. Hammer
   We are grateful to James F. Crow for stimulating suggestions. We thank Uzi Nur, John Jaenike, Brian Charlesworth, Norman Johnson, Parul Doshi, Avis James, and Mike Palopoli for helpful comments. C.-I. W. is supported by an NIH grant (GM39902) and a National Down Syndrome Society Research Scholar Fellowship.

Chapter 11   Masatoshi Nei and Austin L. Hughes
This study was supported by research grants from the National Institutes of Health and the National Science Foundation.

Chapter 12   Philip W. Hedrick, William Klitz, Wendy P. Robinson, Mary K. Kuhner and Glenys Thomson
This research was supported by NIH Grants HD12731 and GM35326.

Chapter 13   Ross C. Hardison
Work described from this laboratory was supported by a Public Health Service Grant DK27635 and an R.C.D.A. DK01589.

Bibliography

Numbers in parentheses at the end of each reference indicate the chapter or chapters in which the work is cited.

Achenbach-Richter, L. and C. R. Woese (1988) The ribosomal gene spacer region in archaebacteria. Syst. Appl. Microbiol. 10: 211–214. (1)

Achenbach-Richter, L., K. O. Stetter and C. R. Woese (1987a) A possible biochemical missing link among archaebacteria. Nature 327: 348–349. (1)

Achenbach-Richter, L., R. Gupta, K. O. Stetter and C. R. Woese (1987b) Were the original eubacteria thermophiles? Syst. Appl. Microbiol. 9: 34–39. (1)

Achenbach-Richter, L., R. Gupta, W. Zillig and C. R. Woese (1988) Rooting the archaebacterial tree: The pivotal role of *Thermococcus celer* in archaebacterial evolution. Syst. Appl. Microbiol. 10: 231–240. (1)

Aguadé, M., N. Miyashita and C. H. Langley (1989a) Restriction-map variation at the *Zeste-tko* region in natural populations of *Drosophila melanogaster*. Mol. Biol. Evol. 6: 123–130. (8, 10)

Aguadé, M., N. Miyashita and C. H. Langley (1989b) Reduced variation in the *yellow-achaete-scute* region in natural populations of *Drosophila melanogaster*. Genetics 122: 607–615. (8, 10)

Air, G. M. (1981) Sequence relationships among the hemagglutinin genes of 12 subtypes of influenza A virus. Proc. Natl. Acad. Sci. USA 78: 7639–7643. (5)

Ajioka, J. W. and W. F. Eanes (1989) The accumulation of P-elements on the tip of the X chromosome in populations of *Drosophila melanogaster*. Genet. Res. 53: 1–6. (8)

Albert, E. D., M. P. Baur and W. R. Mayr (eds.) (1984) *Histocompatibility Testing 1984*. Springer-Verlag, Berlin. (12)

Alizon, M., S. Wain-Hobson, L. Montagnier and P. Sonigo (1986) Genetic variability of the AIDS virus: Nucleotide sequence analysis of two isolates from African patients. Cell 46: 63–74. (5)

Alper, C. A., Z. L. Awdeh, D. D. Raum and E. J. Yunis (1985) Possible human analogs of the murine *T/t* complex. Exp. Clin. Immunogenet. 2: 125–136. (12)

Altwegg, M., F. W. Hickman-Brenner and J. J. Farmer III (1989) Ribosomal RNA gene restriction patterns provide increased sensitivity for typing *Salmonella typhi* strains. J. Infect. Dis. 160: 145–149. (2)

Amann, R., W. Ludwig and K.-H. Schleifer (1988) β-Subunit of ATP-synthase: A useful marker for studying the phylogenetic relationship of eubacteria. J. Gen. Microbiol. 134: 2815–2821. (1)

Antoine, M. and J. Niessing (1984) Intron-less globin genes in the insect *Chironomus thummi thummi*. Nature 310: 795–798. (13)

Appleby, C. A., J. D. Tjepkema and M. J. Trinick (1983) Hemoglobin in a nonleguminous plant, *Parasponia*: Possible genetic origin and function in nitrogen fixation. Science 220: 951–953. (13)

Aquadro, C. F. (1989) Contrasting levels of DNA sequence variation in *Drosophila* species revealed by "six-cutter" restriction map surveys. In: *Molecular Evolution*, M. Clegg and S. O'Brien (eds.). UCLA Symposium on Molecular and Cellular Biology, New Series, Vol. 122. Liss, New York. (10)

Aquadro, C. F., S. F. Desse, M. M. Bland, C. H. Langley and C. C. Laurie-Ahlberg (1986) Molecular population genetics of the alcohol dehydrogenase gene region of *Drosophila melanogaster*. Genetics 114: 1165–1190. (8, 9, 10)

Aquadro, C. F., K. M. Lado and W. A. Noon (1988) The *rosy* region of *Drosophila melanogaster* and *Drosophila simulans*. I. Contrasting levels of naturally occurring DNA restriction map variation and divergence. Genetics 119: 875–888. (8, 10)

Arbeit, R. D., M. Arthur, R. Dunn, C. Kim, R. K. Selander and R. Goldstein (1990) Resolution of recent evolutionary divergence among *Escherichia coli* from related lineages: The application of pulsed field electrophoresis to molecular epidemiology. J. Infect. Dis. 161: 230–235. (2)

Argos, P. and M. G. Rossman (1979) Structural comparisons of heme binding proteins. Biochemistry 18: 4951–4960. (13)

Arnold, J., M. A. Asmussen and J. C. Avise (1988) An epistatic mating system model can produce permanent cytonuclear disequilibria in a hybrid zone. Proc. Natl. Acad. Sci. USA 85: 1893–1896. (6)

Artymiuk, P. J., C. C. F. Blake and A. E. Sippel (1981) Genes pieced together—exons delineate homologous structures of diverged lysozymes. Nature 290: 287–288. (3)

Asmussen, M. A., J. Arnold and J. C. Avise (1987) Definition and properties of disequilibrium statistics for associations between nuclear and cytoplasmic genotypes. Genetics 115: 755–768. (6)

Asmussen, M. A., J. Arnold and J. C. Avise (1989) The effects of assortative mating and migration on cytonuclear associations in hybrid zones. Genetics 122: 923–934. (6)

Avise, J. C., J. E. Neigel and J. Arnold (1984) Demographic influences on mitochondrial DNA lineage survivorship in animal populations. J. Mol. Evol. 20: 99–105. (6)

Avise, J. C., R. M. Ball and J. Arnold (1988) Current versus historical population sizes in vertebrate species with high gene flow: A comparison based on mitochondrial DNA lineages and inbreeding theory for neutral mutations. Mol. Biol. Evol. 5: 331–344. (6)

Awdeh, Z. L., D. Raum, E. J. Yunis and C. A. Alper (1983) Extended HLA/complement allele haplotypes: Evidence for T/t-like complex in man. Proc. Natl. Acad. Sci. USA 80: 259–263. (12)

Bachmann, B. J. (1987) Linkage map of *Escherichia coli* K-12, Edition 7. In: Escherichia coli *and* Salmonella typhimurium: *Cellular and Molecular Biology*, F. C. Neidhardt, J. L. Ingraham, K. B. Low, B. Magasanik, M. Schaechter and H. E. Umbarger (eds.). American Society for Microbiology, Washington, D.C. pp. 807–876. (3)

Bailey, D. W. and H. I. Kohn (1965) Inherited histocompatibility changes in progeny of irradiated and unirradiated inbred mice. Genet. Res. 6: 330–340. (11)

Baine, W. B., J. J. Farmer III, E. J. Gangarosa, G. T. Hermann, C. Thornsberry and P. A. Rice (1977) Typhoid fever in the United States associated with the 1972–1973 epidemic in Mexico. J. Infect Dis. 135: 649–653. (2)

Balch, W. E., G. E. Fox, L. J. Magrum, C. R. Woese and R. S. Wolfe (1979) Methanogens: Reevaluation of a unique biological group. Microbiol. Rev. 43: 260–296. (1)

Balner, H., W. V. Vreeswijk, J. H. Roger and J. D'Amaro (1978) The major histocompatibility complex of chimpanzees: Identification of several new antigens controlled by the A and B loci of ChLA. Tissue Antigens 12: 1–18. (12)

Banks, J. A. and C. W. Birky, Jr. (1985) Chloroplast DNA diversity is low in a wild plant, *Lupinus texensis*. Proc. Natl. Acad. Sci. USA 82: 6950–6954. (7)

Barker, R. M. and D. C. Old (1989) The usefulness of biotyping in studying the epidemiology and phylogeny of salmonellae. J. Med. Microbiol. 29: 81–88. (2)

Barker, R. M., G. M. Kearney, P. Nicholson, A. L. Blair, R. C. Porter and P. B. Crichton (1988) Types of *Salmonella paratyphi* B and their phylogenetic significance. J. Med. Microbiol. 26: 285–293. (2)

Barre-Sinoussi, F., J. C. Chermann, F. Rey, M. T. Nugeyre, S. Chamaret, J. Gruest, C. Dauguet, C. Axler-Blin, F. Vezinet-Brun, C. Rouzioux, W. Rozenbaum and L. Montagnier (1983) Isolation of a T-lymphotropic retrovirus from a patient at risk for acquired immune deficiency syndrome (AIDS). Science 220: 868–871. (5)

Barton, N. and J. S. Jones (1983) Mitochondrial DNA: New clues about evolution. Nature 306: 317–318. (6)

Baur, M. P., M. Neugebauer and E. D. Albert (1984) References Tables of two-locus haplotype frequencies for all MHC marker loci. In: *Histocompatibility Testing 1984*, E. D. Albert, M. P. Baur and W. R. Mayr (eds.). Springer-Verlag, Berlin, pp. 677–755. (12)

Beckenbach, A. T. (1978) The "Sex-Ratio" trait in *Drosophila pseudoobscura*: Fertility relations of males and meiotic drive. Am. Nat. 112: 97–117. (9)

Beckenbach, A .T. (1983) Fitness analysis of the "Sex-Ratio" polymorphism in experimental populations of *Drosophila pseudoobscura*. Am. Nat. 121: 630–648. (9)

Beckenbach, A. T., J. W. Curtsinger and D. Policansky (1982) Fruitless experiments with fruitflies: The "sex-ratio" chromosomes of *D. pseudoobscura*. Drosophila Information Service 58: 22. (9)

Beech, R. N. and A. J. Leigh Brown (1989) Insertion–deletion variation at the *yellow-achaete-scute* region in two natural populations of *Drosophila melanogaster*. Genet. Res. 53: 7–15. (8 ,10)

Beltran, P., J. M. Musser, R. Helmuth, J. J. Farmer III, W. M. Frerichs, I. K. Wachsmuth, K. Ferris, A. C. McWhorter, J. G. Wells, A. Cravioto and R. K. Selander (1988) Toward a population genetic analysis of *Salmonella*: Genetic diversity and relationships among strains of serotypes *S. choleraesuis*, *S. derby*, *S. dublin*, *S. enteritidis*, *S. heidelberg*, *S. infantis*, *S. newport*, and *S. typhimurium*. Proc. Natl. Acad. Sci. USA 85: 7753–7757. (2)

Berg, D. E. and M. M. Howe (eds.) (1989) *Mobile DNA*. American Society of Microbioliology, Washington, D.C. (8)

Bernardi, G. (1989) The isochore organization of the human genome. Annu. Rev. Genet. 23: 637–661. (13)

Bernardi, G., B. Olofsson, J. Filipski, M. Zerial, J. Salinas, G. Cuny, M. Meunier-Rotival, and F. Rodier (1985) The mosaic genome of warm-blooded vertebrates. Science 228: 953–958. (13)

Berthou, F., C. Mathieu and F. Vedel (1983) Chloroplast and mitochondrial DNA variation as indicator of phylogenetic relationships in the genus *Coffea* L. Theor. Appl. Genet. 65: 77–84. (7)

Berzofsky, J. A., L. K. Richman and D. J. Killion (1979) Distinct *H*-2-linked *Ir* genes control both antibody and T cell responses to different determinants on the same antigen, myoglobin. Proc. Natl. Acad. Sci. USA 76: 4046–4050. (11)

Bhatti, A. R. (1973) Variation of alkaline phosphatase isoenzymes in *Escherichia coli* and *Serratia marcescens*. FEBS Lett. 32: 81–83. (3)

Bhatti, A. R. (1975) Some distinctive characteristics of the alkaline phosphatase of *Serratia marcescens*. Can. J. Biochem. Cell Biol. 53: 819–822. (3)

Bhatti, A. R. and J. Done (1973) Antigenic independence of the two types of alkaline phosphatases from *Serratia marcescens* strain 211. Life Sci. 13: 1421–1428. (3)

Bird, C. R., B. Koller, A. D. Auffret, A. K. Huttly, C. J. Howe, T. A. Dyer and J. C. Gray (1985) The wheat chloroplast gene for CF$_0$ subunit I of ATP synthase contains a large intron. EMBO J. 4: 1381–1388. (7)

Birky, C. W., Jr. (1978) Transmission genetics of mitochondria and chloroplasts. Annu. Rev. Genet. 12: 471–512. (6)

Birky, C. W., Jr. (1983a) The partitioning of cytoplasmic organelles at cell division. Int. Rev. Cytol. Supplement 15: 49–89. (6)

Birky, C. W., Jr. (1983b) Relaxed cellular controls and organelle heredity. Science 222: 468–475. (6)

Birky, C. W., Jr. (1988) Evolution and variation in plant chloroplast and mitochondrial genomes. In: *Plant Evolutionary Biology*. L. D. Gottlieb and S. K. Jain (eds.). Chapman and Hall, London, pp. 23–53. (6)

Birky, C. W., Jr. and R. V. Skavaril (1976) Maintenance of genetic homogeneity in systems with multiple genomes. Genet. Res. 27: 249–265. (6)

Birky, C. W., Jr. and J. B. Walsh (1988) Effects of linkage on rates of molecular evolution. Proc. Natl. Acad. Sci. USA 85: 6414–6418. (6)

Birky, C. W., Jr., T. Maruyama and P. Fuerst (1983) An approach to population and evolutionary genetic theory for genes in mitochondria and chloroplasts, and some results. Genetics 103: 513–527. (6)

Birky, C. W., Jr., P. Fuerst and T. Maruyama (1989) Organelle gene diversity under migration, mutation, and drift: Equilibrium expectations, approach to equilibrium, effects of heteroplasmic cells, and comparison to nuclear genes. Genetics 121: 613–627. (6)

Bishop, J. M. (1983) Cellular oncogenes and retroviruses. Annu. Rev. Biochem. 52: 301–354. (5)

Bjorkman, P. J., M. A. Saper, B. Samraoui, W. S. Bennett, J. L. Strominger and D. C. Wiley (1987a) Structure of the human class I histocompatibility antigen, HLA-A2. Nature 329: 506–512. (11)

Bjorkman, P. J., M. A. Saper, B. Samraoui, W. S. Bennett, J. L. Strominger and D. C. Wiley (1987b) The foreign antigen binding site and T cell recognition regions of class I histocompatibility antigens. Nature 329: 512–518. (11, 12)

Black, F. L. and F. M. Salzano (1981) Evidence for heterosis in the HLA system. Am. J. Hum. Genet. 33: 894–899. (12)

Blackman, R. K., R. Grimaila, M. M. D.Koehler and W. M. Gelbart (1987) Mobilization of hobo elements residing within the decapentaplegic gene complex: Suggestion of a new hybrid dysgenesis system in *Drosophila melanogaster*. Cell 49: 497–505. (8)

Blake, C. C. F. (1978) Do genes-in-pieces imply proteins-in-pieces? Nature 273: 267. (13)

Blake, C. C. F. (1981) Exons and the structure, function and evolution of haemoglobin. Nature 291: 616. (13)

Blanchetot, A., V. Wilson and A. J. Jeffreys (1983) The seal myoglobin gene: An unusually long globin gene. Nature 301: 732–734. (13)

Blanden, R. V. (1980) How do immune response genes work? Immunol. Today 1: 33–36. (12)

Bodmer, M. and M. Ashburner (1984) Conservation and change in the DNA sequences coding for alcohol dehydrogenase in sibling species of Drosophila. Nature 309: 425–430. (10)

Bodmer, W. (1972) Evolutionary significance of the HL-A system. Nature 237: 139–145. (11, 12)

Bodmer, W. (1989) HLA 1987. In: Immunobiology of HLA, Vol. II, Immuno-genetics and Histocompatibility, B. Dupont (ed.). Springer-Verlag, New York, pp. 1–9. (11)

Bodmer, W. and J. G. Bodmer (1989) Statistics and population genetics of the HLA system. In: Mathematical Evolutionary Theory, M. Feldman (ed.), Princeton Univ. Press, Princeton, NJ, pp. 315–334. (12)

Boeke, J. D. (1988) Retrotransposons. In: RNA Genetics, Vol. II, Retroviruses, Viroids and RNA Recombination, E. Domingo, J. J. Holland, P. Ahlquist (eds.). CRC Press, Boca Raton, FL, pp. 59–103. (8)

Bonhomme, F., J. Catalan, J. Britton-Davidian, V. M. Chapman, K. Moriwaki, E. Nevo and L. Thaler (1984) Biochemical diversity and evolution in the genus Mus. Biochem. Genet. 22: 275–303. (9)

Bonnard, G., F. Michel, J. H. Weil and A. Steinmetz (1984) Nucleotide sequence of the split tRNA$^{Leu}_{UAA}$ gene from Vicia faba chloroplasts: Evidence for structural homologies of the chloroplast tRNA^{Leu} intron with the intron from the autosplicable Tetrahymena ribosomal RNA precursor. Mol. Gen. Genet. 194: 330–336. (7)

Bosch, M.C., A. Termijtelen, J. J. VanRood and M. J. Giphart (1989) Evidence for hot spot for generalized gene conversion in the second exon of MHC genes. In: Immunobiology of HLA, Vol. II, Histocompatibility Testing 1987, B. Dupont (ed.). Springer-Verlag, New York, pp. 266–269. (12)

Bowman, C. M., G. Bonnard and T. A. Dyer (1983) Chloroplast DNA variation between species of Triticum and Aegilops. Location of the variation on the chloroplast genome and its relevance to the inheritance and classification of the cytoplasm. Theor. Appl. Genet. 65: 247–262. (7)

Boyce, T. M., M. E. Zwick and C. F. Aquadro (1989) Mitochondrial DNA in the bark weevils: Size, structure, and heteroplasmy. Genetics 123: 825–836. (6)

Brahmbhatt, H. N., P. Wyk, N. B. Quigley and P. R. Reeves (1988) Complete physical map of the rfb gene cluster encoding biosynthetic enzymes for the O antigen of Salmonella typhimurium LT2. J. Bacteriol. 170: 98–102. (2)

Bregliano, J.–C. and M. G. Kidwell (1983) Hybrid dysgenesis determinants. In: Mobile Genetic Elements, J. A. Shapiro (ed.). Academic Press, New York, pp. 363–410. (8)

Brenner, S. (1988) The molecular evolution of genes and proteins: A tale of two serines. Nature 334: 528–530. (3)

Brisson, N. and D. P. S. Verma (1982) Soybean leghemoglobin gene family: Normal, pseudo, and truncated genes. Proc. Natl. Acad. Sci. USA 79: 4055–4059. (13)

Brittnacher, J. G. and B. Ganetzky (1983) On the components of segregation distortion in Drosophila melanogaster II. Deletion mapping and dosage analysis of the SD locus. Genetics 103: 659–673. (9)

Brookfield, J. F. Y. (1986) Population biology of transposable elements. Phil. Trans. R. Soc. London Ser. B 312: 217–226. (8)

Brookfield, J. F. Y., E. Montgomery and C. H. Langley (1984) Apparent absence of transposable elements related to the P elements of D. melanogaster in other species of Drosophila. Nature 310: 330–332. (8)

Brooks, B. W., R. G. E. Murray, J. L. Johnson, E. Stackebrandt, C. R. Woese and G. E. Fox (1980) Red-pigmented micrococci: A basis for taxonomy. Int. J. Syst. Bacteriol. 30: 627–646. (1)

Brosius, J., J. L. Palmer, J. P. Kennedy and H. F. Noller (1978) Complete nucleotide sequence of a 16S ribosomal RNA gene from Escherichia coli. Proc. Natl. Acad. Sci. USA 75: 4801–4805. (1)

Brown, A. H. D. (1979) Enzyme polymorphism in plant populations. Theor. Pop. Biol. 15: 1–42. (8)

Brown, D. D., P. C. Wensink and E. Jordan (1972) A comparison of the ribosomal DNAs of Xenopus laevis and Xenopus mulleri: The evolution of tandem genes. J. Mol. Biol. 63: 57–73. (9)

Brown, J. H., T. Jardetzky, M. A. Saper, B. Samraoui, P. J. Bjorkman and D. C. Wiley (1988) A hypothetical model of the foreign antigen binding site of Class II histocompatibility molecules. Nature 332: 845–850. (11)

Brown, J., J. A. Cebra-Thomas, J. D. Bleil, P. M. Wassarman and L. M. Silver (1989) A premature acrosome reaction is programmed by mouse t haplotypes during sperm differentiation and could play a role in transmission ratio distortion. Development 106: 769–773. (9)

Buchanan, R. E. (1925) *General Systematic Bacteriology*. Williams & Wilkins, Baltimore. (1)

Bucheton, A., R. Paro, H. M. Sang, A. Pelisson and D. J. Finnegan (1984) The molecular basis of I–R hybrid dysgenesis in *Drosophila melanogaster*: Identification, cloning and properties of the I factor. Cell 38: 153–163. (8)

Burgin, A. B., K. Parodos, D. J. Lane and N. R. Pace (1990) The excision of intervening sequences from *Salmonella* 23S ribosomal RNA. Cell 60: 405–414. (2)

Burton, F. H., D. D. Loeb, S. F. Chao, C. A. Hutchison and M. H. Edgell (1985) Transposition of a long member of the L1 major interspersed DNA family into the mouse β globin gene locus. Nucleic Acids Res. 13: 5071–5084. (13)

Butler-Ransohoff, J. E., D. A. Kendall and E. T. Kaiser (1988) Use of site-directed mutagenesis to elucidate the role of arginine-166 in the catalytic mechanism of alkaline phosphatase. Proc. Natl. Acad. Sci. USA 85: 4276–4278. (3)

Buus, S., S. Colon, C. Smith, J. H. Freed, C. Miles and H. M. Grey (1986) Interaction between a "processed" ovalbumin peptide and Ia molecule. Proc. Natl. Acad. Sci. USA 83: 3968–3971. (11)

Cairns, J., J. Overbaugh and S. Miller (1988) The origin of mutants. Nature 335: 142–145. (4)

Cameron, J. R., E. Y. Loh and R. W. Davis (1979) Evidence for transposition of dispersed repetitive DNA families in yeast. Cell 16: 739–751. (8)

Cammarano, P., A. Teichner and P. Londei (1986) Intralineage heterogeneity of archaebacterial ribosomes: Evidence for two physicochemically distinct ribosome classes within the third ur-kingdom. Syst. Appl. Microbiol. 7: 137–145. (1)

Campbell, A. (1983) Transposons and their evolutionary significance. In: *Evolution of Genes and Proteins*, M. Nei and R. K. Koehn (eds.). Sinauer Associates, Sunderland, Mass., pp. 258–279. (8)

Campbell, A. (1990) Epilogue. In: *The Bacterial Chromosome*, K. Drlica and M. Riley (eds.). American Society for Microbiology, Washington, D.C., pp. 459–464. (2)

Cann, R. L., M. Stoneking and A. C. Wilson (1987) Mitochondrial DNA and human evolution. Nature 325: 31–36. (9)

Canters, G. W. (1987) Cloning and sequencing of the azurin gene from *Pseudomonas aeruginosa*. Prot. Eng. 1: 243. (3)

Carlile, M. J. and J. J. Skehel (eds.) (1974) Evolution in the microbial world. Cambridge Univ. Press, Cambridge. (2)

Caspari, E., G. S. Watson and W. Smith (1966) The influence of cytoplasmic pollen sterility on gene exchange between populations. Genetics 53: 741–746. (6)

Cattolico, R. A. (1986) Chloroplast evolution in algae and land plants. Trends Ecol. Evol. 1: 64–67. (7)

Cech, T. R. (1986) The generality of self-splicing RNA: Relationship to nuclear mRNA splicing. Cell 44: 207–210. (7)

Celis, E. and R. W. Karr (1989) Presentation of an immunodominant T-cell epitope of hepatitis B surface antigen by the HLA–DPw4 molecule. J. Virol. 63: 747–752. (11)

Chada, K., J. Magram and F. Costantini (1986) An embryonic pattern of expression of a human fetal globin gene in transgenic mice. Nature 319: 685–689. (13)

Chaidaroglou, A., D. J. Brezinski, S. A. Middleton and E. R. Kantrowitz (1988) Function of arginine-166 in the active site of *Escherichia coli* alkaline phosphatase. Biochemistry 27: 8338–8343. (3)

Chakrabarti, L., M. Guyader, M. Alizon, M. D. Daniel, R. C. Desrosiers, P. Tiollais and P. Sonigo (1987) Sequence of simian immunodeficiency virus from macaque and its relationship to other human and simian retroviruses. Nature 328: 543–547. (5)

Chalker, R. B. and M. J. Blaser (1988) A review of human salmonellosis: III. Magnitude of *Salmonella* infection in the United States. Rev. Infect. Dis. 10: 111–124. (2)

Chang, M., D. W. Essar and I. P. Crawford (1990) Diverse regulation of the tryptophan genes in fluorescent pseudomonads. In: *Pseudomonas: Biotransformations, Pathogenesis, and Evolving Biotechnology*, S. Silver (ed.). American Society for Microbiology, Washington, D.C., pp. 292–302. (4)

Chao, L., C. Vargas, B. B. Spear and E. C. Cox (1983) Transposable elements as mutator genes in evolution. Nature 303: 633–635. (8)

Chapman, R. W., J. C. Stephens, R. A. Lansman and J. C. Avise (1982) Models of mitochondrial DNA transmission genetics and evolution in higher eucaryotes. Genet. Res. 40: 41–57. (6)

Chardon, P., M. Vaiman, M. Kirszenbaum, C. Geffrotin, C. Renard and D. Cohen (1985) Restriction fragment length polymorphism of the major histocompatibility complex of the pig. Immunogenetics 21: 161–171. (11)

Charlesworth, B. (1985) The population genetics of transposable elements. In: *Population Genetics and Molecular Evolution*, T. Ohta and K. Aoki (eds.). Springer-Verlag, Berlin, pp. 213–232. (8)

Charlesworth, B. (1988) The maintenance of transposable elements in natural populations. In: *Plant Transposable Elements*, O. J. Nelson (ed.). Plenum Press, New York, pp. 189–212. (8)

Charlesworth, B. and D. Charlesworth (1983) The population dynamics of transposable elements. Genet. Res. 42: 1–27. (8)

Charlesworth, B. and D. L. Hartl (1978) Population dynamics of the segregation distorter polymorphism of *Drosophila melanogaster*. Genetics 89: 171–192. (9)

Charlesworth, B. and C. H. Langley (1986) The evolution of self-regulated transposition of transposable elements. Genetics 112: 359–383. (8)

Charlesworth, B. and C. H. Langley (1989) The population genetics of *Drosophila* transposable elements. Ann. Rev. Genet. 23: 251–287. (8)

Charlesworth, B. and A. Lapid (1989) A study of ten families of transposable elements on X chromosomes from a population of *Drosophila melanogaster*. Genet. Res. 54: 113–125. (8)

Charlesworth, B., C. H. Langley and W. Stephan (1986) The evolution of restricted recombination and the accumulation of repeated DNA sequences. Genetics 112: 947–962. (8)

Charnay, P., R. Treisman, P. Mellon, M. Chao, R. Axel and T. Maniatis (1984) Differences in human α- and β-globin gene expression in mouse erythroleukemia cells: The role of intragenic sequences. Cell 38: 251–263. (13)

Chatton. E. (1937) *Titres et Travoux Scientifiques*. Sete, Sottano. (1)

Cheng, J.-F. and R. C. Hardison (1988) The rabbit α-like globin gene cluster is polymorphic both in the sizes of *Bam*HI fragments and in the numbers of duplicated sets of genes. Mol. Biol. Evol. 5: 486–498. (13)

Cheng, J.-F., R. Printz, T. Callaghan, D. Shuey and R. C. Hardison (1984) The rabbit C family of short, interspersed repeats: Nucleotide sequence determination and transcriptional analysis. J. Mol. Biol. 176: 1–20. (13)

Cheng, J.-F., L. Raid and R. C. Hardison (1986) Isolation and nucleotide sequence of the rabbit globin gene cluster $\phi\zeta$-α1-$\phi\alpha$. J. Biol. Chem. 261: 839–848. (13)

Cheng, J.-F., L. Raid and R. C. Hardison (1987) Block duplications of a ζ-ζ-αθ gene set in the rabbit α-like globin gene cluster. J. Biol. Chem. 262: 5414–5421. (13)

Cheng, J.-F., D. E. Krane and R. C. Hardison (1988) Nucleotide sequence and expression of rabbit globin genes ζ1, ζ2, and ζ3: Pseudogenes generated by block duplications are transcriptionally competent. J. Biol. Chem. 263: 9981–9993. (13)

Cheung, W. Y. and N. S. Scott (1989) A contiguous sequence in spinach nuclear DNA is homologous to three separated sequences in chloroplast DNA. Theoret. Appl. Genet. 77: 625–633. (7)

Chiu, I.-M., A. Yaniv, J. E. Dahlberg, A. Gazit, S. F. Skuntz, S. R. Tronick and S. A. Aaronson (1985) Nucleotide sequence evidence for relationship of AIDS retrovirus to lentivirus. Nature 317: 366–368. (5)

Chiu, W.-L. and B. B. Sears (1985) Recombination between chloroplast DNAs does not occur in sexual crosses of *Oenothera*. Mol. Gen. Genet. 198: 525–528. (7)

Chou, P. Y. and G. D. Fasman (1978) Empirical prediction of protein conformation. Annu. Rev. Biochem. 47: 251–276. (3)

Chovnick, A., W. Gelbart and M. McCarron (1977) Organization of the Rosy locus in *Drosophila melanogaster*. Cell 11: 1–10. (10)

Christopher, D. A., J. C. Cushman, C. A. Price and R. B. Hallick (1988) Organization of ribosomal protein genes rpl23, rpl2, rps19, rpl22 and rps3 on the *Euglena gracilis* chloroplast genome. Curr. Genet. 14: 275–286. (7)

Cifelli, R. L. and J. G. Eaton (1987) Marsupial from the earliest Late Cretaceous of western US. Nature 325: 520–522. (13)

Clark, A. G. (1984) Natural selection with nuclear and cytoplasmic transmission. I. A deterministic model. Genetics 107: 679–701. (6)

Clark, A. G. (1985) Natural selection with nuclear and cytoplasmic transmission. II. Tests with *Drosophila* from diverse populations. Genetics 111: 97–112. (6)

Clark, A. G. (1988) Deterministic theory of heteroplasmy. Evolution 42: 621–626. (6)

Clark, A. G. and E. M. S. Lyckegaard (1988) Natural selection with nuclear and cytoplasmic transmission. III. Joint analysis of segregation and mtDNA in *Drosophila melanogaster*. Genetics 118: 471–481. (6)

Clarke, A. G. (1971) The effects of maternal pre-immunization on pregnancy in the mouse. J. Reprod. Fert. 24: 369–375. (11)

Clarke, B. and D. R. S. Kirby (1966) Maintenance of histocompatibility polymorphisms. Nature 211: 999–1000. (11)

Clegg, J. B. (1987) Can the product of the θ gene be a real globin?. Nature 329: 465–466. (13)

Clegg, J. B., S. E. Y. Goodbourn and M. Braend (1984) Genetic organization of the polymorphic equine α globin locus and sequence of the BII α1 gene. Nucleic Acids Res. 12: 7847–7858. (13)

Clegg, M. T. (1990) Dating the monocot–dicot divergence. Trends Ecol. Evol. 5: 1–2. (7)

Clegg, M. T., A. H. D. Brown and P. R. Whitfeld (1984a) Chloroplast DNA diversity in wild and cultivated barley: Implications for genetic conservation. Genet. Res. 43: 339–343. (7)

Clegg, M. T., J. R. Y. Rawson and K. Thomas (1984b) Chloroplast DNA variation in pearl millet and related species. Genetics 106: 449–461. (7)

Clegg, M. T., K. Ritland and G. Zurawski (1986) Processes of chloroplast DNA evolution. In: *Evolutionary Processes and Theory*, S. Karlin and E. Nevo (eds.). Academic Press, New York, pp. 275–294. (7)

Close, P. S., R. C. Shoemaker and P. Keim (1989) Distribution of restriction site polymorphism within the chloroplast genome of the genus *Glycine*, subgenus *Soja*. Theor. Appl. Genet. 77: 768–776. (7)

Coates, D. and C. A. Cullis (1987) Chloroplast DNA variability among *Linum* species. Am. J. Bot. 74: 260–268. (7)

Cobbs, G. (1977) Multiple insemination and male sexual selection in natural populations of *Drosophila pseudoobscura*. Am. Nat. 111: 641–656. (9)

Coffin, J., A. Haase, J. A. Levy, L. Motagnier, S. Oroszlan, N. Teich, H. Temin, K. Toyoshima, H. Varmus, P. Vogt and R. Weiss (1986) Human immunodeficiency viruses. Science 232: 697. (5)

Cohn, V. (1985) Organization and evolution of the alcohol dehydrogenase gene in *Drosophila*. Ph.D. thesis, University of Michigan microfilms. (10)

Collins, F. S. and S. M. Weissman (1984) The molecular genetics of human hemoglobin. Prog. Nucleic Acids Res. Mol. Biol. 31: 315–462. (13)

Condit, R. (1990) The evolution of transposable elements: conditions for establishment in bacterial populations. Evolution 44: 347–359. (2)

Costagliola, D., J.-Y. Mary, N. Brouard, A. Laporte and A.-J. Valleron (1989) Incubation time for AIDS from French transfusion-associated cases. Nature 338: 768–769. (5)

Coyne, J. A. (1976) Lack of genic similarity between two sibling species of *Drosophila* as revealed by varied techniques. Genetics 84: 593–607. (11)

Coyne, J. A. (1989) A test of meiotic drive in fixing a pericentric inversion. Genetics 123: 241–243. (9)

Coyne, J. A. and M. Kreitman. (1986) Evolutionary genetics of two sibling species, *Drosophila simulans* and *Drosophila sechellia*. Evolution 40: 673–691. (10)

Craik, C. S., S. R. Buchman and S. Beychok (1980) Characterization of globin domains: Heme binding to the central exon product. Proc. Natl. Acad. Sci. USA 77: 1384–1388. (13)

Craik, C. S., S. Sprang, R. Fletterick and W. J. Rutter (1982) Intron–exon splice junctions map at protein surfaces. Nature 299: 180–182. (3)

Craven, P. C., D. C. Mackel, W. B. Baine, W. H. Barker, E. J. Gangarosa, M. Goldfield, H. Rosenfeld, R. Altman, G. Lachapelle, J. W. Davies and R. C. Swanson (1975) International outbreak of *Salmonella eastbourni* infection traced to contaminated chocolate. Lancet April 5: 7888–793. (2)

Crawford, I. P. (1982) Nucleotide sequences and bacterial evolution. In: *Perspectives on Evolution*, R. Milkman (ed.). Sinauer Associates, Sunderland, MA, pp. 148–163. (4)

Crawford, I. P. (1989) Evolution of a biosynthetic pathway: The tryptophan paradigm. Annu. Rev. Microbiol. 43: 567–600. (4)

Crawford, I. P. and L. Eberly (1986) Structure and regulation of the anthranilate synthase genes in *Pseudomonas aeruginosa*: I. Sequence of *trpG* encoding the glutamine amidotransferase subunit. Mol. Biol. Evol. 3: 436–448. (4)

Crawford, I. P., M. Clarke, M. van Cleemput and C. Yanofsky (1987) Crucial role of the connecting region joining the two functional domains of yeast tryptophan synthetase. J. Biol. Chem. 262: 239–244. (4)

Crosa, J. H., D. J. Brenner, W. H. Ewing and S. Falkow (1973) Molecular relationships among the salmonellae. J. Bacteriol. 115: 307–315.

Crow, J. F. (1979) Genes that violate Mendel's rules. Sci. Am. 240: 134–146. (9)

Crow, J. F. (1988) The ultraselfish gene. Genetics 118: 389–391. (9)

Crow, J. F. and K. Aoki (1984) Group selection for a polygenic behavioral trait: Estimating the degree of population subdivision. Proc. Natl. Acad. Sci. USA 81: 6073–6077. (6)

Crow, J. F. and M. Kimura (1970) *An Introduction to Population Genetics Theory*. Burgess, Minneapolis, MN. (9)

Curtsinger, J. W. and M. W. Feldman (1980) Experimental and theoretical analysis of the "Sex-ratio" polymorphism in *Drosophila pseudoobscura*. Genetics 94: 445–466. (9)

Czelusniak, J., M. Goodman, D. Hewett-Emmett, M. L. Weiss, P. J. Venta and R. E. Tashian (1982) Phylogenetic origins and adaptive evolution of avian and mammalian haemoglobin genes. Nature 298: 297–300. (13)

Damian, R. T. (1964) Molecular mimicry: Antigen sharing by parasite and host and its consequences. Am. Nat. 98: 129–149. (11)

Damian, R. T. (1987) Molecular mimicry revisited. Parasit. Today 3: 236–266. (11)

Daniels, E. M., R. Schneerson, W. M. Egan, S. C. Szu and J. B. Robbins (1989) Characterization of the *Salmonella paratyphi* C Vi polysaccharide. Infect. Immun. 57: 3159–3164. (2)

Daniels, S.B., L.D. Strausbaugh, L. Ehrman, and R. Armstrong (1984) Sequences homologous to P elements occur in *Drosophila paulistorum*. Proc. Natl. Acad. Sci. USA 81: 6794–6797. (8)

Davis, P. S., M. W. Shen and B. H. Judd (1987) Asymmetrical pairings of transposons in and proximal to the white locus of *Drosophila* account for four classes of regularly occurring exchange products. Proc. Natl. Acad. Sci. USA 84: 174–178. (8)

Dawkins, R. (1976) *The Selfish Gene*. Oxford University Press, Oxford. (9)

Dayhoff, M. O., L. T. Hunt, P. J. McLaughlin and D. D. Jones (1972) Gene duplications in evolution: The globins. In: *Atlas of Protein Sequence and Structure*, Vol. 5, M. O. Dayhoff (ed.). National Biomedical Research Foundation, Silver Spring, MD, pp. 17–30. (13)

DeBonte, L. R., B. F. Matthews and K. G. Wilson (1984) Variation of plastid and mitochondrial DNAs in the genus *Daucus*. Am. J. Bot. 71: 932–940. (7)

De Carvalho, A. B., A. A. Peixoto and L. B. Klaczko (1989) Sex-ratio in *Drosophila mediopunctata*. Heredity 62: 425–428. (9)

Demers, G. W., M. J. Matunis and R. C. Hardison (1989) The L1 family of long interspersed repetitive DNA in rabbits: Sequence, copy number, conserved open reading frames, and similarity to keratin. J. Mol. Evol. 29: 3–19. (13)

Deng, X.-W., R. A. Wing and W. Gruissem (1989) The chloroplast genome exists in multimeric forms. Proc. Natl. Acad. Sci. USA 86: 4156–4160 (7)

Denny, T. P., M. N. Gilmour and R. K. Selander (1988) Genetic diversity and relationships of two pathovars of *Pseudomonas syringae*. J. Gen. Microbiol. 134: 1949–1960.

Deno, H. and M. Sugiura (1984) Chloroplast tRNA^{Gly} gene contains a long intron in the D stem: Nucleotide sequences of tobacco chloroplast genes for tRNA^{Gly}(UCC) and tRNA^{Arg}(UCU). Proc. Natl. Acad. Sci. USA 81: 405–408. (7)

Deno, H., A. Kato, K. Shinozaki and M. Sugiura (1982) Nucleotide sequences of tobacco chloroplast genes for elongator tRNA^{Met} and tRNA^{Val}(UAC): The tRNA^{Val}(UAC) gene contains a long intron. Nucleic Acids Res. 10: 7511–7520. (7)

Di Nocera, P. P. and I. B. Dawid (1983) Interdigitated arrangement of two oligo(A)-terminated DNA sequences in *Drosophila*. Nucleic Acids Res. 11: 5475–5482. (8)

Dibb, N. J. and A. J. Newman (1989) Evidence that introns arose at proto-splice sites. EMBO J. 8: 2015–2021. (7)

Dickerson, R. E. and I. Geis (1983) *Hemoglobin: Structure, Function, Evolution, and Pathology*. Benjamin/Cummings, Menlo Park, CA, pp. 20–63. (13)

Dodgson, J. B., K. C. McCune, D. J. Rusling, A. Krust and J. D. Engel (1981) Adult chicken α-globin genes, αA and αD: No anemic shock α-globin exists in domestic chickens. Proc. Natl. Acad. Sci. USA 78: 5998–6002. (13)

Dodgson, J. B., S. J. Stadt, O.-R. Choi, M. Dolan, H. D. Fischer and J. D. Engel (1983) The nucleotide sequence of the embryonic chicken β-type globin genes. J. Biol. Chem. 258: 12685–12692. (13)

Doebley, J., W. Renfroe and A. Blanton (1987) Restriction site variation in the *Zea* chloroplast genome. Genetics 117: 139–147. (7)

Doebley, J., M. L. Durbin, E. M. Golenberg, M. T. Clegg and D. P. Ma (1990) Evolutionary analysis of the large subunit of carboxylase (*rbcL*) nucleotide sequence among the grasses (Gramineae). Evolution 44: 1097–1108. (7)

Doherty, P. C. and R. M. Zinkernagel (1975) Enhanced immunological surveillance in mice heterozygous at the H-2 gene complex. Nature 256: 50–52. (11)

Dolan, M., B. J. Sugarman, J. B. Dodgson and J. D. Engel (1981) Chromosomal arrangement of the chicken β-type globin genes. Cell 24: 669–677. (13)

Doolittle, R. F. (1986) *Of URFS and ORFS: A Primer on How to Analyze Derived Amino Acid Sequences.* University Science Books, Mill Valley, CA. (4)

Doolittle, R. F. (1987) The origin and function of intervening sequences in DNA: A review. Am. Nat. 130: 915–928. (7)

Doolittle, R. F. (1989) The simian–human connection. Nature 339: 338–339. (5)

Doolittle, R. F., D.-F. Feng, M. S. Johnson and M. A. McClure (1989) Origins and evolutionary relationships of retroviruses. Quart. Rev. Biol. 64: 1–30. (5, 7)

Doolittle, W. F. and C. Sapienza (1980) Selfish genes, the phenotype paradigm and genome evolution. Nature 284: 601–603. (8, 9)

Doshi, P., S. Kaushal and C.-I. Wu. (1991) Molecular analysis of the *Responder* satellite DNA in *Drosophila melanogaster*. I. DNA tending, nucleosome structure of *Rsp*-binding proteins. Submitted. (9)

Downie, S. R., R. G. Olmstead, G. Surawski, D. E. Soltis, P. S. Soltis, J. C. Watson and J. D. Palmer (1990) Loss of the *rpl2* intron demarcates five major lineages of dicotyledons: Molecular and phylogenetic implications. Evolution (submitted). (7)

Dowsett, A. P. and M. W. Young (1982) Differing levels of dispersed repetitive DNA among closely related species of *Drosophila*. Proc. Natl. Acad. Sci. USA 79: 4570–4574. (8)

Drake, J. W. (1969) Comparative rates of spontaneous mutation. Nature 221: 1132. (3)

DuBose, R. F., D. E. Dykhuizen and D. L. Hartl (1987) Function and fitness studies of alkaline phosphatase in *E. coli* by in vitro mutagenesis. Genetics 116: s43. (3)

DuBose, R. F., D. E. Dykhuizen and D. L. Hartl (1988) Genetic exchange among natural isolates of bacteria: Recombination within the *phoA* gene of *Escherichia coli*. Proc. Natl. Acad. Sci. USA 85: 7036–7040. (2, 3)

Duesberg, P. H (1989) Human immunodeficiency virus and acquired immunodeficiency syndrome: Correlation but not causation. Proc. Natl. Acad. Sci. USA 86: 755–764. (5)

Dupont, B. (ed.) (1989) Immunology of HLA. Springer-Verlag, New York. (12)

Dykhuizen, D. E. and D. L. Hartl (1983) Selection in chemostats. Microbiol. Rev. 47: 150–168. (3)

Eanes, W. F., C. Wesley, J. Hey, D. Houle and J. W. Ajioka (1988) The fitness consequences of P element insertion in *Drosophila melanogaster*. Genet. Res. 52: 17–26. (8)

Eanes, W. F., J. W. Ajioka, J. Hey and C. Wesley (1989a) Restriction map variation associated with the G6PD polymorphism in natural populations of *Drosophila melanogaster*. Mol. Biol. Evol. 6: 384–398. (8, 10)

Eanes, W. F., J. Labate and J. W. Ajioka (1989b) Restriction map variation with the *yellow-achaete-scute* region in five populations of *Drosophila melanogaster*. Mol. Biol. Evol. 6: 492–502. (8, 10)

Eardly, B. D., L. A. Materon, N. H. Smith, D. A. Johnson, M. D. Rumbaugh and R. K. Selander (1990) Genetic structure of natural populations of the nitrogen-fixing bacterium *Rhizobium meliloti*. Appl. Environ. Microbiol. 56: 187–194. (2)

Eaton, W. A. (1980) The relationship between coding sequences and function in hemoglobin. Nature 284: 183–185. (13)

Edelman, R. and M. M. Levine (1986) Summary of an international workshop on typhoid fever. Rev. Infect. Dis. 8: 329–349. (2)

Edidin, M. (1986) Major histocompatibility complex haplotypes and the cell physiology of peptide hormones. Hum. Immunol. 15: 357–365. (12)

Edwards, A. W. F. (1961) The population genetics of "sex-ratio" in *Drosophila pseudoobscura*. Heredity 16: 291–304. (9)

Edwards, P. R., A. C. McWhorter and G. W. Douglas (1962). A culture of *Salmonella infantis* of complex antigenic composition. J. Bacteriol. 84: 95–98. (2)

Efstratiadis, A., J. W. Posakony, T. Maniatis, R. L. Lawn, C. O'Connell, R. A. Spritz, J. K. DeRiel, B. Forget, S. M. Weissman, J. L. Slightom, A. E. Blechl, O. Smithies, F. E. Baralle, C. C. Shoulders, and N. J. Proudfoot (1980) The structure and evolution of the human β-globin gene family. Cell 21: 653–668. (13)

Eggleston, W. B., D. M. Johnson-Schlitz and W. R. Engels (1988) P–M hybrid dysgenesis does not mobilize other transposable element families in *D. melanogaster*. Nature 331: 368–370. (8)

Egid, K. and J. L. Brown (1989) The major histocompatibility complex and female mating preferences in mice. Anim. Behav. 38.3: 548–549. (11)

Eicher, E. M., K. W. Hutchison, S. J. Phillips, P. K. Tucker and B. K. Lee (1989) A repeated segment on the mouse Y chromosome is composed of retroviral-related, Y-enriched and Y-specific sequences. Genetics 122: 181–192. (8)

Eisenberg, D., W. Wilcox, S. M. Eshita, P. M. Pryciak, S. P. Ho and W. F. DeGrado (1986) The design, synthesis, and crystallization of an α-helical peptide. Prot. Struct. Funct. Genet. 1: 16–22. (3)

Endow, S. A. and D. J. Komma (1986) One-step and stepwise magnification of a *bobbed lethal* chromosome in *Drosophila melanogaster*. Genetics 114: 511–523. (9)

Engel, J. D., D. J. Rusling, K. C. McCune and J. B. Dodgson (1983) Unusual structure of the chicken embryonic α-globin gene, π'. Proc. Natl. Acad. Sci. USA 80: 1392–1396. (13)

Engels, W. R. (1981a) Estimating genetic divergence and genetic variability with restriction endonucleases. Proc. Natl. Acad. Sci. USA. 78: 6329–6333. (10)

Engels, W. R. (1981b) Hybrid dysgenesis in *Drosophila* and the stochastic loss hypothesis. Cold Spring Harbor Symp. Quant. Biol. 45: 561–565. (8)

Engels, W. R. (1986) On the evolution and population genetics of hybrid-dysgenesis-causing transposable elements in *Drosophila*. Phil. Trans. R. Soc. London Ser. B 312: 205–215. (8)

Engels, W .R. (1989) P elements in *Drosophila*. In: *Mobile DNA*, D. E. Berg and M. M. Howe (eds.). American Society for Microbiology, Washington D.C., pp. 437–484. (8)

Ennis, P. D., J. Zemmour, R. D. Sulter and P. Parham (1990) Rapid cloning of HLA-A, B cDNA using the polymerase chain reaction. Proc. Natl. Acad. Sci. USA 87: 2833–2837. (12)

Enssle, K.-H., H. Wagner and B. Fleischer (1987) Human mumps virus-specific cytotoxic T lymphocytes: Quantitative analysis of HLA restriction. Human Immunol. 18: 135–149. (11)

Erickson, L. R., N. A. Straus and W. D. Beversdorf (1983) Restriction patterns reveal origins of chloroplast genomes in *Brassica* amphiploids. Theor. Appl. Genet. 65: 201–206. (7)

Eshel, I. (1985) Evolutionary genetic stability of Mendelian segregation and the role of free recombination in the chromosomal system. Am. Nat. 125: 412–420. (9)

Essar, D. W., L. Eberly, C.-Y. Han and I. P. Crawford (1990a) DNA sequence and characterization of four early genes of the tryptophan pathway in *Pseudomonas aeruginosa*. J. Bacteriol. 172: 853–866. (4)

Essar, D. W., L. Eberly and I. P. Crawford (1990b) Evolutionary differences in chromosomal location of four early genes of the tryptophan pathway in fluorescent pseudomonads: DNA sequence and characterization of *P. putida trpE* and *trpGDC*. J. Bacteriol. 172: 867–883. (4)

Essar, D. W., L. Eberly, A. Hadero and I. P. Crawford (1990c) Identification and characterization of the genes for a second anthranilate synthase in *Pseudomonas aeruginosa*: Interchangeability of the two anthranilate synthases and evolutionary implications. J. Bacteriol. 172: 884–900. (4)

Essex, M. and P. J. Kanki (1988) The origin of the AIDS virus. Sci. Am. 259: 64–71. (5)

Ewens, W. J. (1977) Population genetics theory in relation to the neutralist–selectionist controversy. Adv. Hum. Genet. 8: 67–134. (10)

Ewens, W. J. (1972) The sampling theory of selectively neutral alleles. Theor. Pop. Biol. 3: 87–112. (12)

Ewens, W. J. (1989) Population genetics theory—the past and the future. In *Mathematical and Statistical Problems of Evolutionary Theory*, S. Lessard (ed.). Kluwer Academic Publications, Dordrecht. (10)

Ewing, W. J. (1986) *Edwards and Ewing's Identification of Enterobacteriaceae*, 4th Ed. Elsevier, New York. (2)

Fan, W., M. Kasahara, J. Gutknecht, D. Klein, W. E. Mayer, M. Jonker and J. Klein (1989) Shared class II MHC polymorphisms between humans and chimpanzees. Hum. Immunol. 26: 107–121. (11)

Farace, M. G., B. A. Brown, G. Raschella, J. Alexander, R. Gambari, A. Fantoni, S. C. Hardies and C. A. Hutchison (1984) The mouse βh1 gene codes for the z chain of embryonic hemoglobin. J. Biol. Chem. 259: 7123–7128. (13)

Feller, W. (1957) *An Introduction to Probability Theory and Its Applications*, Vol. I, 2nd Ed. Wiley, New York. (3)

Felsenstein, J. (1982) Numerical methods for inferring evolutionary trees. Q. Rev. Biol. 57: 379–404. (1)

Ferris, S. D., R. D. Sage, E. M. Prager, U. Ritte and A. C. Wilson (1983) Mitochondrial DNA evolution in mice. Genetics 105: 681–721. (9)

Fierer, J. (1983) Invasive *Salmonella dublin* infections associated with drinking raw milk. Western J. Med. 138: 665–669. (2)

Figueroa, F., M. Golubic, D. Nizetic and J. Klein (1985) Evolution of mouse major histocompatibility complex genes borne by t chromosomes. Proc. Natl. Acad. Sci. USA 82: 2819–2823. (12)

Figueroa, F., H. Tichy, R. J. Berry and J. Klein (1986) MHC polymorphism in island populations of mice. Curr. Top. Microbiol. Immunol. 127: 100–105. (11)

Figueroa, F., E. Gunther and J. Klein (1988) MHC polymorphism pre–dating speciation. Nature 335: 265–267. (11)

Finnegan, D.J. and D. H. Fawcett (1986) Transposable elements in *Drosophila melanogaster*. Oxf. Surv. Eukar. Genes 3: 1–62. (8)

Fischel-Ghodsian, N., R. D. Nicholls and D. R. Higgs (1987) Unusual features of CpG-rich (HTF) islands in the human α globin complex: Association with non-functional pseudogenes and presence within the 3′ portion of the ζ gene. Nucleic Acids Res. 15: 9215–9225. (13)

Fisher, J. A. and T. Maniatis (1988) *Drosophila Adh*: A promoter element expands the tissue specificity of an enhancer. Cell 53: 451–461. (10)

Fitch, D. H. A., C. Mainone, J. L. Slightom and M. Goodman (1988) The spider monkey ψη-globin gene and surrounding sequences: Recent or ancient insertions of LINEs and SINEs? Genomics 3: 237–255. (13)

Fitch, W. M. (1971) Toward defining the course of evolution: Minimum change for a specified tree topology. Syst. Zool. 20: 406–416. (1)

Fitch, W. M. and E. Margoliash (1967) Construction of phylogenetic trees. Science 155: 279–284. (1)

Fitch, W.M. and E. Margoliash (1970) The usefulness of amino acid and nucleotide sequences in evolutionary studies. In *Evolutionary Biology*, Vol. IV, Th. Dobzhansky, M. K. Hecht, and W. C. Steere (eds.). Plenum, New York. pp. 67–109. (4)

Forrester, W. C., S. Takegawa, T. Papayannopoulou, G. Stamatoyannopoulos and M. Groudine (1987) Evidence for a locus activating region: The formation of developmentally stable hypersensitive sites in globin–expressing hybrids. Nucleic Acids Res. 15: 10159–10177. (13)

Fox, G. E., E. Stackebrandt, R. B. Hespell, J. Gibson, J. Maniloff, T. A. Dyer, R. S. Wolfe, W. E. Balch, R. Tanner, L. Magrum, L. B. Zablen, R. Blakemore, R. Gupta, L. Bonen, B. J. Lewis, D. A. Stahl, K. R. Luehrsen, K. N. Chen and C. R. Woese (1980) The phylogeny of prokaryotes. Science 209: 457–463. (1)

Franchini, G., C. Gurgo, H.-G. Guo, R. C. Gallo, E. Collalti, K. A. Fargnoli, L. F. Hall, F. Wong-Staal and M. S. Reitz, Jr. (1987) Sequence of simian immunodeficiency virus and its relationship to the human immunodeficiency viruses. Nature 328: 539–543. (5)

Frank, S. A. (1989) The evolutionary dynamics of cytoplasmic male sterility. Am. Nat. 133: 345–376. (6)

Frankel, G., S. M. C. Newton, G. K. Schoolnik and B. A. D. Stocker (1989) Intragenic recombination in a flagellin gene: characterization of the H1:j gene of *Salmonella typhi*. EMBO J. 8: 3149–3152. (2)

Fritsch, E. F., R. M. Lawn and T. Maniatis (1980) Molecular cloning and characterization of the human β-like globin gene cluster. Cell 19: 959–972. (13)

Fukasawa, M., T. Miura, A. Hasegawa, S. Morikawa, H. Tsujimoto, K. Miki, T. Kitamura and M. Hayami (1988) Sequence of simian immunodeficiency virus from African green monkey, a new member of the HIV/SIV group. Nature 333: 457–461. (5)

Gallo, R. C. and L. Montagnier (1988) AIDS in 1988. Sci. Am. 259: 41–48. (5)

Ganetzky, B. (1977) On the components of segregation distortion in *Drosophila melanogaster*. Genetics 86: 321–355. (9)

Garner, K. J. and J. B. Lingrel (1988) Structural organization of the β-globin locus of B-haplotype sheep. Mol. Biol. Evol. 5: 134–140. (13)

Garner, K. J. and J. B. Lingrel (1989) A comparison of the βA- and βB- globin gene clusters of sheep. J. Mol. Evol. 28: 175–184. (13)

Garrett, T. P., M. A. Saper, P. J. Bjorkman, J. L. Stominger and D. C. Wiley (1989) Specificity pockets for the side chains of peptide antigens in HLA-A68. Nature 340: 692–696. (12)

Gepts, P. and M. T. Clegg (1989) Genetic diversity in pearl millet (*Pennisetum glaucum* [L.] R. Br.) at the DNA sequence level. J. Hered. 80: 203–208. (7)

Gest, H. and J. L. Favinger (1983) *Heliobacterium chlorum*, an anoxygenic brownish-green photosynthetic bacterium containing a "new" form of bacteriochlorophyll. Arch. Microbiol. 136: 11–16. (1)

Ghosh, S. S., S. C. Bock, S. E. Rokita and E. T. Kaiser (1986) Modification of the active site of alkaline phosphatase by site-directed mutagenesis. Science 231: 145–148. (3)

Gibson, J., W. Ludwig, E. Stackebrandt and C. R. Woese (1985) The phylogeny of the green photosynthetic bacteria: Absence of a close relationship between *Chlorobium* and *Chloroflexus*. Syst. Appl. Microbiol. 6: 152–156. (1)

Gilbert, W. (1978) Why genes in pieces? Nature 271: 501. (3, 13)

Gilbert, W. (1986) The RNA world. Nature 319: 618. (7)

Gilbert, W. (1987) The exon theory of genes. Cold Spring Harbor Symp. Quant. Biol. 52: 901–905. (7)

Gill, T. J., III. (1983) Immunogenetics of spontaneous abortions in humans. Transplantation 35: 1–6. (11, 12)

Gillespie, J. H. (1986a) Natural selection and the molecular clock. Mol. Biol. Evol. 3: 138–155. (10)

Gillespie, J. H. (1986b) Variability of evolutionary rates of DNA. Genetics 113: 1077–1091. (10)

Gilmour, M. N., T. S. Whittam, M. Killian and R. K. Selander (1987) Genetic relationships among the oral streptococci. J. Bacteriol. 169: 5247–5257. (2)

Gillham, N. W. (1978) *Organelle Heredity*. Raven Press, New York. (6)

Ginzburg, L. R., P. M. Bingham and S. Yoo (1984) On the theory of speciation induced by transposable elements. Genetics 107: 331–341. (8)

Giovannoni, S. J., S. Turner, G. J. Olsen, S. Barns, D. J. Lane and N. R. Pace (1988) Evolutionary relationships among cyanobacteria and green chloroplasts. J. Bacteriol. 170: 3584–3592. (1)

Gō, M. (1981) Correlation of DNA exonic regions with protein structural units in haemoglobin. Nature 291: 90–92. (3, 13)

Gō, M. (1983) Modular structural units, exons, and function in chicken lysozyme. Proc. Natl. Acad. Sci. USA 80: 1964–1968. (3)

Gō, M. (1985) Protein structures and split genes. Adv. Biophys. 19: 91–131. (3)

Goedert, J. J. and W. A. Blattner (1985) The epidemiology of AIDS and related conditions. In: *AIDS: Etiology, Diagnosis, Treatment, and Prevention*, V. T. DeVita, Jr., S. Hellman, and S. A. Rosenberg (eds.). Lippincott, Philadelphia, pp. 1–30. (5)

Gojobori, T. and S. Yokoyama (1985) Rates of evolution of the retroviral oncogene of Moloney murine sarcoma virus and of its cellular homologues. Proc. Natl. Acad. Sci. USA 82: 4198–4201. (5)

Gojobori, T. and S. Yokoyama (1987) Molecular evolutionary rates of oncogenes. J. Mol. Evol. 26: 148–156. (5)

Goldberg, M. L., J.-Y. Sheen, W. J. Gehring and M. M. Green (1983) Unequal crossing-over associated with asymmetrical synapsis between nomadic elements in the *Drosophila melanogaster* genome. Proc. Natl. Acad. Sci. USA 80: 5017–5021. (8)

Goldstein, F. W., J. C. Chumpitaz, J. M. Guevara, B. Papadopoulous, J. F. Acar and J. F. Vieu (1986) Plasmid-mediated resistance to multiple antibiotics in *Salmonella typhi*. J. Infect. Dis. 153: 261–266. (2)

Goncharoff, P. and B. P. Nichols (1984) Nucleotide sequence of *Escherichia coli pabB* indicates a common evolutionary origin of *p*-aminobenzoate synthetase and anthranilate synthetase. J. Bacteriol. 159: 57–62. (4)

Goncharoff, P. and B. P. Nichols (1988) Evolution of aminobenzoate synthases: Nucleotide sequences of *Salmonella typhimurium* and *Klebsiella aerogenes pabB*. Mol. Biol. Evol. 5: 531–548. (4)

Gonda, M. A., F. Wong-Staal, R. C. Gallo, J. E. Clements, O. Narayan and R. V. Gilden (1985) Sequence homology and morphologic similarity of HTLV-III and Visna virus, a pathogenic lentivirus. Science 227: 173–177. (5)

Goodfellow, M. and R. G. Board (eds.) (1980) *Microbial Classification and Identification*. Academic Press, London. (2)

Goodman, M. (1981) Decoding the pattern of protein evolution. Prog. Biophys. Mol. Biol. 38: 105–164. (13)

Goodman, M., B. F. Koop, J. Czelusniak, M. L. Weiss and J. L. Slightom (1984) The η-globin gene: Its long evolutionary history in the β-globin gene families of mammals. J. Mol. Biol. 180: 803–824. (13)

Goodman, M., J. Czelusniak, B. F. Koop, D. A. Tagle and J. L. Slightom (1987) Globins: A case study in molecular phylogeny. Cold Spring Harbor Symp. Quant. Biol. 52: 875–890. (13)

Goodman, M., J. Pedwaydon, J. Czelusniak, T. Suzuki, T. Gotoh, L. Moens, F. Shishikura and D. Walz (1988) An evolutionary tree for invertebrate globin sequences. J. Mol. Evol. 27: 236–249. (13)

Gordon, K. H. J., E. J. Crouse, H. J. Bohnert and R. G. Herrmann (1982) Physical mapping of differences in chloroplast DNA of the five wild-type plastomes in *Oenothera* subsection *Euoenothera*. Theor. Appl. Genet. 61: 373–384. (7)

Gouy, M. and W.-H. Li (1989) Phylogenetic analysis based on rRNA sequences supports the archaebacterial rather than the eocyte tree. Nature 339: 145–147. (1)

Govindaraju, D. R., D. B. Wagner, G. P. Smith and B. P. Dancik (1988) Chloroplast DNA variation within individual trees of a *Pinus banksiana–Pinus contorta* sympatric region. Can. J. For. Res. 18: 1347–1350. (7)

Govindaraju, D. R., B. P. Dancik and D. B. Wagner (1989) Novel chloroplast DNA polymorphism in a sympatric region of two pines. J. Evol. Biol. 2: 49–59. (7)

Green, M. M. (1980) Transposable elements in *Drosophila* and other Diptera. Annu. Rev. Genet. 14: 109–120. (8)

Gregorius, H.-R. (1986) Polymorphisms for purely cytoplasmically inherited traits in bisexual plants. Genetics 112: 385–392. (6)

Gregorius, H.-R. and M. D. Ross (1984) Selection with gene–cytoplasm interactions. I. Maintenance of cytoplasm polymorphisms. Genetics 107: 165–178. (6)

Grosveld, F., G. B. van Assendelft, D. Greaves and G. Kollias (1987) Position-independent high-level expression of the human β-globin gene in transgenic mice. Cell 51: 975–985. (13)

Gruissem, W. (1989) Chloroplast RNA: Transcription and processing. In: *The Biochemistry of Plants: A Comprehensive Treatise*, Vol. 15, A. Marcus (ed.). Academic Press, New York, pp. 151–191. (7)

Gummere, G. R., P. J. McCormick and D. Bennett (1986) The influence of genetic background and the homologous chromosome 17 on *t*-haplotype transmission ratio distortion in mice. Genetics 114: 235–245. (9)

Gunsalus, I. C., C. F. Gunsalus, A. M. Chakrabarty, S. Sikes and I. P. Crawford (1968) Fine structure mapping of the tryptophan genes in *Pseudomonas putida*. Genetics 60: 419–435. (4)

Gutell, R. R., B. Weiser, C. R. Woese and H. F. Noller (1985) Comparative anatomy of 16S-like ribosomal RNA. Progress Nucleic Acids Res. Mol. Biol. 32: 155–216. (1)

Guyader, M., M. Emerman, P. Sonigo, F. Clavel, L. Montagnier and M. Alizon (1987) Genome organization and transactivation of the human immunodeficiency virus type 2. Nature 326: 662–669.

Gyllensten, U. B. and H. A. Erlich. (1988) Generation of single-stranded DNA by the polymerase chain reaction and its application to direct sequencing of the HLA-DQA locus. Proc. Natl. Acad. Sci. USA. 85: 7652–7656. (10)

Gyllensten, U. B. and H. A. Ehrlich (1989) Ancient roots for polymorphism at the HLA-DQα locus in primates. Proc. Natl. Acad. Sci. USA 86: 9986–9990. (12)

Gyllensten, U., D. Wharton and A. C. Wilson (1985) Maternal inheritance of mitochondrial DNA during backcrossing of two species of mice. J. Hered. 76: 321–324. (6)

Haeckel, E. (1866) *Generelle Morphologie der Organismen*. Verlag Georg Reimer, Berlin. (1)

Hahn, B. H., G. M. Shaw, M. E. Taylor, R. R. Redfield, P. D. Markham, S. Z. Salahuddin, F.

Wong-Staal, R. C. Gallo, E. S. Parks and W. P. Parks (1986) Genetic variation in HTLV-III/LAV over time in patients with AIDS or at risk for AIDS. Science 232: 1548–1553. (5)

Haigh, J. (1978) The accumulation of deleterious genes in a population— Muller's ratchet. Theor. Pop. Biol. 14: 251–267. (8)

Haldane, J. B. S (1956) The estimation of viabilities. J. Genet. 54: 294–296. (12)

Hall, B. G. (1988) Adaptive evolution that requires multiple spontaneous mutations. I. Mutations involving an insertion sequence. Genetics 120: 887–897. (4)

Hamilton, B. L., A. Hamilton and M. S. Hamilton (1985) Maternal-fetal disparity at multiple minor histocompatibility loci affects the weight of the feto-placental unit in mice. J. Reprod. Immunol. 8: 257–261. (11)

Hamilton, W. D. (1967) Extraordinary sex ratios. Science 156: 477–488. (9)

Hammer, M. F., J. Schimenti and L. M. Silver (1989) Evolution of mouse chromosome 17 and the origin of inversions associated with t haplotypes. Proc. Natl. Acad. Sci. USA 86: 3261–3265. (9)

Hansen, H. E., S. E. Larsen, L. P. Ryder and L. S. Nielsen (1979) HLA-A, B haplotype frequencies in 5,202 unrelated Danes by a maximum-likelihood method of gene counting. Tissue Antigens 13: 143–153. (12)

Hardies, S. C., M. H. Edgell and C. A. Hutchison (1984) Evolution of the mammalian β-globin gene cluster. J. Biol. Chem. 259: 3748–3756. (13)

Hardison, R. C. (1984) Comparison of the β-like globin gene families of rabbits and humans indicates that the gene cluster 5'-ε-γ-δ-β-3' predates the mammalian radiation. Mol. Biol. Evol. 1: 390–410. (13)

Hardison, R. C. and R. Gelinas (1986) Assignment of orthologous relationships among mammalian α-globin genes by examining flanking regions reveals a rapid rate of evolution. Mol. Biol. Evol. 3: 243–261. (13)

Hardison, R. C. and J. B. Margot (1984) Rabbit globin pseudogene ψ β2 is a hybrid of δ- and β-globin gene sequences. Mol. Biol. Evol. 1: 302–316. (13)

Harris, E. H. (1989) The Chlamydomonas Sourcebook. Academic Press, New York. (6)

Harris, S., P. A. Barrie, M. L. Weiss and A. J. Jeffreys (1984) The primate ψ β1 gene: An ancient β-globin pseudogene. J. Mol. Biol. 180: 785–801. (13)

Harris, S., J. R. Thackeray, A. J. Jeffreys and M. L. Weiss (1986) Nucleotide sequence analysis of the lemur β-globin gene family: Evidence for major rate fluctuations in globin polypeptide evolution. Mol. Biol. Evol. 3: 465–484. (13)

Hartl, D. L. (1977) Mechanism of a case of genetic coadaptation in populations of Drosophila melanogaster. Proc. Natl. Acad. Sci. USA 74: 324–328. (9)

Hartl, D. L. (1989) Evolving theories of enzyme evolution. Genetics 122: 1–6. (3)

Hartl, D. L. and D. E. Dykhuizen (1984) The population genetics of Escherichia coli. Annu. Rev. Genet. 18: 31–68. (8, 2)

Hartl, D. L. and D. E. Dykhuizen (1985) The neutral theory and the molecular basis of preadaptation, In: Population Genetics and Molecular Evolution, T. Ohta and K. Aoki (eds.) Japan Scientific Societies Press, Tokyo/Springer-Verlag, Berlin, pp. 107–124. (2)

Hartl, D. L. and Y. Hiraizumi (1976) Segregation distortion. In: The Genetics and Biology of Drosophila, Vol. 1b, M. Ashburner and E. Novitski (eds). Academic Press, New York, pp. 615–666. (9)

Hartl, D. L. and S. A. Sawyer (1988) Why do unrelated insertion sequences occur together in the genome of Escherichia coli? Genetics 118: 537–541. (8)

Hartl, D. L. and S. A. Sawyer (1990) Inference of selection and recombination from nucleotide sequence data. Theor. Pop. Biol. (in press) (3)

Hartl, D. L., M. Medhora, L. Green and D. E. Dykhuizen (1986) The evolution of DNA sequences in Escherichia coli. Phil. Trans. R. Soc. London Ser. B 312: 191–204. (8)

Haseltine, W. A. and F. Wong-Staal (1988) The molecular biology of the AIDS virus. Sci. Am. 259: 52–62. (5)

Hayashida, H. and T. Miyata (1983) Unusual evolutionary conservation and frequent DNA segment exchange in class I genes of the major histocompatibility complex. Proc. Natl. Acad. Sci. USA 80: 2671–2675. (11)

Hayashida, H., H. Toh, R. Kikuno and T. Miyata (1985) Evolution of influenza virus genes. Mol. Biol. Evol. 2: 289–303. (5)

Hedges, R. W., A. E. Jacob and I. P. Crawford (1977) Wide ranging plasmid bearing the Pseudomonas aeruginosa tryptophan synthase genes. Nature 267: 283–284. (4)

Hedrick, P. W. (1987) Gametic disequilibrium measures: Proceed with caution. Genetics 117: 331–341. (12)

Hedrick, P. W. (1988a) Inference of recombinational hotspots using gametic disequilibrium values. Heredity 60: 435–438. (12)

Hedrick, P. W. (1988b) Can segregation distortion influence gametic disequilibrium? Genet. Res. 52: 237–242. (12)

Hedrick, P. W. (1988c) HLA-sharing, recurrent spontaneous abortion, and the genetic hypothesis. Genetics 118: 199–204. (12)

Hedrick, P. W. (1990) Evolution at HLA: Possible explanations for observed deficiencies of homozygotes. Hum. Hered. 40: 213–220. (12)

Hedrick, P. W. and G. Thomson (1983) Evidence for balancing selection at HLA. Genetics 104: 449–456. (11, 12)

Hedrick, P. W. and G. Thomson (1986) A two locus neutrality test: Applications to humans, *E. coli*, and lodgepole pine. Genetics 112: 135–156. (12)

Hedrick, P. W. and G. Thomson (1988) Maternal–fetal interactions and the maintenance of HLA polymorphism. Genetics 119: 205–212. (11, 12)

Hedrick, P. W., S. Jain and L. Holden (1978) Multilocus systems in evolution. Evol. Biol. 11: 101–184. (12)

Heinhorst, S. and J. M. Shively (1983) Encoding of both subunits of ribulose-1,5-bisphosphate carboxylase by organelle genome of *Cyanophora paradoxa*. Nature 304: 373–374. (7)

Hespell, R. B., B. J. Paster, T. J. Macke and C. R. Woese (1984) The origin and phylogeny of the bdellovibrios. Syst. Appl. Microbiol. 5: 196–203. (1)

Hey, J. (1989) The transposable portion of the genome of *Drosophila algonquin* is very different from that in *D. melanogaster*. Mol. Biol. Evol. 6: 66–79. (8)

Hickey, D. A. (1982) Selfish DNA: A sexually-transmitted nuclear parasite. Genetics 101: 519–531. (8)

Higuchi, R. G. and H. Ochman. (1989) Production of single–stranded DNA templates by exonuclease digestion following the polymerase chain reaction. Nucleic Acids Res. 17: 5865. (10)

Hill, A., S. C. Hardies, S. J. Phillips, M. G. Davis, C. A. Hutchison and M. H. Edgell (1984) Two mouse early embryonic β-globin gene sequence: Evolution of the nonadult β-globins. J. Biol. Chem. 259: 3739–3747. (13)

Hill, W. G. (1975) Linkage disequilibrium among multiple neutral alleles produced by mutation in finite population. Theor. Pop. Biol. 8: 117–126. (12)

Hill, W. G. and A. Robertson (1966) The effect of linkage on limits to artificial selection. Genet. Res. 8: 269–294. (6)

Hill, W. G. and A. Robertson (1968) Linkage disequilibrium in finite populations. Theor. Appl. Genet. 38: 226–231. (12)

Hilu, K. W. (1988) Identification of the "A" genome of finger millet using chloroplast DNA. Genetics 118: 163–167. (7)

Hiraizumi, Y. and A. M. Thomas (1984) Suppressor systems of Segregation Distorter (*SD*) chromosomes in natural populations of *Drosophila melanogaster*. Genetics 106: 279–292. (9)

Hiraizumi, Y., D. W. Martin and I. A. Eckstrand (1980) A modified model of segregation distortion in *Drosophila melanogaster*. Genetics 95: 693–706. (9)

Hiratsuka, J., H. Shimada, R. Whittier, T. Ishibashi, M. Sakamoto, M. Mori, C. Kondo, Y. Honji, C.-R. Sun, B.-Y. Meng, Y.-Q. Li, A. Kanno, Y. Nishizawa, A. Hirai, K. Shinozaki and M. Sugiura (1989) The complete sequence of the rice (*Oryza sativa*) chloroplast genome: Intermolecular recombination between distinct tRNA genes accounts for a major plastid DNA inversion during the evolution of the cereals. Mol. Gen. Genet. 217: 185–194. (6, 7)

Hirsch, V., N. Riedel and J. I. Mullins (1987) The genome organization of STLV-3 is similar to that of the AIDS virus except for a truncated transmembrane protein. Cell 49: 307–319. (5)

Hirsch, V. M., R. A. Olmsted, M. Murphey-Corb, R. H. Purcell and P. R. Johnson (1989) An African primate lentivirus (SIV$_{SM}$) closely related to HIV–2. Nature 339: 389–392. (5)

Ho, S. P. and W. F. DeGrado (1987) Design of a four-helix bundle protein: Synthesis of peptides which self-associate into a helical protein. J. Am. Chem. Soc. 109: 6751–6758. (3)

Hoffman, T. A., C. J. Ruiz, G. W. Counts, J. M. Sachs and J. L. Nitzkin (1975) Waterborne typhoid fever in Dade County, Florida. Am. J. Med. 59: 481–487. (2)

Holland, J., K. Spindler, F. Horodyski, E. Grabau, S. Nichol and S. VandePol (1982) Rapid evolution of RNA genomes. Science 215: 1577–1585. (5)

Holmes, N. and P. Parham (1985) Exon shuffling in vivo can generate novel HLA class I molecules. EMBO J. 4: 2849–2854. (11, 12)

Holwerda, B. C., S. Jana and W. L. Crosby (1986) Chloroplast and mitochondrial DNA variation in *Hordeum vulgare* and *Hordeum spontaneum*. Genetics 114: 1271–1291. (7)

Hook, E. W. (1979) *Salmonella* species (including typhoid fever) In: *Principles and Practices of Infectious Diseases*, 2nd Ed., G. L. Mandell, R. G. Douglas Jr., and J. E. Bennett (eds.). Wiley, New York., pp. 1256–1269. (2)

Hornick, R. B. (1985) Selective primary health care: strategies for control of disease in the developing world. XX. Typhoid fever. Rev. Infect. Dis. 7: 536–546. (2)

Hosaka, K. and R. E. Hanneman, Jr. (1988a) The origin of the cultivated tetraploid potato based on chloroplast DNA. Theor. Appl. Genet. 76: 172–176. (7)

Hosaka, K. and R. E. Hanneman, Jr. (1988b) Origin of chloroplast DNA diversity in the Andean potatoes. Theor. Appl. Genet. 76: 333–340. (7)

Hosbach, H. A., T. Wyler and R. Weber (1983) The *Xenopus laevis* globin gene family: Chromosomal arrangement and gene structure. Cell 32: 45–53. (13)

Howard, J. C. (1987) MHC organization of the rat: Evolutionary considerations. In: *Evolution and Vertebrate Immunity*, G. Kelsoe and D. H. Schulze (eds.). University of Texas Press, Austin, pp. 397–411. (11)

Hsieh, C.-H. and J. D. Griffith (1988) The terminus of SV40 DNA replication and transcription contains a sharp sequence-directed curve. Cell 52: 535–544. (9)

Hsu, S.-L., J. Marks, J.-P. Shaw, M. Tam, D. R. Higgs, C. C. Shen and C.-K. J. Shen (1988) Structure and expression of the human θ1 globin gene. Nature 331: 94–96. (13)

Huang, X., R. Hardison and W. Miller (1990) A space-efficient algorithm for local similarities. Computer Applications in the Biosciences (in press). (13)

Huber, R., C. R. Woese, T. A. Langworthy, H. Fricke and K. O. Stetter (1989) *Thermosipho africanus*, gen. nov., represents a new genus of thermophilic eubacteria within the "Thermotogales". Syst. Appl. Microbiol. 12: 32–37. (1)

Hüdepohl, U., W.-D. Reiter and W. Zillig (1990) In vitro transcription of two ribosomal RNA genes of the archaebacterium *Sulfolobus* sp. B12 indicates a factor requirement for specific initiation. Proc. Natl. Acad. Sci. USA 87: 5851–5855. (1)

Hudson, R. R. (1982) Estimating genetic variability with restriction endonucleases. Genetics 100: 711–719. (10)

Hudson, R. R (1983) Properties of a neutral allele model with intragenic recombination. Theor. Pop. Biol. 23: 183–201. (10, 12)

Hudson, R. R. (1991) Gene genealogies and the coalescent process. In *Oxford Surveys in Evolutionary Biology*, P. Harvey and L. Partridge (eds.) Oxford University Press, Oxford (in press). (10)

Hudson, R. R. and N. L. Kaplan (1988) The coalescent process in models with selection and recombination. Genetics 120: 831–840. (10)

Hudson, R. R., M. Kreitman and M. Aguadé (1987) A test of neutral molecular evolution based on nucleotide data. Genetics 116: 153–159. (10)

Hughes, A. L. and M. Nei (1988) Pattern of nucleotide substitution at major histocompatibility complex class I loci reveals overdominant selection. Nature 335: 167–170. (11, 12)

Hughes, A. L. and M. Nei (1989a) Nucleotide substitution at major histocompatibility complex class II loci: Evidence for overdominant selection. Proc. Natl. Acad. Sci. USA 86: 958–962. (11, 12)

Hughes, A. L. and M. Nei (1989b) Ancient interlocus exon exchange in the history of the *HLA-A* locus. Genetics 122: 681–686. (11, 12)

Hughes, A. L. and M. Nei (1989c) Evolution of the major histocompatibility complex: Independent origin of nonclassical class I genes in different groups of mammals. Mol. Biol. Evol. 6: 559–579 (11)

Humphries, R. K., T. Ley, P. Turner, A. D. Moulton and A. W. Neinhuis (1982) Differences in human α, β and δ-globin gene expression in monkey kidney cells. Cell 30: 173–183. (13)

Hunt, J. A., J. G. Bishop III and H. L. Carson (1984) Chromosomal mapping of a middle-repetitive DNA sequence in a cluster of five species of Hawaiian *Drosophila*. Proc. Natl. Acad. Sci. USA 81: 7146–7150. (8)

Hutchison, C. A., III, S. C. Hardies, R. W. Padgett, S. Weaver and M. H. Edgell (1984) The mouse globin pseudogene βh3 is descended from a premammalian δ-globin gene. J. Biol. Chem. 259: 12881–12889. (13)

Hwu, H. R., J. W. Roberts, E. H. Davidson and R. J. Britten (1986) Insertion and/or deletion of many repeated DNA sequences in human and higher ape evolution. Proc. Natl. Acad. Sci. USA 83: 3875–3879. (9)

Hyldig-Nielsen, J. J., E. O. Jensen, K. Paludan, O. Wiborg, R. Garrett, P. Jorgensen and K. A. Marcker (1982) The primary structures of two leghemoglobin genes from soybean. Nucleic Acids Res. 10: 689–701. (13)

Iwabe, N., K. Kuma, M. Hasegawa, S. Osawa and T. Miyata (1989) Evolutionary relationship of archaebacteria, eubacteria, and eukaryotes inferred from phylogenetic trees of duplicated genes. Proc. Natl. Acad. Sci. USA 86: 9355–9359. (1)

Jahn, C. L., C. A. Hutchison, S. A. Phillips, S. Weaver, N. L. Haigwood, C. F. Voliva and M. H. Edgell (1980) DNA sequence organization of the β-globin complex of the BALB/c mouse. Cell 21: 159–168. (13)

James, A. C. and J. Jaenike (1990) "Sex Ratio" meiotic drive in Drosophila testacla. Genetics (in press). (9)

James, D. A. (1965) Effects of antigenic dissimilarity between mother and foetus on placental size in mice. Nature 205: 613–614. (11)

James, D. A. (1967) Some effects of immunological factors on gestation in mice. J. Reprod. Fertil. 14: 265–275. (11)

James, T. C. and S. C. R. Elgin (1986) Identification of a nonhistone chromosomal protein associated with heterochromatin in Drosophila melanogaster and its gene. Mol. Cell. Biol. 6: 3862–3872. (9)

Jann, K. and B. Jann (1984) Structure and biosynthesis of O-antigen. In: Handbook of Endotoxin, Vol. 1, E. Rietschel (ed.). Elsevier Scientific Publishers, Amsterdam, pp. 138–186. (2)

Jansen, R. K. and J. D. Palmer (1987) A chloroplast DNA inversion marks an ancient evolutionary split in the sunflower family (Asteraceae). Proc. Natl. Acad. Sci. USA 84: 5818–5822. (7)

Jansen, R. K. and J. D. Palmer (1988) Phylogenetic implications of chloroplast DNA restriction site variation in the Mutisieae (Asteraceae). Am. J. Bot. 75 753–766. (7)

Jeffreys, A. J. (1982) Evolution of globin genes. In Genome Evolution, G. A. Dover and R. B. Flavell (eds.). Academic Press, London, pp. 157–176. (13)

Jeffreys, A. J. and R. A. Flavell (1977) The rabbit β-globin gene contains a large insert in the coding sequence. Cell 12: 1097–1108. (13)

Jeffreys, A. J., V. Wilson, A. Blanchetot, P. Weller, A. G. van Kessel, N. Spurr, E. Solomon and P. Goodfellow (1984) The human myoglobin gene: A third dispersed globin locus in the human genome. Nucleic Acids Res. 12: 3235–3243. (13)

Jeffreys, A. J., V. Wilson, D. Wood, J. P. Simons, R. M. Kay and J. G. Williams (1980) Linkage of adult α- and β-globin genes in X. laevis and gene duplication by tetraploidization. Cell 21: 555–564. (13)

Jeffreys, A., P. Barrie, S. Harris, D. Fawcett, Z. Nugent and C. Boyd (1982) Isolation and sequence analysis of a hybrid δ-globin pseudogene from the brown lemur. J. Mol. Biol. 156: 487–503. (13)

Jeffreys, A. J., V. Wilson and S. L. Thein (1985) Hypervariable "minisatellite" regions in human DNA. Nature 314: 67–73. (9)

Jensen, E. O., K. Paludan, J. J. H. Hyldig-Nielsen, P. Jorgensen and K. A. Marcker (1981) The structure of a chromosomal leghaemoglobin gene from soybean. Nature 291: 677–679. (13)

Jhiang, S., J. Garey, and A. Riggs (1988) Exon–intron organization in genes of earthworm and vertebrate globins. Science 240: 334–336. (13)

Jin, L. and M. Nei (1990) Limitations of the evolutionary parsimony method of phylogenetic anaylsis. Mol. Biol. Evol. 7: 82–102. (1)

John, B. and G. L. G. Miklos (1988) The Eukaryote Genome in Development and Evolution. Allen and Unwin, London. (8)

Johnson, E. M., B. Krauskopf and L. S. Baron (1965) Genetic mapping of Vi and somatic antigenic determinants in Salmonella. J. Bacteriol. 90: 302–308. (2)

Johnson, L. B. and J. D. Palmer (1989) Heteroplasmy of chloroplast DNA in Medicago. Plant Mol. Biol. 12: 3–11. (6, 7)

Joyce, G. F. (1989) RNA evolution and the origins of life. Nature 338: 217–224. (7)

Joys, T. M. (1976) Identification of an antibody binding site in the phase-1 flagellar protein of Salmonella typhimurium. Microbios 15: 221–228. (2)

310   Bibliography

Joys, T. M. (1985) The covalent structure of the phase-1 flagellar filament protein of *Salmonella typhimurium* and its comparison with other flagellins. J. Biol. Chem. 260: 15758–15761. (2)

Joys, T. M., J. F. Martin, H. L. Wilson and V. Rankis (1974) Differences in the primary structure of the phase-1 flagellins of two strains of *Salmonella typhimurium*. Biochim. Biophys. Acta 351: 301–305. (2)

Jukes, T. H. and C. R. Cantor (1969) Evolution of protein molecules. In *Mammalian Protein Metabolism*, H. N. Munro (ed.). Academic Press, New York, pp. 21–32. (1)

Jung, A., A. E. Sippel, M. Grez and G. Schutz (1980) Exons encode functional and structural units of chicken lysozyme. Proc. Natl. Acad. Sci. USA 77: 5759–5763. (3)

Kaine, B. P. 1987. Intron-containing tRNA genes of *Sulfolobus solfataricus*. J. Mol. Evol. 25: 248–254. (7)

Kane, J. F. (1977) Regulation of a common amidotransferase subunit. J. Bacteriol. 132: 419–425. (4)

Kane, J. F., W. M. Holmes and R. A. Jensen (1972) Metabolic interlock: The dual function of a folate pathway gene as an extra-operonic gene of tryptophan biosynthesis. J. Biol. Chem. 247: 1587–1596. (4)

Kanki, P. J., J. Alroy and M. Essex (1985) Isolation of T-lymphotropic retrovirus related to HTLV-III/LAV from wild-caught African green monkeys. Science 230: 951–954. (5)

Kaplan, J. B., W. K. Merkel and B. P. Nichols (1985b) Evolution of glutamine amidotransferase genes: Nucleotide sequences of the *pabA* genes from *Salmonella typhimurium*, *Klebsiella aerogenes* and *Serratia marcescens*. J. Mol. Biol. 183: 327–340. (4)

Kaplan, N. L. and J. F. Y. Brookfield (1983) Transposable elements in Mendelian populations. III. Statistical results. Genetics 104: 485–495. (8)

Kaplan, N., T. Darden and C. H. Langley (1985a) Evolution and extinction of transposable elements in Mendelian populations. Genetics 109: 459–480. (8)

Kaplan, N. L., R. R. Hudson and C. H. Langley (1989) The "hitch-hiking" effect revisited. Genetics 123: 887–899. (10)

Kaplan, N. L., T. Darden and R. R. Hudson. (1988) The coalescent process in models with selection. Genetics 120: 819–829. (10)

Kapperud, G., J. Lassen, K. Dommarsnes, B.-E. Kristiansen, D. A. Caugant, E. Ask and M. Jahkola (1989) Comparison of epidemiological marker methods for identification of *Salmonella typhimurium* isolates from an outbreak caused by contaminated chocolate. J. Clin. Microbiol. 27: 2019–2024. (2)

Karlsson, S. and A. W. Nienhuis (1985) Developmental regulation of human globin genes. Annu. Rev. Biochem. 54: 1071–1108. (13)

Kato, K., J. A. Trapain, J. Allopenna, B. Dupont and S. Y. Yang (1989) Molecular analysis of the serologically defined HLA-Aw19 antigens: A genetically distinct family of HLA-A antigens comprising A29, A31, A32 and Aw33, but probably not A30. J. Immunol. (in press) (12)

Kauffmann, F. (1966) *The Bacteriology of Enterobacteriaceae*. Williams & Wilkins, Baltimore. (2)

Kawamura, M., P. S. Keim, Y. Goto, H. Zalkin and R. L. Heinrikson (1978) Anthranihilate synthetase component II from *Pseudomonas putida*: Covalent structure and identification of the cysteine residue involved in catalysis. J. Biol. Chem. 253: 4659–4665. (4)

Kazazian, H. H., C. Wong, H. Youssoufian, A. F. Scott, D. G. Phillips and S. E. Antonarakis (1988) Haemophilia A resulting from de novo insertion of L1 sequences represents a novel mechanism for mutation in man. Nature 332: 164–166. (13)

Keilin, D. (1953) Occurrence of haemoglobin in yeast and the supposed stabilization of the oxygenated cytochrome oxidase. Nature 172: 390–393. (13)

Keller, M. and F. Michel (1985) The introns of the *Euglena gracilis* chloroplast gene which codes for the 32-kDa protein of photosystem II. Evidence for structural homologies with class II introns. FEBS Letters 179: 69–73. (7)

Kessel, M., and F. Klink (1980) Archaebacterial elongation factor is ADP–ribosylated by diphtheria toxin. Nature 287: 250–251. (1)

Kestler, H. W., III, Y. Li, Y. M. Naidu, C. V. Butler, M. F. Ochs, G. Jaenel, N. W. King, M. D. Daniel and R. C. Desrosiers (1988) Comparison of simian immunodeficiency virus isolates. Nature 331: 619–621. (5)

Kidwell, M. G. (1983) Evolution of hybrid dysgenesis determinants in *Drosophila melanogaster*. Proc. Natl. Acad. Sci. USA 80: 1655–1659. (8)

Kimura, M. (1983) *The Neutral Theory of Molecular Evolution*. Cambridge Univ. Press, Cambridge. (1, 2, 10)

Kimura, M. (1986) Variability of evolutionary rates of DNA. Genetics 113: 1077: 1091. (10)

Kimura, M. and J. F. Crow (1964) The number of alleles that can be maintained in a finite population. Genetics 49: 725–738. (11)

Kimura, M. and T. Ohta (1971) *Theoretical Aspects of Population Genetics*. Princeton Univ. Press, Princeton, NJ. (7)

Kingman, J. F. C. (1980) *Mathematics of Genetic Diversity*. CBMS-NSF Regional Conference Series in Applied Mathematics, Vol. 34. Society for Industrial and Applied Mathematics, Philadelphia. (10)

Kingman, J. F. C. (1982a) The coalescent. Stochast. Proc. Appl. 13: 235–248. (10)

Kingman, J. F. C. (1982b) On the genealogy of large populations. J. Appl. Prob. 19A: 27–43. (10)

Kirk, J. T. O. and R. A. E. Tilney-Bassett (1978) *The Plastids*. Elsevier/North-Holland Biomedical Press, Amsterdam. (6)

Klein, J. (1978) H-2 mutations: Their genetics and effect on immune functions. Adv. Immunol. 26: 55–146. (12)

Klein, J. (1986) *Natural History of the Major Histocompatibility Complex*. Wiley, New York. (9, 11, 12)

Klein, J. (1987) Origin of major histocompatibility complex polymorphism: The trans-species hypothesis. Hum. Immunol. 19: 155–162. (11, 12)

Klein, J. and F. Figueroa (1986) Evolution of the major histocompatibility complex. CRC Crit. Rev. Immunol. 6: 295–386. (11)

Klein, J., M. Kasahara, J. Gutknecht and F. Figueroa (1989) Origin and function of *Mhc* polymorphism. Hum. Immunol. 26: 107–121. (11)

Klekowski, E. J., Jr. (1988) *Mutation, Developmental Selection, and Plant Evolution*. Columbia Univ. Press, New York. (6)

Klitz, W. (1988) Inheritance of insulin-dependent diabetes. Nature 333: 402–403. (12)

Klitz, W. and G. Thomson (1987) Disequilibrium pattern analysis. II. Application to Danish HLA A and B locus data. Genetics 116: 633–643. (12)

Klitz, W., G. Thomson and M. P. Baur (1986) Contrasting evolutionary histories among tightly linked HLA loci. Am. J. Hum. Genet. 39: 340–349 (11, 12)

Klitz, W., S. K. Lo, M. Neugebauer, M. P. Baur, E. D. Albert and G. Thomson (1987) A comprehensive search for segregation distortion in HLA. Hum. Immun. 18: 163–180. (12)

Klitz, W., G. Thomson, J. Barbosa and S. S. Rich (1991) Transmission ratio analysis. Am. J. Hum. Genet. (submitted). (12)

Kluyver, A. J., and C. B. van Niel (1936) Prospects for a natural system of classification of bacteria. Zentralb. Bacteriol. Parasit. Infektion. II. 94: 369–403. (1)

Kobori, J. A., A. Winoto, J. McNicholas and L. Hood (1984) Molecular characterization of the recombination region of six murine major histocompatibility complex (MHC) I-region recombinants. J. Mol. Cell. Immunol. 1: 125–131. (12)

Koch, W., K. Edwards and H. Kössel (1981) Sequencing of the 16S–23S spacer in a ribosomal RNA operon of *Zea mays* chloroplast DNA reveals two split tRNA genes. Cell 25: 203–213. (7)

Koeberl, D. D., C. D. K. Bottema, J. Buerstedde and S. Sommer (1989) Functionally important regions of the factor IX gene have a low rate of polymorphism and a high rate of mutation in the dinucleotide CpG. Am. J. Hum. Genet. 45: 448–457. (11)

Kohchi, T., H. Shirai, H. Fukuzawa, T. Sano, T. Komano, K. Umesono, H. Inokuchi, H. Ozeki and K. Ohyama (1988) Structure and organization of *Marchantia polymorpha* chloroplast genome. IV. Inverted repeat and small single copy regions. J. Mol. Biol. 203: 353–372. (7)

Koop, B. F. and M. Goodman (1988) Evolutionary and developmental aspects of two hemoglobin β-chain genes (εM and βM) of opossum. Proc. Natl. Acad. Sci. USA 85: 3893–3897. (13)

Krainer, A. R. and T. Maniatis (1988) RNA splicing. In: Transcription and Splicing, B. D. Hames and D. M. Glover (eds.). IRL Press, Oxford, pp. 131–206. (7)

Krane, D., J.-F. Cheng, and R. Hardison (1991) Subfamily relationships and clustering of rabbit C repeats. Mol. Biol. Evol. (in press). (13)

Krebbers, E., A. Steinmetz and L. Bogorad (1984) DNA sequences for the *Zea mays* tRNA genes tV-UAC and tS-UGA: tV-UAC contains a large intron. Plant Mol. Biol. 3: 13–20. (7)

Kreitman, M. (1983) Nucleotide polymorphism at the alcohol dehydrogenase locus of *Drosophila melanogaster*. Nature 304: 412–417. (10, 11)

Kreitman, M. and M. Aguadé (1986a) Genetic uniformity in two populations of *Drosophila melanogaster* as revealed by filter hybridization of four-nucleotide-recognizing restriction enzyme digests. Proc. Natl. Acad. Sci. USA 83: 3562–3566. (10)

Kreitman, M. and M. Aguadé (1986b) Excess polymorphism at the *Adh* locus in *Drosophila melanogaster*. Genetics 114: 93–110. (10)

Kreitman, M. and R. R. Hudson (1991) Inferring the evolutionary histories of the *Adh* and *Adh-dup* loci in *Drosophila melanogaster* from patterns of polymorphism and divergence. Genetics (in press). (10)

Kreitman, M. and L. L. Landweber. (1989) A strategy for producing single-stranded DNA in the polymerase chain reaction: A direct method for genomic sequencing. Gene Anal. Techn. 6: 84–88. (10)

Krieg, N. R. and J. G. Holt (eds.) (1984) *Bergey's Manual of Systematic Bacteriology*, Vol. 1. Williams & Wilkins, Baltimore. (2)

Kroll, J. S. and E. R. Moxon (1990) Capsulation in distantly related strains of *Haemophilus influenzae* type b: Genetic drift and gene transfer at the capsulation locus. J. Bacteriol. 172: 1374.–1379. (2)

Krystal, M., D. Buonagurio, J. F. Young and P. Palese (1983) Sequential mutations in the NS genes of influenza virus field strains. J. Virol. 45: 547–554. (5)

Kuhner, M., S. Watts, W. Klitz, G. Thomson and R. S. Goodenow (1991) Gene conversion in the evolution of both the *H-2* and *Qa* Class I genes of the murine major histocompatibility complex. Genetics (in press). (12)

Kung, S. D., Y. S. Zhu and G. F. Shen (1982) *Nicotiana* chloroplast genome III. Chloroplast DNA evolution. Theor. Appl. Genet. 61: 73–79. (7)

Kutsukake, K. and T. Iino (1980) A *trans*-acting factor mediates inversion of a specific DNA segment in flagellar phase variation of *Salmonella*. Nature 284: 479–481. (2)

Kwakman, J. H., D. Konings, H. J. Pel and L. A. Grivell (1989) Structure–function relationships in a self-splicing group II intron: A large part of domain II of the mitochondrial intron aI5 is not essential for self–splicing. Nucleic Acids Res. 17: 4205–4216. (7)

Lacy, E., R. C. Hardison, D. Quon and T. Maniatis (1979) The linkage arrangement of four rabbit β-like globin genes. Cell 18: 1273–1283. (13)

Lader, E., H.-S. Ha, M. O'Neill, K. Artzt and D. Bennett (1989) tctex-1: A candidate gene family for a mouse *t* complex sterility locus. Cell 58: 969–979. (9)

Lake, J. A. (1987) A rate-independent technique for analysis of nucleic acid sequences: Evolutionary parsimony. Mol. Biol. Evol. 4: 167–191. (1)

Lake, J. A. (1988) Origin of the eukaryotic nucleus determined by rate-invariant analysis of rRNA sequences. Nature 331: 184–186. (1)

Lambowitz, A. M. (1989) Infectious introns. Cell 56: 323–326. (7)

Langley, C. H. and C. F. Aquadro (1987) Restriction-map variation in natural populations of *Drosophila melanogaster*: White-locus region. Mol. Biol. Evol. 4: 651–663. (8, 9, 11)

Langley, C. H., J. F. Y. Brookfield and N. Kaplan (1983) Transposable elements in Mendelian populations. I. A theory. Genetics 104: 457–471. (8)

Langley, C. H., E. Montgomery, R. Hudson, N. Kaplan and B. Charlesworth (1988a) On the role of unequal exchange in the containment of transposable element copy number. Genet. Res. 52: 223–235. (8)

Langley, C. H., A. E. Shrimpton, T. Yamazaki, N. Miyashita, Y. Matsuo and C. F. Aquadro (1988b) Naturally occurring variation in the restriction map of the *Amy* region of *Drosophila melanogaster*. Genetics 119: 619–629. (8, 10)

Lansman, R. A., J. C. Avise and M. D. Huettel (1983) Critical experimental test of the possibility of "paternal leakage" of mitochondrial DNA. Proc. Natl. Acad. Sci. USA 80: 1969–1971. (6)

Lauer, J., C.-K. J. Shen and T. Maniatis (1980) The chromosomal arrangement of human α-like globin genes: Sequence homology and α-globin gene deletions. Cell 20: 119–130. (13)

Lawlor, D. A., F. E. Ward, P. D. Ennis, A. P. Jackson and P. Parham (1988) HLA-A and B polymorphisms predate the divergence of humans and chimpanzeees. Nature 335: 268–271. (11, 12)

Lawlor, D. A., J. Zemmour, P. D. Ennis and P. Parham (1990) Evolution of class I MHC genes and proteins: From natural selection to thymic selection. Annu. Rev. Immunol. (in press) (11)

Lawn, R. M., A. Efstratiadis, C. O'Connell and T. Maniatis (1980) The nucleotide sequence of the human β-globin gene. Cell 21: 647–651. (13)

Lawrence, J. G., D. E. Dykhuizen, R. F. DuBose and D. L. Hartl (1989) Phylogenetic analysis using insertion sequence fingerprinting in *Escherichia coli*. Mol. Biol. Evol. 6: 1–14.

Leder, A., H. Miller, D. Hamer, J. G. Seidman, B. Norman, M. Sullivan and P. Leder (1978) Comparison of cloned mouse α- and β-globin genes: Conservation of intervening sequence locations and extragenic homology. Proc. Natl. Acad. Sci. USA 75: 6187–6191. (13)

Leder, A., D. Swan, F. Ruddle, P. D'Eustachio and P. Leder (1981) Dispersion of α-like globin genes of the mouse to three different chromosomes. Nature 293: 196–200. (13)

Lederberg, J. and T. Iino (1956) Phase variation in *Salmonella*. Genetics 41: 743–757. (2)

Lee, J. S., G. G. Brown and D. P. S. Verma (1983) Chromosomal arrangement of leghemoglobin genes in soybean. Nucleic Acids Res. 11: 5541–5552. (13)

Lefevre, G. (1976) A photographic representation of the polytene chromosomes of *Drosophila melanogaster* salivary glands. In: *The Genetics and Biology of Drosophila*. Vol. 1a, M. Ashburner and E. Novitski (eds.). Academic Press, Orlando, FL, pp. 31–66. (8)

Leffers, H., J. Kjems, L. Ostergaard, N. Larsen and R. A. Garrett (1987) Evolutionary relationships amongst archaebacteria: A comparative study of 23S ribosomal RNA of a sulfur-dependent extreme thermophile, an extreme halophile, and a thermophilic methanogen. J. Mol. Biol. 195: 43–61. (1)

Lehvaslaiho, H., A. Saura and J. Lokki (1987) Chloroplast DNA variation in the grass tribe Festuceae. Theoret. Appl. Genet. 74: 298–302. (7)

Leigh Brown, A. J. (1983) Variation at the 87A heat shock locus in *Drosophila melanogaster*. Proc. Natl. Acad. Sci. USA 80: 5350–5354. (8, 10)

Leigh Brown, A. J. and J. E. Moss (1987) Transposition of the I element and *copia* in a natural population of *Drosophila melanogaster*. Genet. Res. 49: 121–128. (8)

Lemeunier, F., J. R. David, L. Tsacas and M. Ashburner (1986) The melanogaster species group. In: *The Genetics and Biology of Drosophila*, Vol. 3e, M. Ashburner, H. L. Carson and J. N. Thompson Jr. (eds.). Academic Press, London, pp. 147–265. (10)

Le Minor, L. (1984) *Salmonella* Lignieres 1900. In: *Bergey's Manual of Systematic Bacteriology*, Vol. 1, N. R. Krieg and J. G. Holt (eds.). Williams & Wilkins, Baltimore, pp. 427–458. (2)

Lenington, S. and K. Egid (1985) Female discrimination of male odors correlated with male genotype at the T locus: A response to T-locus or H-2-locus variability? Behav. Genet. 15: 53–67. (9)

Lenington, S. and I. L. Heisler (1987) Effect of behavior on transmission ratio distortion of t-haplotypes in wild house mice. Am. Nat. (9)

Leung, S. O., N. J. Proudfoot and E. Whitelaw (1987) The gene for θ-globin is transcribed in human fetal erythroid tissues. Nature 329: 551–554. (13)

Levin, B. R. (1981) Periodic selection, infectious gene exchange and the genetic structure of *E. coli* populations. Genetics 99: 1–23. (2)

Levin, B. R. (1988) The evolution of sex in bacteria. In: *The Evolution of Sex*, R. E. Michod and B. R. Levin (eds.). Sinauer Associates, Sunderland, MA, pp. 194–211. (2)

Levin, B. R., M. L. Petras and D. I. Rasmussen (1969) The effect of migration on the maintenance of a lethal polymorphism in the house mouse. Am. Nat. 103: 647–661. (9)

Levinger, L. and A. Varshavsky (1982) Protein D1 preferentially binds A+T-rich DNA *in vitro* and is a component of *Drosophila melanogaster* nucleosomes containing A+T-rich satellite DNA. Proc. Natl. Acad. Sci. USA 79: 7152–7156. (9)

Levinthal, C., E. R. Signer and K. Fetherolf (1962) Reactivation and hybridization of reduced alkaline phosphatase. Proc. Natl. Acad. Sci. USA 48: 1230–1237. (3)

Levy, J. A., A. D. Hoffman, S. M. Kramer, J. A. Landis, J. M. Shimabukuro and L. S. Oshiro (1984) Isolation of lymphocytopathic retroviruses from San Francisco patients with AIDS. Science 225: 840–842. (5)

Lewis, E. B (1951) Pseudoallelism and gene evolution. Cold Spring Harbor Symp. Quant. Biol. 16: 159–174. (4)

Lewontin, R. C. (1985) Population genetics. Annu. Rev. Genet. 19: 81–102. (10)

Lewontin, R. C. (1962) Interdeme selection controlling a polymorphism in the house mouse. Am. Nat. 96: 65–78. (9)

Lewontin, R. C. (1964) The interaction of selection and linkage. I. General considerations; heterotic models. Genetics 49: 49–67. (12)

Lewontin, R. C. (1988) On measures of gametic disequilibrium. Genetics 120: 849–852. (12)

Lewontin, R. C. and L. C. Dunn (1960) The evolutionary dynamics of a polymorphism in the house mouse. Genetics 45: 705–722. (9)

Li, W.-H. and M. Tanimura (1987) The molecular clock runs more slowly in man than in apes and monkeys. Nature 326: 93–96. (11)

Li, W.-H., C.-C. Luo and C.-I. Wu (1985a) Evolution of DNA sequences. In: *Molecular Evolutionary Genetics*, R. J. MacIntyre (ed.). Plenum Press, New York, pp. 1–94. (8)

Li, W.-H., C.-I. Wu and C.-C. Luo (1985b) A new method for estimating synonymous and nonsynonymous rates of nucleotide substitution considering the relative likelihood of nucleotide and codon changes. Mol. Biol. Evol. 2: 150–174. (3)

Li, W.-H., M. Tanimura and P. M. Sharp (1988) Rates and dates of divergence between AIDS virus nucleotide sequences. Mol. Biol. Evol. 5: 313–330. (5)

Liebhaber, S. A., M. J. Goossens and Y. W. Kan (1980) Cloning and complete nucleotide sequence of human 5'–α-globin gene. Proc. Natl. Acad. Sci. USA 77: 7054–7058. (13)

Lindsley, D. L. and E. H. Grell (1969) Spermiogenesis without chromosomes in *Drosophila melanogaster*. Genetics 61: 69–77. (9)

Lindsley, D. L. and L. Sandler (1977) The genetic analysis of meiosis in female *Drosophila*. Phil. Trans. R. Soc. London Ser. B 277: 295–312. (8)

Lingrel, J. B., T. M. Townes, S. G. Shapiro, S. M. Wernke, P. A. Liberator and A. G. Menon (1985) Structural organization of the α- and β-globin loci of the goat. In: *Experimental Approaches for the Study of Hemoglobin Switching*, G. Stamatoyannopoulos and A. W. Nienhuis (eds.). Liss, New York, pp. 67–79. (13)

Loeb, D. D., R. W. Padgett, S. C. Hardies, W. R. Shehee, M. B. Comer, M. H. Edgell and C. A. Hutchison (1986) The sequence of a large L1Md element reveals a tandemly repeated 5' end and several features found in retrotransposons. Mol. Cell. Biol. 6: 168–182. (13)

Longenecker, B. M. and T. R. Mosmann (1981) Structure and properties of the major histocompatibility complex of the chicken. Speculations on the advantages and evolution of polymorphism. Immunogenetics 13: 1–23. (11)

Longenecker, B. M., F. Pazderka, J. S. Gavora, J. L. Spencer and R. F. Ruth (1976) Lymphoma induced by Herpesvirus: Resistance associated with a major histocompatibility gene. Immunogenetics 3: 401–407. (11)

Lopez de Castro, J. A. (1989) HLA-B27 and HLA-A2 subtypes: Structure, evolution and function. Immunol. Today 10: 239–246. (12)

Lopez de Castro, J. A., J. L. Strominger, D. M. Strong and H. T. Orr (1982) Structure of crossreactive human histocompatibility antigens HLA-A28 and HLA-A2: Possible implications for the generation of HLA polymorphism. Proc. Natl. Acad. Sci. USA 79: 3813–3817. (11)

Lynch, M. (1988) Estimation of relatedness by DNA fingerprinting. Mol. Biol. Evol. 5: 584–599. (9)

Lyon, M. (1989) The genetic basis of transmission ratio distortion and male sterility due to the *t* complex. Am. Nat. (in press) (9)

Lyon, M. F. and J. Zenthon (1987) Differences in or near the responder region of complete and partial mouse *t*-haplotypes. Genet. Res. 50: 29–34. (9)

Lyttle, T. W. (1977) Experimental population genetics of meiotic drive systems I. Pseudo-Y chromosomal drive as a means of eliminating cage populations of *Drosophila melanogaster*. Genetics 86: 413–445. (9)

Lyttle, T. W. (1989) The effect of novel chromosome position and variable dose on the genetic behavior of the *Responder* (*Rsp*) element of the *Segregation Distorter* (*SD*) system of *Drosophila melanogaster*. Genetics 121: 751–763. (9)

Lyttle, T. W., J. G. Brittnacher and B. Ganetzky (1986) Detection of *Rsp* and modifier variation in the meiotic drive system *Segregation Distorter* (*SD*) of *Drosophila melanogaster*. Genetics 114: 183–202. (9)

Maeda, N., C.-I. Wu, J. Bliska and J. Reneke (1988) Molecular evolution of intergenic DNA in higher primates: Pattern of DNA changes, molecular clock, and evolution of repetitive sequences. Mol. Biol. Evol. 5: 1–20. (13)

Mager, D. L., P. S. Henthorn and O. Smithies (1985) A Chinese Gγ+Aγ δ β0 thalassemia deletion: Comparison to other deletions in the human β-globin gene and sequence analysis of the breakpoints. Nucleic Acids Res. 13: 6559–6575. (13)

Mahan, M. J., A. M. Segall and J. R. Roth (1990) Recombination events that rearrange the chromosome: Barriers to inversion. In: *The Bacterial Chromosome*, K. Drlica and M. Riley (eds.). American Society for Microbiology, Washington, D.C., pp. 341–349. (2)

Makela, P. H. and B. A. D. Stocker (1984) Genetics of lipopolysaccharide. In: *Handbook of*

Endotoxin, Vol. 1., E. T. Rietschel (ed.). Elsevier Science Publishers, Amsterdam, pp. 75–92. (2)

Maniatis, T., S. Goodbourn and J. Fischer (1987) Regulation of inducible and tissue-specific gene expression. Science 236: 1237–1245. (13)

Mann, J. M., J. Chin, P. Piot and T. Quinn (1988) The international epidemiology of AIDS. Sci. Am. 259: 82–89. (5)

Margot, J. B., G. W. Demers and R. C. Hardison (1989) Complete nucleotide sequence of the rabbit β-like globin gene cluster: Analysis of intergenic sequences and comparison with the human β-like globin gene cluster. J. Mol. Biol. 205: 15–40. (13)

Mark, D. F., A. Wang and C. Levenson (1987) Site-specific mutagenesis to modify the human tumor necrosis factor gene. Methods Enzymol. 154: 403–414. (3)

Marks, J., J.-P. Shaw and C.-K. J. Shen (1986) Sequence organization and genomic complexity of primate θ1 globin gene, a novel α-globin-like gene. Nature 321: 785–788. (13)

Martin, G., D. Wiernasz and P. Schedl (1983) Evolution of *Drosophila* repetitive-dispersed DNA. J. Mol. Evol. 19: 203–213. (8)

Martin, S. L., K. A. Vincent and A. C. Wilson (1983) Rise and fall of the δ-globin gene. J. Mol. Biol. 164: 513–528. (13)

Martin, W., A. Gierl and H. Saedler (1989) Molecular evidence for pre-Cretaceous angiosperm origins. Nature 339: 46–48. (7)

Martinez, C., L. del Rio, A. Portela, E. Domingo and J. Ortin (1983) Evolution of the influenza virus neuraminidase gene during drift of the N2 subtype. Virology 130: 539–545. (5)

Maruyama, T. (1982) Stochastic integrals and their application to population genetics. In: *Molecular Evolution, Protein Polymorphism and the Neutral Theory*, M. Kimura (ed.). Japan Scientific Societies Press, Tokyo, pp. 151–166. (12)

Maruyama, T. and M. Kimura (1980) Genetic variability and effective population size when local extinction and recolonization of subpopulations are frequent. Proc. Natl. Acad. Sci. USA 77: 6710–6714. (2)

Maruyama, T. and M. Nei (1981) Genetic variability maintained by mutation and overdominant selection in finite populations. Genetics 98: 441–459. (11)

Maruyama, T. and C. W. Birky, Jr. (1991) Effects of periodic selection on gene diversity in organelle genomes and other systems without recombination. Genetics (in press). (6)

Matsui, K., K. Sano and E. Ohtsubo (1986) Complete nucleotide and deduced amino acid sequences of the *Brevibacterium lactofermentum* tryptophan operon. Nucleic Acids Res. 14: 10113–10114. (4)

Matthews, B. W. and S. J. Remington (1974) The three dimensional structure of the lysozyme from bacteriophage T4. Proc. Natl. Acad. Sci. USA 71: 4178–4182. (3)

Mayer, L. W. (1988) Use of plasmid profiles in epidemiologic surveillance of disease outbreaks and in tracing the transmission of antibiotic resistance. Clin. Microbiol. Rev. 1: 228–243. (2)

Mayer, W. E., M. Jonker, D. Klein, P. Ivanyi, G. van Seventer and J. Klein (1988) Nucleotide sequences of chimpanzee MHC class I alleles: Evidence for trans-species mode of evolution. EMBO J. 7: 2765–2774. (11, 12)

Maynard Smith, J. (1987) On the equality of origin and fixation times in genetics. J. Theor. Biol. 128: 247–252. (6)

McConnell, T. J., W. S. Talbot, R. A. McIndoe and E. K. Wakeland (1988) The origin of MHC class II gene polymorphism within the genus *Mus*. Nature 332: 651–654. (11)

McGuire, K. L., W. R. Duncan and P. W. Tucker (1985) Syrian hamster DNA shows limited polymorphism at class I-like loci. Immunogenetics 22: 257–268. (11)

Meagher, R. B., S. Berry-Lowe and K. Rice (1989) Molecular evolution of the small subunit of ribulose bisphosphate carboxylase: Nucleotide substitution and gene conversion. Genetics 123: 845–863. (7)

Medley, G. F., R. M. Anderson, D. R. Cox and L. Billard (1987) Incubation period of AIDS in patients infected via blood transfusion. Nature 328: 719–721. (5)

Metzlaff, M., T. Borner and R. Hagemann (1981) Variations of chloroplast DNAs in the genus *Pelargonium* and their biparental inheritance. Theor. Appl. Genet. 60: 37–41. (7)

Michel, F. and B. Dujon (1983) Conservation of RNA secondary structures in two intron families including mitochondrial-, chloroplast-, and nuclear-encoded members. EMBO J. 2: 33–38. (7)

Michel, F. and A. Jacquier (1987) Long-range intron-exon and intron-intron pairings involved in self-splicing of class II catalytic introns. Cold Spring Harbor Symp. Quant. Biol. 52: 201–212. (7)

Michel, F. and B. F. Lang (1985) Mitochondrial class II introns encode proteins related to the reverse transcriptases of retroviruses. Nature 316: 641–643. (7)

Michel, F., K. Umesono and H. Ozeki (1989) Comparative and functional anatomy of group II catalytic introns—a review. Gene 82: 5–30. (7)

Miklos, G. L. G. (1985) Localized highly repetitive DNA sequences in vertebrate and invertebrate genomes. In: *Molecular Evolutionary Genetics*, R. J. MacIntyre (ed.). Plenum Press, New York, pp. 241–321. (9)

Miklos, G. L. G., M. J. Healy, P. Pain, A. J. Howells and R. J. Russell (1984) Molecular and genetic studies on the euchromatin-heterochromatin transition region of the X chromosome of *Drosophila melanogaster*. 1. A cloned entry point near to the uncoordinated (*unc*) locus. Chromosoma 89: 218–227. (8)

Miklos, G. L. G., M.-T. Yamamoto, J. Davies and V. Pirrotta (1988) Microcloning reveals a high frequency of repetitive sequences characteristic of chromosome 4 and the β-heterochromatin of *Drosophila melanogaster*. Proc. Natl. Acad. Sci. USA 85: 2051–2055. (8)

Miles, E. W., R. Bauerle and S. A. Ahmed (1987) Tryptophan synthase from *Escherichia coli* and *Salmonella typhimurium*. Methods Enzymol. 142: 398–414. (4)

Milkman, R. (1973) Electrophoretic variation in *Escherichia coli* from natural sources. Science 182: 1024–1026. (2)

Milo, J., A. Levy, G. Ladizinsky and D. Palevitch (1988) Phylogenetic and genetic studies in *Papaver* section *Oxytona*: Cytogenetics, isozyme analysis and chloroplast DNA variation. Theor. Appl. Genet. 75: 795–802. (7)

Miyamoto, M. M., J. L. Slightom and M. Goodman (1987) Phylogenetic relations of humans and African apes from DNA sequences in the ψη-globin region. Science 238: 369–373. (11)

Miyashita, N. and C. H. Langley (1988) Molecular and phenotypic variation of the *white* locus region in *Drosophila melanogaster*. Genetics 120: 199–212. (8, 10)

Miyata, T. and T. Yasunaga (1980) Molecular evolution of mRNA: A method for estimating evolutionary rates of synonymous and amino acid substitutions from homologous nucleotide sequences and its application. J. Mol. Evol. 16: 23–36. (5)

Moloney, J. B. (1966) A virus-induced rhabdomyosarcoma of mice. Natl. Cancer Inst. Monogr. 22: 139–142. (5)

Monnat, R. J., Jr., C. L. Maxwell and L. A. Loeb (1985) Nucleotide sequence preservation of human Leukemic mitochondrial DNA. Cancer Res. 45: 1809–1814. (6)

Montgomery, E. A. and C. H. Langley (1983) Transposable elements in Mendelian populations. II. Distribution of three *copia*-like elements in a natural population of *Drosophila melanogaster*. Genetics 104: 473–483. (8)

Montgomery, E., B. Charlesworth and C. H. Langley (1987) A test for the role of natural selection in the stabilization of transposable element copy number in a population of *Drosophila melanogaster*. Genet. Res. 49: 31–41. (8)

Moon, E., T.-H. Kao and R. Wu (1987) Rice chloroplast DNA molecules are heterogeneous as revealed by DNA sequences of a cluster of genes. Nucleic Acids Res. 15: 611–630. (7)

Muesing, M. A., D. H. Smith, C. D. Cabradilla, C. V. Benton, L. A. Lasky and D. J. Capon (1985) Nucleic acid structure and expression of the human AIDS/lymphadenopathy retrovirus. Nature 313: 450–458. (5)

Mukai, T. and O. Yamaguchi (1974) The genetic structure of natural populations of *Drosophila*. XI. Genetic variability in a local population. Genetics 76: 339–366. (8)

Mukai, T., L. E. Mettler and S. I. Chigusa (1971) Linkage disequilibrium in a local population of *Drosophila melanogaster*. Proc. Natl. Acad. Sci. USA 68: 1065–1069. (9)

Muller, H. J. (1964) The relation of recombination to mutational advance. Mutat. Res. 1: 2–9. (8)

Murai, K. and K. Tsunewaki (1986) Molecular basis of genetic diversity among cytoplasms of *Triticum* and *Aegilops* species. IV. ctDNA variation in *Ae. triuncialis*. Heredity 57: 335–339. (7)

Murray, R. G. E. (1962) Fine structure and taxonomy of bacteria. In: *Microbial Classification*, G. C. Ainsworth and P. H. A. Sneath (eds.). Soc. Gen. Microbiol. Symp. 12, pp. 119–144. (1)

Murray, R. G. E. (1974) A place for bacteria in the living world. In: *Bergey's Manual of Determinative Bacteriology*, 8th ed., R. E. Buchanan and N. E. Gibbons (eds.). Williams & Wilkins, Baltimore, pp. 4–9. (1)

Musser, J. M. and 27 others. (1990) Global genetic structure and molecular epidemiology of encapsulated *Haemophilus influenzae*. Rev. Infect. Dis. 12: 75–111. (2)

Nadeau, J. H. (1983) Absence of detectable gametic disequilibrium between the *t*-complex and linked allozyme-encoding loci in house mice. Genet. Res. 42: 323–333. (12)

Nadeau, J. H. (1986) A glyoxalase-1 variant associated with the *t*-complex in house mice. Genetics 113: 91–99. (12)

Nathenson, S. G., J. Geliebter, G. M. Pfaffenbach and R. A. Zeff (1986) Murine major histocompatibility complex class-I mutants: Molecular analysis and structure-function implications. Annu. Rev. Immunol. 4: 471–502. (12)

Neale, D. B. and R. R. Sederoff (1989) Paternal inheritance of chloroplast DNA and maternal inheritance of mitochondrial DNA in loblolly pine. Theor. Appl. Genet. 77: 212–216. (7)

Neale, D. B., N. C. Wheeler. and R. W. Allard (1986) Paternal inheritance of chloroplast DNA in Douglas-fir. Can. J. For. Res. 16: 1152–1154. (7)

Neale, D. B., M. A. Saghai-Maroof, R. W. Allard, Q. Zhang and R. A. Jorgensen (1988) Chloroplast DNA diversity in populations of wild and cultivated barley. Genetics 120: 1105–1110. (7)

Nei, M. (1987) *Molecular Evolutionary Genetics*. Columbia Univ. Press, New York. (2, 10, 11, 13)

Nei, M. and T. Gojobori (1986) Simple methods for estimating the numbers of synonymous and nonsynonymous nucleotide substitutions. Mol. Biol. Evol. 3: 418–426. (11)

Nei, M. and D. Graur (1984) Extent of protein polymorphism and the neutral mutation theory. Evol. Biol. 17: 73–118. (2, 11)

Nei, M. and L. Jin (1989) Variances of the average numbers of nucleotide substitutions within and between populations. Mol. Biol. Evol. 6: 290–300. (10, 11)

Nei, M. and F. Tajima (1981) DNA polymorphism detectable by restriction endonucleases. Genetics 97: 145–163. (10)

Neigel, J. E. and J. C. Avise (1986) Phylogenetic relationships of mitochondrial DNA under various demographic models of speciation. In: *Evolutionary Processes and Theory*, S. Karlin and E. Nevo (eds.). Academic Press, New York, pp. 515–534. (6)

Nevers, P. and H. Saedler (1977) Transposable genetic elements as agents of gene instability and chromosomal rearrangements. Nature 268: 109–115. (8)

Nevo, E., A. Beiles and R. Ben Shlomo (1984) The evolutionary significance of genetic diversity: ecological, demographic and life history correlates. In: *Evolutionary dynamics of genetic diversity*, G. S. Mani (ed.), Springer-Verlag, Berlin, pp. 13–213. (2)

Nicholls, R. D., N. Fischel-Ghodsian and D. R. Higgs (1987) Recombination at the human α-globin gene cluster: Sequence features and topological constraints. Cell 49: 369–378. (13)

Nichols, B. P., A. M. Seibold and S. Z. Doktor (1989) *para*-Aminobenzoate synthesis from chorismate occurs in two steps. J. Biol. Chem. 264: 8597–8601. (4)

Nicoghosian, K., M. Bigras, D. Sankoff and R. Cedergren (1987) Archetypical features in tRNA families. J. Mol. Evol. 26: 341–346. (7)

Nierzwicki-Bauer, S. A., S. E. Curtis and R. Haselkorn (1984) Cotranscription of genes encoding the small and large subunits of ribulose-1,5-bisphosphate carboxylase in the cyanobacterium *Anabaena* 7120. Proc. Natl. Acad. Sci. USA 81: 5961–5965. (7)

Nizetic, D., F. Figueroa, E. Nevo and J. Klein (1985) Major histocompatibility complex of the mole-rat. Immunogenetics 22: 55–67. (11)

Nizetic, D., M. Stevanovic, B. Soldatovic, I. Savic and R. Crkvenjakov (1988) Limited polymorphism of both classes of MHC genes in four different species of the Balkan mole rat. Immunogenetics 28: 91–98. (11)

Noll, K. (1989) Chromosome map of the thermophilic archaebacterium *Thermococus celer*. J. Bacteriol. 171: 6720–6725. (1)

Norris, G. E., B. F. Anderson and E. N. Baker (1983) Structure of Azurin from *Alcaligenes denitrificans* at 2.5 Å resolution. J. Mol. Biol. 165: 501–521. (3)

Nur, U. and B. L. H. Brett (1987) Control of meiotic drive of B chromosomes in the mealybug, *Pseudococcus affinis* (*obscurus*). Genetics 115: 499–510. (9)

Nur, U. and C.-I Wu (1991) Delayed spermatid maturation and meiotic drive in *Drosophila melanogaster*. Genetics (submitted). (9)

Nur, U., J. H. Werren, D. G. Eickbush, W. D. Burke and T. H. Eickbush (1988) A "Selfish" B chromosome that enhances its transmission by eliminating the paternal genome. Science 240: 512–514. (9)

O'Brien, S. J. (1986) Molecular genetics in the domestic cat and its relatives. Trends Genet. 2: 137–142. (13)

O'Brien, S. J., D. E. Wildt, D. Goldman, C. R. Merril and M. Bush (1983) The cheetah is depauperate in genetic variation. Science 221: 459–462. (11)

O'Brien, S. J., M. E. Roelke, L. Marker, A. Newman, C. A. Winkler, D. Meltzer, L. Colly, J. F. Evermann, M. Bush and D. E. Wildt (1985) Genetic basis for species vulnerability in the cheetah. Science 227: 1428–1434. (11)

Ochman, H. and R. K. Selander (1984) Standard reference strains of *Escherichia coli* from natural populations. J. Bacteriol. 157: 690–693. (3)

Ochman, H. and A. C. Wilson (1987) Evolution in bacteria: Evidence for a universal substitution rate in cellular genomes. J. Mol. Evol. 26: 74–86. (2, 3)

Ochman, H., T. S. Whittam, D. A. Caugant and R. K. Selander (1983) Enzyme polymorphism and genetic population structure in *Escherichia coli* and *Shigella*. J. Gen. Microbiol. 129: 2715–2726. (2)

Ogihara, Y. and K. Tsunewaki. (1988) Diversity and evolution of chloroplast DNA in *Triticum* and *Aegilops* as revealed by restriction fragment analysis. Theor. Appl. Genet. 76: 321–332. (7)

Ohta, T. (1977) On the gene conversion model as a mechanism for maintenance of homogeneity in systems with multiple genomes. Genet. Res. 30: 89–91. (6)

Ohta, T. (1982) Allelic and nonallelic homology of a supergene family. Proc. Natl. Acad. Sci. USA 79: 3251–3254. (11)

Ohta, T. and M. Kimura (1969) Linkage disequilibrium due to random genetic drift. Genet. Res. 13: 47–55. (12)

Ohyama, K., H. Fukazawa, T. Kohchi, H. Shirai, T. Sano, S. Sano, K. Umesono, Y. Shiki, M. Takeuchi, Z. Chang, S. Aota, H. Inokuchi and H. Ozeki (1986) Complete nucleotide sequence of liverwort *Marchantia polymorpha* chloroplast DNA. Plant Mol. Biol. Rep. 4: 148–175. (7)

Oksenberg, J. R., E. Persitz, A. Amar and C. Brautbar (1984) Maternal-paternal histocompatibility: Lack of association with habitual abortions. Fertil. Steril. 42: 389–395. (12)

Olivieri, G. and A. Olivieri (1965) Autoradiographic study of nucleic acid synthesis during spermatogenesis in *Drosophila melanogaster*. Mutat. Res. 2: 366–380. (9)

Olsen, G. J. (1988) The earliest phylogenetic branchings: Comparing rRNA-based evolutionary trees inferred with various techniques. Cold Spring Harbor Symp. of Quant. Biol. 52: 825–837. (1)

Olsen, G. J. and C. R. Woese (1989) A brief note concerning archaebacterial phylogeny. Can J. Microbiol. 35: 119–123. (1)

Oren, A., B. J. Paster, and C. R. Woese (1984) *Haloanaerobiaceae*: A new family of moderately halophilic, obligatory anaerobic bacteria. Syst. Appl. Microbiol. 5: 71–80. (1)

Orgel, L. E. and F. H. C. Crick (1980) Selfish DNA: The ultimate parasite. Nature 284: 604–607. (8)

Orkin, S. H., H. H. Kazazian, S. F. Antonarakis, S. Goff, C. D. Boehm, J. P. Sexton, P. G. Waber and P. J. V. Giardina (1982) Linkage of β-thalassaemia mutations and β-globin gene polymorphisms with DNA polymorphisms in human β-globin gene clusters. Nature 296: 627–631. (13)

Ou, J. T., L. S. Baron, F. A. Rubin and D. J. Kopecko (1988) Specific insertion and deletion of insertion sequence 1-like DNA element causes the reversible expression of the virulence capsular antigen Vi of *Citrobacter freundii* in *Escherichia coli*. Proc. Natl. Acad. Sci. USA 85: 4402–4405. (2)

Oyaizu, H. and C. R. Woese (1985) Phylogenetic relationships among the sulfate respiring bacteria, myxobacteria and purple bacteria. Syst. Appl. Microbiol. 6: 257–263. (1)

Oyaizu, H., B. Debrunner-Vossbrinck, L. Mandelco, J. A. Studier, and C. R. Woese (1986) The green non-sulfur bacteria: A deep branching in the eubacterial line of descent. Syst. Appl. Microbiol. 9: 47–53. (1)

Padgett, R. A., P. J. Grabowski, M. M. Konarska, S. Seiler and P. A. Sharp (1986) Splicing of messenger RNA precursors. Annu. Rev. Biochem. 55: 1119–1150. (7)

Palm, J. (1974) Maternal-fetal histoincompatibility in rats: An escape from adversity. Cancer Res. 34: 2061–2065. (11)

Palmer, J. D. (1987) Chloroplast DNA evolution and biosystematic uses of chloroplast DNA variation. Am. Nat. 130: S6–S29. (7)

Palmer, J. D. and D. Zamir (1982) Chloroplast DNA evolution and phylogenetic relationships in *Lycopersicon*. Proc. Natl. Acad. Sci. USA 79: 5006–5010. (7)

Palmer, J. D., C. R. Shields, D. B. Cohen and T. J. Orton (1983) Chloroplast DNA evolution and the origin of amphidiploid *Brassica* species. Theor. Appl. Genet. 65: 181–189. (7)

Palmer, J. D., R. A. Jorgensen and W. F. Thompson (1985) Chloroplast DNA variation and evolution in *Pisum*: Patterns of change and phylogenetic analysis. Genetics 109: 195–213. (7)

Palmer, J. D., R. K. Jansen, H. J. Michaels, M. W. Chase and J. R. Manhart (1988) Chloroplast DNA variation and plant phylogeny. Ann. Missouri Bot. Garden. 75: 1180–1206 (7)

Palmer, J. D., S. L. Baldauf, P. D. Calie and C. W. dePamphillis (1990) Chloroplast gene instability and transfer to the nucleus. In: *Molecular Evolution*, M. T. Clegg and S. J. O'Brien (eds.). Liss, New York, pp. 97–106. (7)

Palmer, M., P. J. Wettstein and J. A. Frelinger (1983) Evidence for extensive polymorphism of class I genes in the rat major histocompatibility complex (*RT1*). Proc. Natl. Acad. Sci. USA 80: 7616–7620. (11)

Pamilo, P., M. Nei and W.-H. Li (1987) Accumulation of mutations in sexual and asexual populations. Genet. Res. 49: 135–146. (8)

Parham, P., C. E. Lomen, D. A. Lawlor, J. P. Ways, N. Holmes, H. L. Coppin, R. D. Salter, A. M. Wan and P. D. Ennis (1988) Nature of polymorphism in HLA-A, -B, and -C molecules. Proc. Natl. Acad. Sci. USA 85: 4005–4009. (12)

Parham, P., D. A. Lawlor, C. E. Lomen and P. D. Ennis (1989) Diversity and diversification of HLA-A, B, C alleles. J. Immunol. 142: 3937–3950. (12)

Partridge, L. (1988) The rare-male effect: What is its evolutionary significance? Phil. Trans. R. Soc. London Ser. B 319: 525–539. (11)

Partridge, L. and T. Halliday (1984) Mating patterns and mate choice. In: *Behavioural Ecology*, 2nd ed., J. R. Krebs and N. B. Davies (eds.). Blackwell, Oxford, pp. 222–250. (11)

Paster, B. J., W. Ludwig, W. G. Weisburg, E. Stackebrandt, R. B. Hespell, C. M. Hahn, H. Reichenbach, K. O. Stetter, and C. R. Woese (1985) A phylogenetic grouping of the bacteroides, cytophagas and certain flavobacteria. System. Appl. Microbiol. 6: 34–42. (1)

Perl-Treves, R. and E. Galun (1985) The *Cucumis plastome*: Physical map, intragenic variation and phylogenetic relationships. Theor. Appl. Genet. 71: 417–429. (7)

Perutz, M. F. (1983) Species adaptation in a protein molecule. Mol. Biol. Evol. 1: 1–28. (13)

Petras, M. L. (1967) Studies of natural populations of *Mus*. II. Polymorphism at the *T* locus. Evolution 21: 466–478. (9)

Pimpinelli, S. and P. Dimitri (1989) Cytogenetic analysis of segregation distortion in *Drosophila melanogaster*: The cytological organization of the Responder (*Rsp*) locus. Genetics 121: 765–772. (9)

Pine, R. and P. C. Huang (1987) An improved method to obtain a large number of mutants in a defined region of DNA. Methods Enzymol. 154: 415–430. (3)

Pinero, D., E. Martinez and R. K. Selander (1988) Genetic diversity and relationships among isolates of *Rhizobium leguminosarum* biovar *phaseoli*. Appl. Environ. Microbiol. 54: 2825–2832. (2)

Piot, P., F. A. Plummer, F. S. Mhalu, J.-L. Lamboray, J. Chin and J. M. Mann (1988) AIDS: An international perspective. Science 239: 573–579. (5)

Plapp, B. V. (1982) Origins of protein and structure and function. In: *Perspectives on Evolution*, R. Milkman (ed.). Sinauer Associates, Sunderland, MA, pp. 129–147. (4)

Plos, K., S. I. Hull, R. A. Hull, B. R. Levin, I. Ørskov, F. Ørskov and C. Svanborg-Edén (1989) Distribution of the P-associated-pilus (*pap*) region among *Escherichia coli* from natural sources: Evidence for horizontal gene transfer. Infect. Immun. 57: 1604–1611. (2)

Popoff, M. Y. and L. Le Minor (1985) Expression of antigenic factor O: 54 is associated with the presence of a plasmid in *Salmonella*. Ann. Inst. Pasteur Microbiol. 136B: 169–179. (2)

Popovic, M., M. G. Sarngadharan, E. Read and R. C. Gallo (1984) Detection, isolation, and continuous production of cytopathic retroviruses (HTLV-III) from patients with AIDS and pre-AIDS. Science 224: 497–500. (5)

Porter, R. D. (1988) Modes of gene transfer in bacteria. In: *Genetic Recombination*, R. Kucherlapati and G. R. Smith (eds.). American Society for Microbiology, Washington, D.C., pp. 1–41 (2)

Powell, J. R. (1983) Interspecific cytoplasmic gene flow in the absence of nuclear gene flow: Evidence from *Drosophila*. Proc. Natl. Acad. Sci. USA 80: 492–495. (9)

Presta, L. G. and G. D. Rose (1988) Helix signals in proteins. Science 240: 1632–1641. (3)

Prévot, A. R. (1940) *Manuel de classification et de détermination des bactéries anaérobies.* Masson et Cie, Paris. (1)

Pringsheim, E. G. (1923) Zur kritik der Bakteriensystematik. Lotos 71: 357–377. (1)

Pringsheim, E. G. (1949) The relationship between bacteria and Myxophyceae. Bacteriol. Rev. 13: 47–98. (1)

Proudfoot, N. J., A. Gil and T. Maniatis (1982) The structure of the human ζ-globin gene and a closely linked, nearly identical pseudogene. Cell 31: 553–563. (13)

Prout, T., J. Bundgaard and S. Bryant (1973) Population genetics of modifiers of meiotic drive. I. The solution of a special case and some general implications. Theor. Pop. Biol. 4: 446–465. (9)

Puhler, G, H. Leffers, F. Gropp, P. Palm, H.-P. Klenk, F. Lottspeich, R. A. Garrett, and W. Zillig (1989) Archaebacterial DNA-dependent RNA polymerases testify to the evolution of the eukaryotic nuclear genome. Proc. Natl. Acad. Sci. USA 86: 4569–4573. (1)

Quigley, F. and J. H. Weil (1985) Organization and sequence of five tRNA genes and of an unidentified reading frame in the wheat chloroplast genome: Evidence for gene rearrangements during the evolution of chloroplast genomes. Curr. Genet. 9: 495–503. (7)

Quigley, F., W. F. Martin and R. Cerff (1988) Intron conservation across the prokaryote-eukaryote boundary: Structure of the nuclear gene for chloroplast glyceraldehyde-3-phosphate dehydrogenase from maize. Proc. Natl. Acad. Sci. USA 85: 2672–2676. (7)

Radic, M. Z., K. Lundgren and B. A. Hamkalo (1987) Curvature of mouse satellite DNA and condensation of heterochromatin. Cell 50: 1101–1108. (9)

Rand, D. M. and R. G. Harrison (1986) Mitochondrial DNA transmission genetics in crickets. Genetics 114: 955–970. (6)

Rand, D. M. and R. G. Harrison (1989) Molecular population genetics of mtDNA size variation in crickets. Genetics 121: 551–569. (6)

Ratner, L., W. Haseltine, R. Patarca, K. J. Livak, B. Starcich, S. F. Josephs, E. R. Doran, J. A. Rafalski, E. A. Whitehorn, K. Baumeister, L. Ivanoff, S. R. Petteway, Jr., M. L. Pearson, J. A. Lautenberger, T. S. Papas, J. Ghrayeb, N. T. Chang, R. C. Gallo and F. Wong-Staal (1985) Complete nucleotide sequence of the AIDS virus, HTLV-III. Nature 313: 277–284. (5)

Reeves, M. W., G. M. Evins, A. A. Heiba, B. D. Plikaytis and J. J. Farmer III (1989) Clonal nature of *Salmonella typhi* and its genetic relatedness to other salmonellae as shown by multilocus enzyme electrophoresis, and proposal of *Salmonella bongori* comb. nov. J. Clin. Microbiol. 27: 313–320. (2)

Regan, L. and W. F. DeGrado (1988) Characterization of a helical protein designed from first principles. Science 241: 976–978. (3)

Reiter, W.-D., P. Palm and S. Yeats (1989) Transfer RNA genes frequently serve as integration sites for prokaryotic genetic elements. Nucleic Acids Res. 17: 1907–1914. (7)

Reith, M. and R. A. Cattolico (1986) Inverted repeat of *Olisthodiscus luteus* chloroplast DNA contains genes for both subunits of ribulose-1,5-bisphosphate carboxylase and the 32,000-dalton Q_B protein: Phylogenetic implications. Proc. Natl. Acad. Sci. USA 83: 8599–8603. (7)

Reuter, G., M. Giarre, J. Farah, J. Gausz, A. Spierer and P. Spierer (1990) Dependence of position-effect variegation in *Drosophila* on dose of a gene encoding an unusual zinc-finger protein. Nature 344: 219–223. (9)

Richardson, J. S (1981) The anatomy and taxonomy of protein structure. Adv. Protein Chem. 34: 167–339. (4)

Richardson, J. S. and D. C. Richardson (1988) Amino acid preferences for specific locations at the ends of a helices. Science 240: 1648–1652. (3)

Rieseberg, L. H., D. E. Soltis and J. D. Palmer (1988) A molecular reexamination of introgression between *Helianthus annuus* and *H. bolanderi* (Compositae). Evolution 42: 227–238. (7)

Riggs, A. (1990) Aspects of the origin and evolution of non-vertebrate hemoglobins. Am. Zool. (in press). (13)

Riley, L. W., G. T. DiFerdinando Jr., T. M. DeMelfi and M. L. Cohen (1983) Evaluation of isolated cases of salmonellosis by plasmid profile analysis: Introduction and transmission of a bacterial clone by precooked roast beef. J. Infect. Dis. 148: 12–17. (2)

Riley, M. A., M. E. Hallas and R. C. Lewontin (1989) Distinguishing the forces controlling genetic variation at the *Xdh* locus in *Drosophila pseudoobscura*. Genetics 123: 359–369. (10)

Riley, M. and S. Krawiec (1987) Evolutionary history of enteric bacteria. In: Escherichia coli *and*

Salmonella typhimurium: *Cellular and molecular biology,* F. C. Neidhardt, J. L. Ingraham, K. B. Low, B. Magasanik, M. Schaechter, and H. E. Umbarger (eds.). American Society for Microbiology, Washington, D.C., pp. 967–981. (2)

Riley, M. and K. E. Sanderson (1990) Comparative genetics of *Escherichia coli* and *Salmonella typhimurium.* In: *The Bacterial Chromosome,* K. Drlica and M. Riley (eds.). American Society for Microbiology, Washington, D.C., pp. 85–95. (2)

Ritland, K. and M. T. Clegg (1987) Evolutionary analysis of plant DNA sequences. Am. Nat. 130: S74–S100. (7)

Robbins, J. D. and J. B. Robbins (1984) Reexamination of the protective role of the capsular polysaccharide (Vi antigen) of *Salmonella typhi.* J. Infect. Dis. 150: 436–449. (2)

Robertson, A. (1961) Inbreeding in artificial selection programmes. Genet. Res. 2: 189–194. (6)

Robinson, W. P. (1989) Population genetic analysis of selection and disease associations at the HLA gene family. Ph.D. Thesis, University of California at Berkeley. (12)

Rogan, P. K., J. Pan and S. M. Weissman (1987) L1 repeat elements in the human ϵ-$^G\gamma$-globin gene intergenic region: Sequence analysis and concerted evolution within this family. Mol. Biol. Evol. 4: 327–342. (13)

Rogers, J. H. (1985) Mouse histocompatibility-related genes are not conserved in other mammals. EMBO J. 4: 749–753. (11)

Rohrbaugh, M. L. and R. C. Hardison (1983) Analysis of rabbit β-like globin gene transcripts during development. J. Mol. Biol. 164: 395–417. (13)

Roninson, I. B. and V. M. Ingram (1982) Gene evolution in the chicken β-globin cluster. Cell 28: 515–521. (13)

Ronsseray, S. and D. Anxolabéhère (1986) Chromosomal distribution of P and I transposable elements in a natural population of *Drosophila melanogaster.* Chromosoma 94: 433–440. (8)

Rose, R. J., L. B. Johnson and R. J. Kemble (1986) Restriction endonuclease studies on the chloroplast and mitochondrial DNAs of alfalfa (*Medicago sativa* L.) protoclones. Plant Mol. Biol. 6: 331–338. (7)

Rosenberg, L. T., D. Cooperman and R. Payne (1983) HLA and mate selection. Immunogenetics 17: 89–93. (11)

Roychoudhury, A. K. and M. Nei (1988) *Human Polymorphic Genes: World Distribution.* Oxford Univ. Press, New York. (11)

Rubin, F. A., D. J. Kopecko, K. F. Noon and L. S. Baron (1985) Development of a DNA probe to detect *Salmonella typhi.* J. Clin. Microbiol. 22: 600–605. (2)

Rubin, G. M. (1983) Dispersed repetitive DNA sequences in *Drosophila.* In: *Mobile Genetic Elements,* J.A. Shapiro (ed.), pp. 329–361. Academic Press, Orlando, FL, pp. 329–361. (8)

Rubin, R. H. and L. Weinstein (1977) *Salmonellosis: Microbiologic, Pathologic, and Clinical Features.* Stratton International Medical Book Corporation, New York. (2)

Ryan, C. A., N. T. Hargrett-Bean and P. A. Blake (1989) *Salmonella typhi* infections in the United States, 1975–1984: Increasing role of foreign travel. Rev. Infect. Dis. 11: 1–8. (2)

Sage, R. D., J. B. Whitney III and A. C. Wilson (1986) Genetic analysis of a hybrid zone between *domesticus* and *musculus* mice (*Mus musculus* Complex): Hemoglobin polymorphisms. Curr. Top. Microbiol. Immunol. 127: 75–85. (9)

Saiki, R. K., D. H. Gelfand, S. Stoffel, S. J. Scharf, R. Higuchi, G. T. Horn, K. B. Mullis and H. A. Erlich (1988) Primer-directed enzymatic amplification of DNA with a thermostable DNA polymerase. Science 239: 487–491. (10)

St. Louis, M. E., D. L. Morse, M. E. Potter, T. M. DeMelfi, J. J. Guzewich, R. V. Tauxe, P. A. Blake and the *Salmonella enteritidis* working group (1988) The emergence of grade A eggs as a major source of *Salmonella enteritidis* infections. J. Am. Med. Assoc. 259: 2103–2107. (2)

Saitou, N. and M. Nei (1986) Polymorphism and evolution of influenza A virus genes. Mol. Biol. Evol. 3: 57–74. (5)

Saitou, N. and M. Nei (1987) The neighbor-joining method: A new method for reconstructing phylogenetic trees. Mol. Biol. Evol. 4: 406–425. (5, 11)

Salts, Y., R. G. Herrmann, N. Peleg, U. Lavi, S. Izhar, R. Frankel and J. S. Beckmann (1984) Physical mapping of plastid DNA variation among eleven *Nicotiana* species. Theor. Appl. Genet. 69: 1–14. (7)

Sanchez-Pescador, R., M. D. Power, P. J. Barr, K. S. Steimer, M. M. Stempien, S. L. Brown-Shimer, W. W. Gee, A. Renard, A. Randolph, J. A. Levy, D. Dina and P. A. Luciw (1985)

Nucleotide sequence and expression of an AIDS-associated retrovirus (ARV-2). Science 227: 484–492. (5)

Sanderson, K. E. and J. R. Roth (1988) Linkage map of *Salmonella typhimurium*, Edition VII. Microbiol. Rev. 52: 485–532. (2)

Sandler, L. and E. Novitski (1957) Meiotic drive as an evolutionary force. Am. Nat. 91: 105–110. (9)

Sandman, K., J. A. Krzycki, B. Dobrinski, R. Lurz, and J. N. Reeve (1990) DNA binding protein HMf, isolated from the hyperthermophilic archaea *Methanothermus fervidus*, is most closely related to histones. Proc. Natl. Acad. Sci. USA 87: 5788–5791. (1)

Sawada, I. and C. W. Schmid (1986) Primate evolution of the a globin gene cluster and its *Alu*-like repeats. J. Mol. Biol. 192: 693–709. (13)

Sawyer, S. (1989) Statistical tests for detecting gene conversion. Mol. Biol. Evol. 6: 526–538. (12)

Sawyer, S. and D. Hartl (1986) Distribution of transposable elements in prokaryotes. Theor. Pop. Biol. 30: 1–16. (8)

Sawyer, S., D. Dykhuizen, R. DuBose, L. Green, T. Mutangadura-Mhlanga and D. Hartl (1987a) Distribution and abundance of insertion sequences among natural isolates of *Escherichia coli*. Genetics 115: 51–63. (2)

Sawyer, S. A., D. E. Dykhuizen and D. L. Hartl (1987b) Confidence interval for the number of selectively neutral amino acid polymorphisms. Proc. Natl. Acad. Sci. USA 84: 6225–6228. (3, 10)

Saxinger, W. C., P. H. Levine, A. G. Dean, G. De The, G. Lange-Wantzin, J. Moghissi, F. Laurent, M. Hoh, M. G. Sarngadharan and R. C. Gallo (1985) Evidence for exposure to HTLV-III in Uganda before 1973. Science 227: 1036–1038. (5)

Schacter, B., L. R. Weitkamp and W. E. Johnson (1984) Parental HLA compatibility, fetal wastage, and neural tube defects: Evidence for a *T/t*-like locus in humans. Am. J. Hum. Genet. 36: 1082–1091. (11)

Schaeffer, S. W. and C. F. Aquadro (1987) Nucleotide sequence of the *Adh* gene region of *Drosophila pseudoobscura*: Evolutionary change and evidence for an ancient gene duplication. Genetics 117: 61–73. (10)

Schaeffer, S. W., C. F. Aquadro and W. W. Anderson (1987) Restriction-map variation in the alcohol dehydrogenase region of *Drosophila pseudoobscura*. Mol. Biol. Evol. 4: 254–265. (8, 10)

Schaeffer, S. W., C. F. Aquadro and C. H. Langley (1988) Restriction-map variation in the *Notch* region of *Drosophila melanogaster*. Mol. Biol. Evol. 5: 30–40. (8, 10, 11)

Schalet, A. (1969) Exchanges at the bobbed locus of *Drosophila melanogaster*. Genetics 63: 133–153. (9)

Schalet, A. and G. Lefevre (1976) The proximal region of the X chromosome. In: *The Genetics and Biology of* Drosophila, Vol. 1b, M. Ashburner and E. Novitski (eds.). Academic Press, Orlando, FL, pp. 848–928. (8)

Schimenti, J. C. and C. H. Duncan (1985) Structure and organization of the bovine β-globin genes. Mol. Biol. Evol. 2: 514–525. (13)

Schimenti, J., L. Vold, D. Socolow and L. M. Silver (1987) An unstable family of large DNA elements in the center of the mouse *t* complex. J. Mol. Biol. 194: 583–594. (9)

Schimenti, J., J. A. Cebra-Thomas, C. L. Decker, S. D. Islam, S. H. Pilder and L. M. Silver (1988) A candidate gene family for the mouse *t* complex responder (*Tcr*) locus responsible for haploid effects on sperm function. Cell 55: 71–78. (9)

Schmid, C. W. and C.-K. J. Shen (1985) The evolution of interspersed repetitive DNA sequences in mammals and other vertebrates. In: *Molecular Evolutionary Genetics*, R. J. MacIntyre (ed.). Plenum Press, New York, pp. 323–358. (8, 13)

Schmidt, G. W. and M. L. Mishkind (1986) The transport of proteins into chloroplasts. Annu. Rev. Biochem. 55: 879–912. (7)

Schnabel, A. and M. A. Asmussen (1989) Definition and properties of disequilibria within nuclear-mitochondrial-chloroplast and other nuclear-dicytoplasmic systems. Genetics 123: 199–215. (6)

Schnabel, R., M. Thomm, R. Gerardy-Schahn, W. Zillig, K. O. Stetter, and J. Huet (1983) Structural homology between different archaebacterial DNA-dependent RNA polymerases analyzed by immunological comparison of their components. EMBO J. 2: 751–755. (1)

Schneider, M. and J. D. Rochaix (1986) Sequence organization of the chloroplast ribosomal spacer of *Chlamydomonas reinhardii*: Uninterrupted tRNA[ile] and tRNA[ala] genes and extensive secondary structure. Plant Mol. Biol. 6: 265–270. (7)

Schnell, G. D. and R. K. Selander (1981) Environmental and morphological correlates of genetic variation in mammals. In: *Mammalian Population Genetics*, M. H. Smith and T. Joule (eds.). University of Georgia Press, Athens, pp. 60–99. (2)

Schötz, F. (1954) Über Plastidenkondurrenz bei *Oenothera*. Planta 43: 183–240. (7)

Schwartz, R. M. and M. O. Dayhoff (1978) Origins of prokaryotes, mitochondria and chloroplasts. Science 199: 395–403. (1)

Segall, M. (1988) HLA and genetics of IDDM. Holism vs. reductionism? Diabetes 37: 1005–1008. (12)

Selander, R. K. and J. M. Musser (1990) The population genetics of bacterial pathogenesis. In: *Molecular Basis of Bacterial Pathogenesis*, B. H. Iglewski and V. L. Clark (eds.). Academic Press, Orlando, FL, pp. 11–36. (2)

Selander, R. K. and N. H. Smith (1990) Molecular population genetics of *Salmonella*. Rev. Med. Microbiol. (in press). (2)

Selander, R. K., D. A. Caugant, H. Ochman, J. M. Musser, M. N. Gilmour and T. S. Whittam (1986) Methods of multilocus enzyme electrophoresis for bacterial population genetics and systematics. Appl. Environ. Microbiol. 51: 873–884. (2)

Selander. R. K., D. A. Caugant and T. S. Whittam (1987) Genetic structure and variation in natural populations of *Escherichia coli*. In: Escherichia coli *and* Salmonella typhimurium: *Cellular and Molecular Biology*, F. D. Neidhardt, J. L. Ingraham, K. B. Low, B. Magasanik, M. Schaechter, and H. E. Umbarger (eds.). American Society for Microbiology, Washington, D.C., pp. 1625–1648. (2)

Selander, R. K., P. Beltran, N. H. Smith, R. M. Barker, P. B. Crichton, D. C. Old, J. M. Musser and T. S. Whittam (1990a) Genetic population structure, clonal phylogeny, and pathogenicity of *Salmonella paratyphi* B. Infect. Immun. 58: 1891–1901. (2)

Selander, R. K., P. Beltran, N. H. Smith, R. Helmuth, F. A. Rubin, D. J. Kopecko, K. Ferris, B. D. Tall, A. Cravioto and J. M. Musser (1990b) Evolutionary genetic relationships of *Salmonella* serovars that cause human typhoid and other enteric fevers. Infect. Immun. 58: 2262–2275. (2)

Serjeantson, S. W. (1989) The reasons for MHC polymorphism in man. Transplant. Proc. 21: 598–601. (12)

Sette, A., S. Buus, S. Colon, C. Miles and H. M. Grey (1988) I-A^d binding peptides derived from unrelated protein antigens share a common structural motif. J. Immunol. 141: 45–48. (11)

Sette, A., S. Buus, E. Appella, J. A. Smith, R. Chesnut, C. Miles, S. M. Colon and H. M. Grey (1989) Prediction of major histocompatibility complex binding regions of protein antigens by sequence pattern analysis. Proc. Natl. Acad. Sci. USA 86: 3296–3300. (11)

Shapiro, J. A. (ed.) (1983) *Mobile Genetic Elements*. Academic Press, Orlando, FL (8)

Shapiro, S. G., E. A. Schon, T. M. Townes and J. B. Lingrel (1983) Sequence and linkage of the goat ε-I and ε-II β-globin genes. J. Mol. Biol. 169: 31–52. (13)

Sharp, P. and W.-H. Li (1986) An evolutionary perspective on synonymous codon usage in unicellular organisms. J. Mol. Evol. 24: 28–38. (10)

Sharp, P. M. and W.-H. Li (1988) Understanding the origins of AIDS viruses. Nature 336: 315. (5)

Sharp, P. M. and W.-H. Li (1989) On the rate of DNA sequence evolution in *Drosophila*. J. Mol. Evol. 28: 398–402. (6)

Sheehy, M. J., S. J. Scharf, J. R. Rowe, M. H. Neme de Gimenez, L. M. Meske, H. A. Ehrlich and B. S. Nepom. (1989) A diabetes-susceptible HLA haplotype is best defined by a combination of HLA-DR and -DQ alleles. J. Clin. Invest. 83: 830–835. (12)

Shehee, W. R., D. D. Loeb, N. B. Adey, F. H. Burton, N. C. Casavant, P. Cole, C. J. Davies, R. A. McGraw, S. A. Schichman, D. M. Severynse, C. F. Voliva, F. W. Weyter, G. B. Wisely, M. H. Edgell and C. A. Hutchison, III (1989) Nucleotide sequence of the BALB/c mouse β-globin complex. J. Mol. Biol. 205: 41–62. (13)

Shen, S.-H., J. L. Slightom and O. Smithies (1981) A history of the human fetal globin gene duplication. Cell 26: 191–203. (13)

Shih, M.-C., P. Heinrich and H. M. Goodman (1988) Intron existence predated the divergence of eukaryotes and prokaryotes. Science 242: 1164–1166. (7)

Shin, H.-S., J. Stavnezer, K. Artzt and D. Bennett (1982) Genetic structure and origin of *t*-haplotypes of mice, analyzed with H-2 cDNA probes. Cell 29: 969–976 (9)

Shinomiya, T. and S. Ina (1989) Genetic comparison of bacteriophage PS17 and *Pseudomonas aeruginosa* R-type pyocin. J. Bacteriol. 171: 2287–2292. (4)

Shinomiya, T., S. Shiga and M. Kageyama (1983) Genetic determinant of pyocin R2 in *Pseudomonas aeruginosa* PAO. I. Localization of the pyocin R2 gene cluster between the *trpCD* and *trpE* genes. Mol. Gen. Genet. 189: 375–381. (4)

Shinozaki, K., H. Deno, M. Sugita, S. Kuramitsu and M. Sugiura (1986a) Intron in the gene for the ribosomal protein S16 of tobacco chloroplast and its conserved boundary sequences. Mol. Gen. Genet. 202: 1–5. (7)

Shinozaki, K., M. Ohme, M. Tanaka, T. Wakasugi, N. Hayashida, T. Matsubayasha, N. Zaita, J. Chunwongse, J. Obokata, K. Yamaguchi-Shinozaki, C. Ohto, K. Torazawa, B. Y. Meng, M. Sugita, H. Deno, T. Kamogashira, K. Yamada, J. Kusuda, F. Takaiwa, Λ. Kata, N. Tohdoh, H. Shimada and M. Sugiura (1986b) The complete nucleotide sequence of the tobacco chloroplast genome. Plant Mol. Biol. Rep. 4: 110–147. (7)

Shyman, S. and S. Weaver (1985) Chromosomal rearrangements associated with LINE elements in the mouse genome. Nucleic Acids Res. 13: 5085–5093. (13)

Signer, E. R., A. Torriani and C. Levinthal (1961) Gene expression in intergeneric merozygotes. Cold Spring Harbor Symp. Quant. Biol. 26: 31–34. (3)

Silver, L. M. (1985) Mouse *t* haplotypes. Annu. Rev. Genet. 19: 179–208. (9, 12)

Silver, L. M., M. Hammer, H. Fox, J. Garrels, M. Bucan, B. Herrmann, A-M. Frischauf, H. Lehrach, H. Winking, F. Figueroa and J. Klein. (1987) Molecular evidence for the rapid propagation of mouse *t* haplotypes from a single, recent, ancestral chromosome. Mol. Biol. Evol. 4: 473–482. (9)

Silverman, M. and M. Simon (1980) Phase variation genetic analysis of switching mutants. Cell 19: 845–854. (2)

Simmons, M. J. and J. F. Crow (1977) Mutations affecting fitness in *Drosophila* populations. Annu. Rev. Genet. 11: 49–78. (8)

Simon, M., M. J. Zieg, M. Silverman, G. Mandel and R. Doolittle (1980) Phase variation: Evolution of a controlling element. Science 209: 1370–1374. (2)

Simon, R., V. Priefer and A. Puhler (1983) A broad host range mobilization system for *in vivo* genetic engineering: Transposon mutagenesis in gram negative bacteria. Bio/Technology 1: 784–791. (4)

Simonson, M., M. Crone, C. Koch and K. Hála (1982) The MHC haplotypes of the chicken. Immunogenetics 16: 513–532. (11)

Singer, M. F. (1982) SINEs and LINEs: Highly repeated short and long interspersed sequences in mammalian genomes. Cell 28: 433–434. (13)

Singer, M. F. and J. Skowronski (1985) Making sense out of LINEs: Long interspersed repeat sequences in mammalian genomes. Trends Biochem. Sci. 10: 119–122. (13)

Singer, D. S., R. Ehrlich, L. Satz, W. Frels, J. Bluestone, R. Hodes and S. Rudikoff (1987). Structure and expression of class I MHC genes in the miniature swine. Vet. Immunol. Immunopathol. 17: 211–221. (11)

Singh, P. B., R. E. Brown and B. Roser (1987) MHC antigens in urine as olfactory recognition cues. Nature 327: 161–164. (11)

Singh, R. S. and L. R. Rhomberg (1987) A comprehensive study of genic variation in natural populations of *Drosophila melanogaster*. I. Estimates of gene flow from rare alleles. Genetics 115: 313–322. (8)

Skibinski, D. O. F. and R. D. Ward (1982) Correlations between heterozygosity and evolutionary rate of proteins. Nature 298: 490–492. (10)

Slatkin, M. (1989) Detecting small amounts of gene flow from phylogenies of alleles. Genetics 121: 609–612. (6, 10)

Slatkin, M. and W. P. Maddison (1989) A cladistic measure of gene flow inferred from the phylogenies of alleles. Genetics 123: 603–613. (10)

Slightom, J. L., A. E. Blechl and O. Smithies (1980) Human fetal $^G\gamma$- and $^A\gamma$-globin genes: Complete nucleotide sequences suggest that DNA can be exchanged between these duplicated genes. Cell 21: 627–638. (13)

Smith, G. R (1985) Site-specific recombination. In: *Genetics of bacteria*, J. Scaife, D. Leach, and A. Galizzi (eds.). Academic Press, London, pp. 147–163. (2)

Smith, J. F. and J. J. Doyle (1986) Chloroplast DNA variation and evolution in the Juglandaceae. Am. J. Bot. 73: 730. (7)

Smith, N. H. and R. K. Selander (1989) Sequence invariance of the antigen-coding central region of the phase 1 flagellar filament gene (*fliC*) among strains of *Salmonella typhimurium*. J. Bacteriol. 172: 603–609. (2)

Smith, N. H., P. Beltran and R. K. Selander (1990) Recombination of *Salmonella* phase 1 flagellin genes generates new serovars. J. Bacteriol. 172: 2209–2216. (2)

Smith, S. E. (1989) Influence of paternal genotype on plastid inheritance in *Medicago sativa*. J. Hered. 80: 214–217. (6)

Smith, T. F., A. Srinivasan, G., Schochetman, M. Marcus and G. Myers (1988) The phylogenetic history of immunodeficiency viruses. Nature 333: 573–575. (5)

Snell, G. D. (1968) The H-2 locus of the mouse: Observations and speculations concerning its comparative genetics and its polymorphism. Folia Biol. 14: 335–358. (11)

Snellings, N. J., E. M. Johnson and L. S. Baron (1977) Genetic basis of Vi antigen expression in *Salmonella paratyphi* C. J. Bacteriol. 13: 57–62. (2)

Snellings, N. J., E. M. Johnson, D. J. Kopecko, H. H. Collins and L. S. Baron (1981) Genetic regulation of variable Vi antigen expression in a strain of *Citrobacter freundii*. J. Bacteriol. 145: 1010–1017. (2)

Soares, M. B., E. Schon, A. Henderson, S. K. Karathanasis, R. Cate, S. Zeitlin, J. Chirgwin and A. Efstratiadis (1985) RNA-mediated gene duplication: The rat preproinsulin I gene is a functional retroposon. Mol. Cell. Biol. 5: 2090–2103. (13)

Solignac, M., J. Génermont, M. Monnerot and J.-C. Mounolou (1987) *Drosophila* mitochondrial genetics: Evolution of heteroplasmy through germ line cell divisions. Genetics 117: 687–696. (6)

Soliman, K., G. Fedak and R. W. Allard (1987) Inheritance of organelle DNA in barley and *Hordeum* × *Secale* intergenic hybrids. Genome 29: 867–872. (7)

Soltis, D. E., P. S. Soltis and B. D. Ness (1989a) Chloroplast-DNA variation and multiple origins of autopolyploidy in *Heuchera micrantha* (Saxifragaceae). Evolution 43: 650–656. (7)

Soltis, D. E., P. S. Soltis, T. A. Ranker and B. D. Ness (1989b) Chloroplast DNA variation in a wild plant, *Tolmiea menziesii*. Genetics 121: 819–826. (7)

Sonea, S. and M. Panisset (1983) *A New Bacteriology*. Jones and Bartlett, Boston, MA. (2)

Sonigo, P., M. Alizon, K. Staskus, D. Klatzmann, S. Cole, O. Danos, E. Retzel, P. Tiollais, A. Haase and S. Wain-Hobson (1985) Nucleotide sequence of the visna lentivirus: Relationship to the AIDS virus. Cell 42: 369–382. (5)

Sowadski, J. M., M. D. Handschumacher, H. M. Krishna Murthy, B. A. Foster and H. W. Wyckoff (1985) Refined structure of alkaline phosphatase from *Escherichia coli* at 2.8 Å resolution. J. Mol. Biol. 186: 417–433. (3)

Spika, J. S. et al. (1987) Chloramphenicol-resistant *Salmonella newport* traced through hamburger to dairy farms. N. Engl. J. Med. 316: 565–570. (2)

Spofford, J. B. (1976) Position effect variegation in *Drosophila*. In: *Genetics and Biology of Drosophila*, Vol. 1c, M. Ashburner and M. Novitski (eds.). Academic Press, New York, pp. 955–1018. (9)

Spradling, A. C. and G. M. Rubin (1981) *Drosophila* genome organization: Conserved and dynamic aspects. Annu. Rev. Genet. 15: 219–264. (8)

Sprinzl, M., T. Hartmann, J. Weber, J. Blank and R. Zeidler (1989) Compilation of tRNA sequences and sequences of tRNA genes. Nucleic Acids Res. 17: r1–r172. (7)

Stackebrandt, E., H. Pohla, R. Kroppenstedt, H. Hippe, and C. R. Woese (1985) 16S rRNA analysis of *Sporomusa*, *Selenomonas*, and *Megasphaera*: on the phylogenetic origin of Gram-positive eubacteria. Arch. Microbiol. 143: 270–276. (1)

Stalker, H. D. (1961) The genetic systems modifying meiotic drive in *Drosophila paramelanica*. Genetics 46: 177–202. (9)

Stanier, R. Y. (1970) Some aspects of the biology of cells and the possible evolutionary significance. In: *Organization and Control in Prokaryotic and Eukaryotic Cells*, H. P. Charles and B. C. J. G. Knight (eds.). Soc. Gen. Microbiol. Symp. 20, pp. 1–38. (1)

Stanier, R. Y. and C. B. van Niel (1941) The main outlines of bacterial classification. J. Bacteriol. 42: 437–466. (1)

Stanier, R. Y. and C. B. van Niel (1962) The concept of a bacterium. Arch. Mikrobiol. 42: 17–35. (1)

Stanier, R. Y., M. Doudoroff, and E. A. Adelberg (1970) *The Microbial World*, 3rd ed. Prentice-Hall, Englewood Cliffs, NJ, p. 529. (1)

Steinemann, M. (1982) Multiple sex chromosomes in *Drosophila miranda*: A system to study the degeneration of a chromosome. Chromosoma 86: 59–76. (8)

Steinmetz, M., K. Minard, S. Horvath, J. McNicholas, J. Srelinger, C. Wake, E. Long, B. Mach

and L. Hood (1982) A molecular map of the immune response region from the major histocompatibility complex of the mouse. Nature 300: 35–42. (12)

Stephan, W. (1986) Recombination and the evolution of satellite DNA. Genet. Res. 47: 167–174.
(8)

Stephan, W. (1989) Molecular genetic variation in the centromeric region of the X chromosome
in three *Drosophila ananassae* populations. II. The *Om(1D)* locus. Mol. Biol. Evol. 6: 624–
635. (8)

Stephan, W. and C. H. Langley (1989) Molecular genetic variation in the centromeric region of
the X chromosome in three *Drosophila ananassae* populations. I. Contrasts between the *vermilion* and *forked* loci. Genetics 121: 89–99. (8)

Stephens, J. C. (1985) Statistical methods of DNA sequence analysis: Detection of intragenic
recombination or gene conversion. Mol. Biol. Evol. 2: 539–556. (3, 12)

Stephens, J. C. and M. Nei (1985) Phylogenetic analysis of polymorphic DNA sequences at the
Adh locus in *Drosophila melanogaster* and its sibling species. J. Mol. Evol. 22: 289–300. (10,
11)

Stewart, W. N. (1983) *Paleobotany and the Evolution of Plants*. Cambridge Univ. Press, Cambridge. (7)

Strauss, S. H., J. D. Palmer, G. T. Howe and A. H. Doerksen (1988) Chloroplast genomes of two
conifers lack a large inverted repeat and are extensively rearranged. Proc. Natl. Acad. Sci. USA
85: 3898–3902 (7)

Streilein, I. W. (1987) Studies on an MHC exhibiting limited polymorphism. In: *Evolution and
Vertebrate Immunity*, G. Kelsoe and D. H. Schulze (eds.). Univ. of Texas Press, Austin, pp.
379–395. (11)

Strobeck, C. (1983) Expected linkage disequilibrium for a neutral locus linked to a chromosomal
arrangement. Genetics 103: 545–555. (10)

Strobel, E., P. Dunsmuir and G. M. Rubin (1979) Polymorphisms in the chromosomal locations
of elements of the 412, copia and 297 dispersed repeated gene families in *Drosophila*. Cell
17: 429–439. (8)

Stocker, B. A. D. (1949) Measurement of rate of mutation of flagellate antigenic phase in *Salmonella
typhimurium*. J. Hyg. 47: 398–412. (2)

Strominger, J. L. (1989) The γδ T cell receptor and class Ib MHC-related proteins: Enigmatic
molecules of immune recognition. Cell 57: 895–898. (11)

Stull, T. L., J. J. LiPuma and T. D. Edlind (1988) A broad-spectrum probe for molecular
epidemiology of bacteria: ribosomal RNA. J. Infect. Dis. 157: 280–286. (2)

Südhof, T. C., J. L. Goldstein, M. S. Brown and D. W. Russell (1985a) The LDL receptor gene:
A mosaic of exons shared with different proteins. Science 228: 815–822. (3)

Südhof, T. C., D. W. Russell, J. L. Goldstein, M. S. Brown, R. Sanchez-Pescador and G. I. Bell
(1985b) Cassette of eight exons shared by genes for LDL receptor and EGF precursor. Science
228: 893–895. (3)

Sved, J. A. (1968) The stability of linked systems of loci with small population size. Genetics 59:
543–563. (12)

Sytsma, K. J. and L. D. Gottlieb (1986) Chloroplast DNA evolution and phylogenetic relationships
in *Clarkia* sect. *Peripetasma* (Onagraceae). Evolution 40: 1248–1261. (7)

Sytsma, K. J. and B. A. Schaal (1985) Phylogenetics of the *Lisianthius skinneri* (Gentianaceae)
species complex in Panama utilizing DNA restriction fragment analysis. Evolution 39: 594–
608. (7)

Syvanen, M. (1984) The evolutionary implications of mobile genetic elements. Annu. Rev. Genet.
18: 271–293. (8)

Szmidt, A. E., T. Alden and J.-E. Hallgren (1987) Paternal inheritance of chloroplast DNA in
Larix. Plant Mol. Biol. 9: 59–64. (7)

Szmidt, A. E., Y. A. El-Kassaby, A. Sigurgeirsson, T. Alden, D. Lindgren and J.-E. Hallgren
(1988) Classifying seedlots of *Picea sitchensis* and *P. glauca* in zones of introgression using
restriction analysis of chloroplast DNA. Theor. Appl. Genet. 76: 841–845. (7)

Tabata, T., S. Kimiko and M. Iwabuchi (1983) The structural organization and DNA sequence of
wheat H4. Nucleic Acid Res. 11: 5865–5875. (10)

Tacket, C. O., L. B. Dominguez, H. J. Fisher and M. L. Cohen (1985) An outbreak of multiple-
drug-resistant *Salmonella enteritis* from raw milk. J. Am. Med. Assoc. 253: 2058–2060. (2)

Tajima, F. (1983) Evolutionary relationship of DNA sequences in finite populations. Genetics 105: 437–460. (10)

Tajima, F. (1989a) Statistical method for testing the neutral mutation hypothesis by DNA polymorphism. Genetics 123: 585–595. (10)

Tajima, F. (1989b) The effect of change in population size on DNA polymorphism. Genetics 123: 597–601. (10)

Takahata, N. (1981) A mathematical study on the distribution of the number of repeated genes per chromosome. Genet. Res. 38: 97–102. (9)

Takahata, N. (1983a) Linkage disequilibrium of extranuclear genes under neutral mutations and random genetic drift. Theor. Pop. Biol. 24: 1–21. (6)

Takahata, N. (1983b) Population genetics of extranuclear genomes under the neutral mutation hypothesis. Genet. Res. 42: 235–255. (6)

Takahata, N. (1984) A model of extranuclear genomes and the substitution rate under within-generation selection. Genet. Res. 44: 109–116. (6)

Takahata, N. (1985) Introgression of extranuclear genomes in finite populations: Nucleo-cytoplasmic incompatibility. Genet. Res. 45: 179–194. (6)

Takahata, N. (1988) The n coalescent in two partially isolated diffusion populations. Genet. Res. 52: 213–222. (10)

Takahata, N. and T. Maruyama (1981) A mathematical model of extranuclear genes and the genetic variability maintained in a finite population. Genet. Res. 37: 291–302. (6)

Takahata, N. and M. Nei (1990) Allelic geneology under overdominant and frequency-dependent selection and polymorphism of major histocompatibility complex loci. Genetics 124: 967–978. (11)

Takahata, N. and S. R. Palumbi (1985) Extranuclear differentiation and gene flow in the finite island model. Genetics 109: 441–457. (6)

Takahata, N. and M. Slatkin (1983) Evolutionary dynamics of extranuclear genes. Genet. Res. 42: 257–265. (6)

Takahata, N. and M. Slatkin (1984) Mitochondrial gene flow. Proc. Natl. Acad. Sci. USA 81: 1764–1767. (6)

Takaiwa, F. and M. Sugiura (1982) Nucleotide sequence of the 16S–23S spacer region in an rRNA gene cluster from tobacco chloroplast DNA. Nucleic Acids Res. 10: 2665–2676. (7)

Talbott, R. L., E. E. Sparger, K. M. Lovelace, W. M. Fitch, N. C. Pedersen, P. A. Luciw and J. H. Elder (1989) Nucleotide sequence and genomic organization of feline immunodeficiency virus. Proc. Natl. Acad. Sci. USA 86: 5743–5747. (5)

Tartof, K. D. (1974) Unequal mitotic sister chromatid exchange as the mechanism of ribosomal RNA gene magnification. Proc. Natl. Acad. Sci. USA 71: 1272–1276. (9)

Tavaré, S. (1984) Line-of-descent and genealogical processes, and their applications in populations genetic models. Theor. Pop. Biol. 26: 119–164. (10)

Taylor, D. E., I. K. Wachsmuth, Y.-H. Shangkuan, E. V. Schmidt, T. J. Barrett, J. S. Schrader, C. S. Scherach, H. B. McGee, A. Feldman and D. J. Brenner (1982) Salmonellosis associated with marijuana: A multi-state outbreak traced by plasmid fingerprinting. N. Engl. J. Med. 306: 1249–1253. (2)

Teich, N. (1982) Taxonomy of retroviruses. In: *RNA Tumor Viruses*, R. Weiss, N. Teich, H. Varmus and J. Coffin (eds.). Cold Spring Harbor Laboratory, Cold Spring Harbor, NY, pp. 25–207. (5)

Temin, R. G., B. Ganetzky, P. A. Powers, T. W. Lyttle, S. Pimpinelli, C.-I. Wu and Y. Hiraizumi (1990) Segregation Distorter (SD) in *Drosophila melanogaster*: Genetic and molecular analyses. Am. Nat. (in press). (9)

Temin, R. G. and M. Marthas (1984) Factors influencing the effect of segregation distortion in natural populations of *Drosophila melanogaster*. Genetics 107: 375–393. (9)

Terasaki, P. I. (ed.) (1980) *Histocompatibility Testing 1980.* University of California, Los Angeles. (12)

Tewarson, S., Z. Zaleska-Rutczynska, F. Figueroa and J. Klein (1983) Polymorphism of Qa and Tla loci of the mouse. Tissue antigens 22: 204–212. (11)

Thomas, M. L., J. H. Harger, D. K. Wagener, B. S. Rabin and T. J. Gill, III. (1985) HLA sharing and spontaneous abortion in humans. Am. J. Obstet. Gynecol. 151: 1053–1058. (12)

Thomson, G. (1988) HLA disease associations: Models for insulin dependent diabetes mellitus and the study of complex human genetic disorders. Annu. Rev. Genet. 22: 31–50. (12)

Thomson, G. J. and M. W. Feldman (1976) Population genetics of modifiers of meiotic drive III.

Equilibrium analysis of a general model for the genetic control of segregation distortion. Theor. Pop. Biol. 10: 10–25. (9)

Thomson, G. and W. Klitz (1987) Disequilibrium pattern analysis. I. Theory. Genetics 116: 623–632. (12)

Thomson, G., W. F. Bodmer and J. Bodmer (1976) The HL-A system as a model for studying the interaction between selection, migration, and linkage. In: *Population Genetics and Ecology*, S. Karlin and E. Nevo, (eds.). Academic Press, New York, pp. 465–498. (12)

Thomson, G., W. P. Robinson, M. K. Kuhner, S. Joe, M. J. MacDonald, J. L. Gottschall, J. Barbosa, S. S. Rich, J. Bertrams, M. P. Baur, J. Partanen, B. A. Tait, E. Schober, W. R. Mayr, J. Ludvigsson, B. Lindblom, N. R. Farid, C. Thompson and I. Deschamps (1988) Genetic heterogeneity, modes of inheritance and risk estimates for a joint study of Caucasians with insulin-dependent diabetes mellitus. Am. J. Hum. Genet. 43: 799–816. (12)

Threlfall, E. J. and J. A. Frost (1990) The identification, typing and fingerprinting of *Salmonella*: Laboratory aspects and epidemiological applications. J. Appl. Bacteriol. 68: 5–16. (2)

Tibayrenc, M., F. Kjellberg and F. J. Ayala (1990) A clonal theory of parasitic protozoa: The population structures of *Entamoeba, Giardia, Trypanosoma* and their medical and taxonomical consequences. Proc. Natl. Acad. Sci. USA 87: 2414–2418. (2)

Tilghman, S., D. Tiemeier, J. Seidman, M. Peterlin, M. Sullivan, J. Maizel and P. Leder (1978a) Intervening sequence of DNA identified in the structural portion of a mouse β-globin gene. Proc. Natl. Acad. Sci. USA 75: 725–729. (13)

Tilghman, S., P. Curtis, D. Tiemeier, P. Leder and C. Weissmann (1978b) The intervening sequence of a mouse β-globin gene is transcribed within the 15S β-globin mRNA precursor. Proc. Natl. Acad. Sci. USA 75: 1309–1313. (13)

Timoney, J. F., J. H. Gillespie, F. W. Scott and J. E. Barlough (1988) *Hagan and Bruner's Microbiology and Infectious Diseases of Domestic Animals*. Comstock, Ithaca, NY. (2)

Timothy, D. H., C. S. Levings III, D. R. Pring, M. F. Conde and J. L. Kermicle (1979) Organelle DNA variation and systematic relationships in the genus *Zea: Teosinte*. Proc. Natl. Acad. Sci. USA 76: 4220–4224. (7)

Tiwari, J. L. and P. I. Terasaki (1985) *HLA and Disease Associations*. Springer-Verlag, New York. (11, 12)

Todd, J. A., J. I. Bell and H. O. McDevitt (1987) HLA-DQβ gene contributes to susceptibility and resistance to insulin-dependent diabetes mellitus. Nature 329: 599–604. (12)

Tokuyasu, K. T., W. J. Peacock and R. W. Hardy (1972) Dynamics of spermiogenesis in *Drosophila melanogaster*. I. Individualization process. Z. Zellforsch. 124: 479–506. (9)

Torriani-Gorini, A., F. G. Rothman, S. Silver, A. Wright and E. Yagil (eds.) (1987) *Phosphate Metabolism and Cellular Regulation in Microorganisms*. American Society for Microbiology, Washington, D.C. (3)

Townes, T. M., M. C. Fitzgerald and J. B. Lingrel (1984) Triplication of a four-gene set during evolution of the goat β-globin gene locus produced three genes now expressed differentially during development. Proc. Natl. Acad. Sci. USA 81: 6589–6593. (13)

Trabuchet, G., Y. Chelbourne, P. Savatier, J. Lachuer, C. Faure, G. Verdier and V. M. Nigon (1987) Recent insertion of an Alu sequence in the β-globin gene cluster of the gorilla. J. Mol. Evolution 25: 288–291. (13)

Treco, D., B. Thomas and N. Arnheim (1985) Recombination hot spot in the human β-globin gene cluster: Meiotic recombination of human DNA fragments in *Saccharomyces cerevisiae*. Mol. Cell Biol. 5: 2029–2038. (13)

Trippa, G. and A. Loverre (1975) A factor on a wild third chromosome (III^{ra}) that modifies the segregation distortion phenomenon in *Drosophila melanogaster*. Genet. Res. 26: 113–125. (9)

Trowsdale, J. (1987) Genetics and polymorphism: Class II antigens. Br. Med. Bull. 43(1): 15–36. (12)

Trowsdale, J., V. Groves and A. Arnason (1989) Limited MHC polymorphism in whales. Immunogenetics 29: 19–24. (11)

Tsunewaki, K. and Y. Ogihara (1983) The molecular basis of genetic diversity among cytoplasms of *Triticum* and *Aegilops* species. II. On the origin of polyploid wheat cytoplasms as suggested by chloroplast DNA restriction fragment patterns. Genetics 104: 155–171. (7)

Tuan, D., W. Soloman, Q. Li and I. London (1985) The "β-like-globin" gene domain in human erythroid cells. Proc. Natl. Acad. Sci. USA 82: 6384–6388. (13)

Tykocinski, M. L., P. N. Marche, E. E. Max and T. J. Kindt (1984) Rabbit class I MHC genes:

cDNA clones define full-length transcripts of an expressed gene and a putative pseudogene. J. Immunol. 133: 2261–2269. (11)

Umesono, K., H. Inokuchi, Y. Shiki, M. Takeuchi, Z. Chang, H. Fukuzawa, T. Kohchi, H. Shirai, K. Ohyama and H. Ozeki (1988) Structure and organization of *Marchantia polymorpha* chloroplast genome. II. Gene organization of the large single copy region from *rps*′12 to *atp*B. J. Mol. Biol. 203: 299–331. (7)

Uyenoyama, M. K. (1985) Quantitative models of hybrid dysgenesis: Rapid evolution under transposition, extrachromosomal inheritance, and fertility selection. Theor. Pop. Biol. 27: 176–201. (8)

Van Delden. W. (1982) The alcohol dehydrogenase polymorphism in *Drosophila melanogaster*. Selection at an enzyme locus. Evol. Biol. 15: 187–222. (10)

Van Delden, W. and A. Kamping (1989) The association between the polymorphisms at the *Adh* and α-*Gpdh* Loci and the In(2L)t inversion in *Drosophila melanogaster* in relation to temperature. Evolution 43(4): 775–793. (10)

Van Eden, W., R. R. P. Devries and J. J. Van Rood (1983) The genetic approach to infectious disease with special emphasis on the MHC. Dis. Markers 1: 221–242. (12)

Vanlerberghe, F., B. Dod, P. Boursot, M. Bellis and F. Bonhomme (1986) Absence of Y-chromosome introgression across the hybrid zone between *Mus musculus domesticus* and *Mus musculus musculus*. Genet. Res. 48: 191–197. (9)

Vanlerberghe, F., P. Boursot, J. T. Nielsen and F. Bonhomme (1988) A steep cline for mitochondrial DNA in Danish mice. Genet. Res. (6)

van Niel, C. B. (1946) The classification and natural relationships of bacteria. Cold Spring Harbor Symp. Quant. Biol. 11: 285–301. (1)

van Niel, C. B. (1955) Classification and taxonomy of the bacteria and bluegreen algae. *A Century of Progress in the Natural Sciences 1853–1953*. California Academy of Sciences, San Francisco, pp. 89–114. (1)

Varmus, H. (1988) Retroviruses. Science 240: 1427–1435. (5)

Vaughn, K. C. (1981) Plastid fusion as an agent to arrest sorting out. Curr. Genet. 3: 243–245. (6)

Verma, N. and P. Reeves (1989) Identification and sequence of *rfb*S and *rfb*E, which determine antigenic specificity of group A and group D salmonellae. J. Bacteriol. 171: 5694–5701. (2)

Von Heijne, G. (1985) Signal sequences: The limits of variation. J. Mol. Biol. 184: 99–105. (3)

Wachsmuth, I. K. (1986) Molecular epidemiology of bacterial infections: Examples of methodology and investigations of outbreaks. Rev. Infect. Dis. 8: 682–692. (2)

Wagner, D. B., G. R. Furnier, M. A. Saghai-Maroof, S. M. Williams, B. P. Dancik and R. W. Allard (1987) Chloroplast DNA polymorphisms in lodgepole and jack pines and their hybrids. Proc. Natl. Acad. Sci. USA 84: 2097–2100. (7)

Wagner, W. H., Jr. (1954) Reticulate evolution in the Appalachian Aspleniums. Evolution 8: 103–118. (4)

Wain-Hobson, S., P. Sonigo, O. Danos, S. Cole and M. Alizon (1985) Nucleotide sequence of the AIDS virus, LAV. Cell 40: 9–17. (5)

Wakabashi, S., H. Matsubara, and O. Webster (1986) Primary sequence of a dimeric bacterial haemoglobin from *Vitreoscillo*. Nature 322: 481–483. (13)

Walsh, J. B. (1983) Role of biased gene conversion in one-locus neutral theory and genome evolution. Genetics 105: 461–468. (6)

Walsh, J. B. (1987) Persistence of tandem arrays: Implications for satellite and simple-sequence DNAs. Genetics 115: 553–567. (8)

Wanntorp, H.-E. (1983) Reticulated cladograms and the identification of hybrid taxa. In: *Advances in Cladistics*, Vol. II, N. I. Platnick and V. A. Funk (eds.), pp. 81–88. (4)

Ward, R. D. and D. O. F. Skibinski (1985) Observed relationships between protein heterozygosity and protein genetic distance and comparisons with neutral expectations. Genet. Res. 45: 315–340. (10)

Waring, R. B. and R. W. Davies (1984) Assessment of a model for intron RNA secondary structure relevant to RNA self-splicing—a review. Gene 28: 277–291. (7)

Warner, C., B. Gerndt, Y. Xu, Y. Bourlet, C. Auffray, S. Lamont and A. Nordskog (1989)

Restriction fragment length polymorphism analysis of major histocompatibility class II genes from inbred chicken lines. Anim. Genet. 20: 225–231. (11)

Watkins, D. I., F. S. Hodi and N. L. Letvin (1988) A primate species with limited major histocompatibility complex class I polymorphism. Proc. Natl. Acad. Sci. USA 85: 7714–7718. (11)

Watson, G. S. and E. Caspari (1960) The behavior of cytoplasmic pollen sterility in populations. Evolution 14: 56–63. (6)

Watterson, G. A. (1975) On the number of segregating sites in genetical models without recombination. Theor. Pop. Biol. 7: 256–276. (10)

Watterson, G. A. (1977) Heterosis or neutrality? Genetics 85: 789–814. (11)

Watterson, G. A. (1978a) An analysis of multi-allelic data. Genetics 88: 171–179. (12)

Watterson, G. A. (1978b) The homozygosity test of neutrality. Genetics 88: 405–417. (12)

Watterson, G. A. (1989) Allele frequencies in multigene families. I. Diffusion equation approach. Theor. Pop. Biol. 35: 142–160. (12)

Watts, S., M. Kuhnel, W. Klitz and R. S. Goodenow (1989) Gene conversion in the evolution of the class I multigene family of the murine major histocompatibility complex. Genetics (Submitted). (12)

Wegmann, T. G. (1984) Foetal protection against abortion: Is it immunosuppression or immunostimulation? Ann. Immunol. 135B: 307–312. (11)

Wei, L.-N. and T. M. Joys (1985) Covalent structure of three phase-1 flagellar filament proteins of *Salmonella*. J. Mol. Biol. 186: 791–803. (2)

Weiner, A. M., P. L. Deininger and A. Efstratiadis (1986) Nonviral retroposons: Genes, pseudogenes, and transposable elements generated by the reverse flow of genetic information. Annu. Rev. Biochem. 55: 631–661. (13)

Weisburg, W. G., Y. Oyaizu, H. Oyaizu, and C. R. Woese (1985a) Natural relationship between bacteroides and flavobacteria. J. Bacteriol. 164: 230–236. (1)

Weisburg, W. G., C. R. Woese, M. E. Dobson, and E. Weiss (1985b) A common origin of rickettsiae and certain plant pathogens. Science 230: 556–558. (1)

Weisburg, W. G., T. P. Hatch, and C. R. Woese (1986) Eubacterial origin of chlamydiae. J. Bacteriol. 167: 570–574. (1)

Weisburg, W. G., M. E. Dobson, J. E. Samuel, G. A. Dasch, L. P. Mallavia, O. Baca, L. Mandelco, J. E. Sechrest, E. Weiss, C. R. Woese (1989a) Phylogenetic diversity of the Rickettsiae. J. Bacteriol. 171: 4202–4206. (1)

Weisburg, W. G., J. G. Tully, D. L. Rose, J. P. Petzel, H. Oyaizu, D. Yang, L. Mandelco, J. Sechrest, T. G. Lawrence, J. Van Etten, J. Maniloff, and C. R. Woese (1989b) Phylogenetic analysis of the mycoplasmas: Basis for their classification. J. Bacteriol. 171: 6455–6467. (1)

Weiss, E. H., A. Mellor, L. Golden, K. Fahrner, E. Simpson, J. Hurst and R. A. Flavell (1983) The structure of a mutant *H-2* gene suggests that the generation of polymorphism in H-2 genes may occur by gene conversion-like events. Nature 301: 671–674. (11)

Werren, J. H., U. Nur and D. Eickbush (1987) An extrachromosomal factor causing loss of paternal chromosomes. Nature 327: 75–76. (9)

Werren, J. H., U. Nur and C.-I. Wu (1988) Selfish genetic elements. Trends Ecol. Evol. 3: 297–302. (9)

White, P. B. (1926) Further studies of the *Salmonella* group. Br. Med. Res. Council Special Rep. Ser. 103: 1–160. (2)

Whittam, T. S., H. Ochman and R. K. Selander (1983) Multilocus genetic structure in natural populations of *Escherichia coli*. Proc. Natl. Acad. Sci. USA 80: 1751–1755. (2)

Willison, K., A. Kelly, K. Dudley, P. Goodfellow, N. Spurr, V. Groves, P. Gorman, D. Shoer, and J. Trowsdale (1987) The human homologue of the mouse *t*-complex gene, TCP1, is located on chromosome 6 but is not near the HCA region. EMBO J. 6(7): 67–74. (12)

Willison, K. R., K. Dudley and J. Potter (1986) Molecular cloning and sequence analysis of a haploid expressed gene encoding *t* complex polypeptide 1. Cell 44: 727–738. (9)

Wilson, A. C., R. L. Cann, S. M. Carr, M. George, U. B. Gyllensten, K. M. Helm-Bychowski, R. G. Higuchi, S. R. Palumbi, E. M. Prager, R. D. Sage and M. Stoneking (1985) Mitochondrial DNA and two perspectives on evolutionary genetics. Biol. J. Linnean Soc. 26: 375–400. (6)

Wilson, A. C., H. Ochman and E. M. Prager (1987) Molecular time scale for evolution. Trends Genet. 3: 241–247. (3)

Wilson, M. A., B. Gaut and M. T. Clegg (1990) Chloroplast DNA evolves slowly in the palm family (Arecaceae). Mol. Biol. Evol. 7: 303–314. (7)

Winkler, C., A. Schultz, S. Cevario and S. J. O'Brien (1989) Genetic characterization of FLA, the cat major histocompatibility complex. Proc. Natl. Acad. Sci. USA 86: 943–947. (11)

Woese, C. R. (1987) Bacterial evolution. Microbiol. Rev. 51: 221–271. (1, 3, 4)

Woese, C. R. and G. E. Fox (1977). Phylogenetic structure of the prokaryotic domain: The primary kingdoms. Proc. Natl. Acad. Sci. USA 74: 5088–5090. (1)

Woese, C. R. and G. J. Olsen (1986) Archaebacterial phylogeny: perspectives on the urkingdoms. Syst. Appl. Microbiol. 7: 161–177. (1)

Woese, C. R., J. Gibson and G. E. Fox (1980) Do genealogical patterns in purple photosynthetic bacteria reflect interspecific gene transfer? Nature 283: 212–214. (1)

Woese, C. R., R. Gutell, R. Gupta, and H. F. Noller (1983) Detailed analysis of the higher-order structure of 16S-like ribosomal ribonucleic acids. Microbiol. Rev. 47: 621–669. (1)

Woese, C. R., R. Gupta, C. M. Hahn, W. Zillig, and J. Tu (1984) The phylogenetic relationships of three sulfur-dependent archaebacteria. Syst. Appl. Microbiol. 5: 97–105. (1)

Woese, C. R., B. A. Debrunner-Vossbrinck, H. Oyaizu, E. Stackebrandt, and L. Ludwig (1985) Gram-positive bacteria: Possible photosynthetic ancestry. Science 229: 762–765. (1)

Woese, C. R., O. Kandler, and M. L. Wheelis (1990a) Towards a natural system of organisms: Proposal for the domains Archaea, Bacteria, and Eucarya. Proc. Natl. Acad. Sci. USA 87: 4576–4579. (1)

Woese, C. R., D. Yang, L. Mandelco, and K. O. Stetter (1990b) The Flexibacter-Flavobacter connection. Syst. Appl. Microbiol. 13 (in press). (1)

Woese, C. R., L. Mandelco, D. Yang, R. Gherna, and M. T. Madigan (1990c) The case for relationship of the flavobacteria and their relatives to the green sulfur bacteria. Syst. Appl. Microbiol. 13 (in press). (1)

Wolf, P. G., P. S. Soltis and D. E. Soltis (1988) Chloroplast DNA variation in Heuchera grossulariifolia (Saxifragaceae). Am. J. Bot. 75(suppl.): 218. (7)

Wolfe, K. H. and P. M. Sharp (1988) Identification of functional open reading frames in chloroplast genomes. Gene 66: 215–222. (7)

Wolfe, K. H., W.-H. Li and P. M. Sharp (1987) Rates of nucleotide substitution vary greatly among plant mitochondrial, chloroplast, and nuclear DNAs. Proc. Natl. Acad. Sci. USA 84: 9054–9058. (6, 7)

Wolfe, K. H., P. M. Sharp and W.-H. Li (1989a) Rates of synonymous substitution in plant nuclear genes. J. Mol. Evol. 29: 208–211. (7)

Wolfe, K. H., M. Gouy, Y.-W. Yang, P. M. Sharp and W.-H. Li (1989b) Date of the monocot-dicot divergence estimated from chloroplast DNA sequence data. Proc. Natl. Acad. Sci. USA 86: 6201–6205. (7)

Woolf, B. (1955) On estimating the relation between blood group and disease. Ann. Eugenics 19: 251–253. (12)

Wright, S. (1931) Evolution in Mendelian populations. Genetics 16: 97–159. (8)

Wright, S. (1966) Polyallelic random drift in relation to evolution. Proc. Natl. Acad. Sci. USA 55: 1074–1081. (11)

Wright, S., A. Rosenthal, R. Flavell and F. Grosveld (1984) DNA sequences required for regulated expression of β-globin genes in murine erythroleukemia cells. Cell 38: 265–273. (13)

Wu, C.-I. (1983a) Virility deficiency and the Sex-Ratio trait in Drosophila pseudoobscura. I. Sperm displacement and sexual selection. Genetics 105: 651–662.

Wu, C.-I. (1983b) The fate of autosomal modifiers of the Sex-Ratio trait in Drosophila and other sex-linked meiotic drive systems. Theor. Pop. Biol. 24: 107–120. (9)

Wu, C.-I. and A. T. Beckenbach (1983) Evidence for extensive genetic differentiation between the Sex-Ratio and the standard arrangement of Drosophila pseudoobscura and D. persimilis and identification of hybrid sterility factors. Genetics 105: 71–86. (9)

Wu, C.-I. and W.-H. Li (1985) Evidence for higher rates of nucleotide substitution in rodents than in man. Proc. Natl. Acad. Sci. USA 82: 1741–1745. (13)

Wu, C.-I., T. W. Lyttle, M.-L. Wu, and G.-F. Lin (1988) Association between a satellite DNA sequence and the responder of Segregation Distorter in D. melanogaster. Cell 54: 179–189. (9)

Wu, C.-I., J. R. True and N.A. Johnson (1989) Fitness reduction associated with the deletion of a satellite DNA array. Nature 341: 248–251. (9)

Wu, T. T. and E. A. Kabat (1970) An analysis of the sequences of the variable regions of Bence

Jones proteins and myeloma light chains and their implications for antibody complementarity. J. Exp. Med. 132: 211–250. (12)

Wyk, P. and P. Reeves (1989) Identification and sequence of the gene for abequose synthase, which confers antigenic specificity on group B salmonellae: Homology with galactose epimerase. J. Bacteriol. 171: 5687–5693. (2)

Xiong, Y. and T. H. Eickbush (1988a) Functional expression of a sequence-specific endonuclease encoded by the retrotransposon R2Bm. Cell 55: 235–246. (9)

Xiong, Y. and T. H. Eickbush (1988b) Similarity of reverse transcriptase-like sequences of viruses, transposable elements, and mitochondrial introns. Mol. Biol. Evol. 5: 675–690. (5, 7, 8)

Yamada, T. and M. Shimaji (1986) Peculiar feature ofthe organization of rRNA genes of the Chlorella chloroplast DNA. Nucleic Acids Res. 14: 3827–3839. (7)

Yamaguchi, O., T. Yamazaki, K. Saigo, T. Mukai and A. Robertson (1987) Distributions of three transposable elements, P, 297 and copia, in natural populations of Drosophila melanogaster. Jpn. J. Genet. 62: 205–216. (8)

Yamazaki, K., E. A. Boyse, V. Mike, H. T. Thaler, B. J. Mathieson, J. Abbott, J. Boyse, Z. A. Zayas and L. Thomas (1976) Control of mating preferences in mice by genes in the major histocompatibility complex. J. Exp. Med. 144: 1324–1335. (11, 12)

Yamazaki, K., M. Yamaguchi, P. W. Andrews, B. Peake and E. A. Boyse (1978) Mating preferences of F₂ segregants of crosses between MHC-congenic mouse strains. Immunogenetics 6: 253–259. (12)

Yamazaki, K., M. Yamaguchi, L. Baranoski, J. Bard, E. A. Boyse and L. Thomas (1979) Recognition among mice: Evidence from the use of a Y-maze differentially scented by congenic mice of different major histocompatibility types. J. Exp. Med. 150: 755–760. (11)

Yamazaki, K., G. K. Beauchamp, C. J. Wysocki, J. Bard, L. Thomas and E. A. Boyse (1983a) Recognition of H-2 types in relation to the blocking of pregnancy in mice. Science 221: 186–188. (12)

Yamazaki, K., G. K. Beauchamp, I. K. Egorov, J. Bard, L. Thomas and E. A. Boyse (1983b) Sensory distinction between H-2^b and H-2^{bm1} mutant mice. Proc. Natl. Acad. Sci. USA 80: 5685–5688. (11)

Yamazaki, K., G. K. Beauchamp, D. Kupniewski, J. Bard, L. Thomas and E. A. Boyse (1988) Familial imprinting determines H-2 selective mating preferences. Science 240: 1331–1332. (12)

Yang, D., Y. Oyaizu, H. Oyaizu, G. J. Olsen, and C. R. Woese (1985) Mitochondrial origins. Proc. Natl. Acad. Sci. USA 82: 4443–4447. (1)

Yannopoulos, G., N. Stamatis, M. Monastirioti, P. Hatzopoulos and C. Louis (1987) hobo is responsible for the induction of hybrid dysgenesis by strains of Drosophila melanogaster bearing the male recombination factor 23.5MRF. Cell 49: 487–495. (8)

Yatskievych, G., D. B. Stein and G. J. Gastony (1988) Chloroplast DNA evolution and systematics of Phanerophlebia (Dryopteridaceae) and related fern genera. Proc. Natl. Acad. Sci. USA 85: 2589–2593. (7)

Yokoyama, S. (1988) Molecular evolution of the human and simian immunodeficiency viruses. Mol. Biol. Evol. 5: 645–659. (5)

Yokoyama, S. and T. Gojobori (1987) Molecular evolution and phylogeny of the human AIDS viruses LAV, HTLV-III and ARV. J. Mol. Evol. 24: 330–336. (5)

Yokoyama, S., E. N. Moriyama and T. Gojobori (1987) Molecular phylogeny of the human immunodeficiency and related retroviruses. Proc. Jpn. Acad. 63: 147–150. (5)

Yokoyama, S., L. Chung and T. Gojobori (1988) Molecular evolution of the human immunodeficiency and related viruses. Mol. Biol. Evol. 5: 237–251. (5)

Young, J. P. W. (1989) The population genetics of bacteria. In: Genetics of Bacterial Diversity, D. A. Hopwood and K. F. Chater (eds.). Academic Press, London, pp. 417–438. (2)

Young, M. W. (1979) Middle repetitive DNA: A fluid component of the Drosophila genome. Proc. Natl. Acad. Sci. USA 76: 6274–6278. (8)

Yuhki, N. and S. J. O'Brien (1988) Molecular characterization and genetic mapping of class I and class II MHC genes of the domestic cat. Immunogenetics 27: 414–425. (11)

Zahn, K. and F. B. Blattner (1985) Sequence-induced DNA curvature at the bacteriophage λ origin of replication. Nature 317: 451–453. (9)

Zeigler, D. R. and D. H. Dean (1987) Genetic map of Bacillus subtilis 168. In: Genetic Maps,

Vol. 4, S. J. O'Brien (ed.). Cold Spring Harbor Laboratory, Cold Spring Harbor, NY, pp. 195–212. (4)

Zinder, N. D. and J. Lederberg (1952) Genetic exchange in *Salmonella*. J. Bacteriol. 64: 679–699. (2)

Zinkernagel, R. M. (1979) Associations between major histocompatibility antigens and susceptibility to disease. Annu. Rev. Microbiol. 33: 201–213. (12)

Zinkernagel, R. M. and P. C. Doherty (1974) Immunological surveillance against altered self components by sensitized T lymphocytes in lymphocytic choriomeningitis. Nature 251: 547–548. (11)

Zoller, M. J. and M. Smith (1983) Oligonucleotide-directed mutagenesis of DNA fragments cloned into M13 vectors. Methods Enzymol. 100: 468–500. (3)

Zubay, G (1985) *Biochemistry*. Addison Wesley, Reading, MA, p. 97. (4)

Zuckerkandl, E. and L. Pauling (1965) Molecules as documents of evolutionary history. J. Theor. Biol. 8: 357–366. (1)

Zuker, M. and P. Stiegler (1981) Optimal computer folding of large RNA sequences using thermodynamics and auxiliary information. Nucleic Acids Res. 9: 133–148. (7)

Zurawski, G. and M. T. Clegg (1984) The barley chloroplast DNA *atpBE*, *trn*M2 and *trn*V1 loci. Nucleic Acids Res. 12: 2549–2559. (7)

Zurawski, G. and M. T. Clegg (1987) Evolution of higher-plant chloroplast DNA-encoded genes: Implications for structure-function and phylogenetic studies. Annu. Rev. Plant Physiol. 38: 391–418. (7)

Zurawski, G., M. T. Clegg and A. H. D. Brown (1984a) The nature of nucleotide sequence divergence between barley and maize chloroplast DNA. Genetics 106: 735–749. (7)

Zurawski, G., W. Bottomley and P. R. Whitfeld (1984b) Junctions of the large single copy region and the inverted repeats in *Spinacea oleracea* and *Nicotiana debneyi* chloroplast DNA: Sequence of the genes for tRNA[His] and the ribosomal proteins S19 and L2. Nucleic Acids Res. 12: 6547–6558. (7)

Zwiebel, L. J., V. H. Cohn, D. R. Wright and G. P. Moore (1982) Evolution of single-copy DNA and the ADH gene in seven Drosophilids. J. Mol. Evol. 19: 62–71. (9)

Index

Abortion, HLA and, 268-269
Acinetobacter calcoaceticus, 93
Acquired immunodeficiency syndrome
    (AIDS), 96-98, 100-101, 111
  AIDS-associated retrovirus, 96,
    99-100, 107
Adaptation
  bacterial *trp* genes and, 79
  globin gene families and, 273, 284
  Salmonella and, 25, 38-47, 56
Alcohol dehydrogenase locus, 235
  causes of variation, 203-209
  selection at DNA level, 205-207, 210,
    221
Algae, chloroplast DNA and, 137, 139,
    144
Alkaline phosphatase, of *E. coli*, 58-59,
    74-76
  amino acid sequences, 62-71
  polymorphisms in sequences, 59-62
  selective constraints, 71-74
Alleles
  enteric fever and, 40, 44, 46
  flagellin genes and, 50, 52-53
  gene conversion and, 265-266
  HLA and, 248-249, 251, 254, 270-271
  host range-genetic diversity
    correlation, 47-48
  MHC loci in mammals, 222, 229-232,
    243, 247
  modes of selection and, 267-269
  multilocus variation and, 258-264
  neutral theory and, 119, 122
  organelle genes and, 114-115, 130,
    132, 134
  polymorphism maintenance and,
    236-242
  recombination of genomes and,
    126-128

  Salmonella, 25-26, 33-35, 55-56
  segregation distortion and, 266-267
  selection at DNA level and, 204, 206,
    210, 212-214, 216, 219
  selection within cells and, 124-125
  single-locus variation and, 254-258
  transposable elements in *Drosophila*
    and, 157, 167
  transspecific polymorphism and,
    234-236
  ultraselfish genes and, 178, 181,
    183-184
Amino acids
  bacterial *trp* genes and, 79, 87
  globin gene families and, 272-273, 277
  HIV and, 99-100, 106, 111
  HLA and, 249-250, 254, 256-258
  MHC loci in mammals and, 225-226,
    234, 240
  ribosomal RNA and, 6
  Salmonella and, 50, 53
  selection at DNA level and, 213, 216
Amino acid sequence, of *E. coli* alkaline
    phosphatase, 59, 74-75
  experimental manipulation of, 62-63
  multiple replacement, 67-71
  polymorphism in, 60-62
  selective constraints, 71-74
  single replacement, 63-67
Anabaena, chloroplast DNA and, 139
Ankylosing spondylitis, 252-253
Anthranilate, bacterial *trp* genes and, 88,
    94-95
Anthranilate synthase, bacterial *trp* genes
    and, 86-87, 89-94
Antibodies
  HIV and, 97
  MHC loci in mammals and, 228, 239

335